NUMERICAL TECHNIQUES FOR STOCHASTIC SYSTEMS

NUMERICAL TECHNIQUES FOR STOCHASTIC SYSTEMS

A collection of papers based on the lectures presented at the conference on Numerical Techniques for Stochastic Systems, held at Gargnano, Italy, September 1979.

edited by

Francesco ARCHETTI

and

Marco CUGIANI

Institute of Mathematics
University of Milano
Italy

1980

NORTH-HOLLAND PUBLISHING COMPANY – AMSTERDAM • NEW YORK • OXFORD

© North-Holland Publishing Company, 1980

All rights reserved. No part of this publication may be reproduced, stored in a retrieval system, or transmitted, in any form or by any means, electronic, mechanical, photocopying, recording or otherwise, without the prior permission of the copyright owner.

ISBN: 0 444 86000 2

Publishers:

NORTH-HOLLAND PUBLISHING COMPANY
AMSTERDAM • NEW YORK • OXFORD

Sole Distributors for the U.S.A. and Canada:

ELSEVIER NORTH-HOLLAND INC.
52 VANDERBILT AVENUE, NEW YORK, N.Y. 10017

Library of Congress Cataloging in Publication Data

Main entry under title:

Numerical techniques for stochastic systems.

1. Stochastic systems--Addresses, essays, lectures.
2. Numerical analysis--Addresses, essays, lectures.
I. Archetti, Francesco, 1946- II. Cugiani, Marco. III. Conference on Numerical Techniques for Stochastic Systems, Gargnano, Italy, 1979.
QA402.N85 519.2 80-14670
ISBN 0-444-86000-2

8100818 / QA 402 N85

PRINTED IN THE NETHERLANDS

PREFACE

This volume collects the invited papers and some of the contributed papers presented at the Conference "Numerical Techniques for Stochastic Systems" held in Gargnano, Italy, in September 1979, with the financial support of the Consiglio Nazionale delle Ricerche and the organizational assistance of the Università di Milano.

The main aim of this volume is the presentation of some recent research results obtained in the interface between numerical methods and stochastic systems.

This research area can be placed under two headings: the first connected with the development of numerical methods for the identification, estimation and modelling of stochastic systems, the second with the applications of ideas from probability and statistics to the design and analysis of numerical methods. Clearly these two areas are strongly correlated and similar tools are often used in both domains of research.

The main aim of the conference was to bring together scholars from each area in order to focus on the most promising aspects of their common methodologies and identify trends of future development.

The editors of these proceedings are confident that the growing necessity of accounting for uncertain factors, while building realistic models or analyzing complex computational procedures, will confirm, in the coming years, the importance of the topics discussed at the conference.

The edtiors are grateful to the speakers for their participation, to the authors for their contributions and to B. Betrò and J.J. Mc Keown also for helping in the running of the conference and the preparation of this volume.

Special thanks are due to L.C.W. Dixon whose advice has been valuable in the definition of the scientific program and the preparation of the proceedings and whose recent volume with G.P. Szegö, entitled "Numerical Optimisation of Dynamic Systems" also contains papers related to those in the first part of this volume.

Finally, thanks are due to Miss Beatrice Midollini and Miss Ornella Nobili who performed the typing of the final master forms for this book.

Milano, February 1980 The Editors

CONTENTS

PART ONE

ESTIMATION, IDENTIFICATION AND MODELLING OF STOCHASTIC SYSTEMS

Numerical Techniques for Stochastic Systems
F. Archetti and M. Cugiani (eds.)
© North-Holland Publishing Company, 1980

NONLINEAR STOCHASTIC SYSTEM IDENTIFICATION VIEWED AS AN OPTIMISATION PROBLEM

by

L.C.W. Dixon and P.J. Lovesey

The Numerical Optimisation Centre
The Hatfield Polytechnic
The United Kingdom

In this paper we review the off line and on line system parameter identification problems, and show that the former can always be expressed as an optimisation problem and that the latter can be so posed when the parameters are time invariant. We discuss the objective when the parameters are time varying and present some new results on the difficulties in solving these problems.

INTRODUCTION

The system parameter identification problem occurs in many different engineering and economic contexts. Because it is of such widespread importance, many different algorithms have been proposed for its solution. In this paper we will restrict our consideration to those approaches that can be viewed as solving the identification problem as if it were an optimisation problem. We will distinguish between two forms of the identification problem, namely the off line and on line problems.

A problem is described as off line if the data is all available before the computation commences and on line if additional data is being received during the computation. The on line problem can again be subdivided into two subclasses dependent on whether the parameters being identified are constant or essentially time varying. We will consider certain implications of posing these three estimation problems as optimisation problems commencing with the off line situation.

Off Line Identification

Let us assume that we are considering a dynamic system of known structure. Further assume that a known input has been applied and that some measurements of the output have been made. From this data we wish to estimate the values of some parameters of the dynamic system.

If we wish these estimates to be unbiassed and to have the minimum possible variance, then by the Cramer-Rao theorem, we should construct the probability of these measurements occuring as a function of the parameters and then maximise this "Likelihood Function". This is known as the maximum likelihood approach and is often time consuming as the likelihood function is frequently highly non linear. However in many circumstances a less non linear problem is obtained if logarithms are taken, so that in practise the problem solved becomes

$$\text{Min} - \log \text{(Likelihood Function)}. \tag{1}$$

Even this however is frequently non linear and time consuming.

Most algorithms for solving the on line problem are adaptations of methods used to solve the off line problem. In the on line situation there is frequently a natural limitation on how much time is available for undertaking the computation to obtain these estimates before they are used to control a system. This natural limitation does not occur in the off line situation where (1) could be solved exactly; but on line the time consuming nature of (1) normally implies a compromise; frequently this involves replacing (1) by a simpler optimisation problem. This in turn may lead to biassed estimates and/or larger variances. These, in turn, then lead to non optimal feed back controls and loss of performance.

The question therefore arises as to how such approximations effect the performance?

One of the simplest estimation problems is the well known single input/single output ARIMA problem. In this the input signal u_j is known and gives rise to a nominal output y_k which is measured as m_k.

The system can be written as

$$y_k + \sum_j^p a_j y_{k-j} = \sum_i^q b_i \, u_{k-i}$$

or an operator notation,

$$(I + A) \; y = Bu \tag{2.1}$$

and the noise relation is

$$m_k = y_k + n_k \tag{2.2}$$

We may express the noise n in terms of the data u, m by

$$(I+A)n = -Bu + (I+A)m.$$

The problem is to estimate the parameters a_j, b_i from this data. To obtain the likelihood function we must make some assumption about the probability distribution of n. A usual assumption is to assume that they are related to a sequence ξ_k of $(0,\sigma)$ random variables by a filter

$$(I+C)(I+A)n = (I+D)\xi. \tag{2.3}$$

This leads to the optimisation problem (Astrom & Bohlin, 1966)

$$\text{Min} \quad F = \sum \xi^2 \tag{2.4}$$

$$\text{s.t } (I+D)\xi = (I+C)\{(I+A)m - Bu\} \tag{2.5}$$

As the objective function is highly nonlinear in the variables $x=(A,B,C,D)^T$ this frequently involves a large number of iterations even of an efficient nonlinear optimisation algorithm. Again as the set of linear equations 2.5 have to be solved for ξ at each function evaluation and as this involves a complete pass through the data set (m, u), this method can be very expensive when applied to large data sets.

In contrast if we take the simple approximation $C = D$ then we obtain the equation error (Astrom & Eykoff, 1972) or ordinary least squares algorithm

$$\begin{aligned} \text{Min } F &= \sum e^2 \\ e &= \{(I+A)m - Bu\}. \end{aligned} \tag{2.6}$$

The objective function is now a simple quadratic in the variables $x=\{A,B\}$ and can therefore be solved in one Newton iteration. These estimates of A and B can be obtained very cheaply but can be exceedingly biassed.

As a compromise between these two extremes, if we set $D=0$ we obtain

<u>Clarke's (1967) Generalised Least Squares Function</u>

$$\begin{aligned} \text{Minimise} \quad F &= \sum e^2 \\ e &= (I+C)\{(I+A)m - Bu\} \end{aligned} \tag{2.7}$$

This is a quartic optimisation problem and therefore an iterative solution is still required. The cost of a function evaluation is greatly reduced compared to problem 2.4/2.5 as we no longer need to solve a set of linear equations for ξ as the e are given directly by 2.7 and also all the information in the data streams m, u can now be precondensed into small information matrices. This therefore leads to a method that is relatively cheap compared to the maximum likelihood method and relatively unbiassed compared to the equation error method.

It is of interest to illustrate this principle of precondensing data streams into relevant information matrices.

Example of Precondensation of Data

Let us consider the simple case where

$A = (a_1, a_2)$, $B = (b_1)$, $C = (c_1)$ then

$$e_k = (m_k+a_1m_{k-1}+a_2m_{k-2}-b_1u_{k-1}) + c_1(m_{k-1}+a_1m_{k-2}+a_2m_{k-3}-b_1u_{k-2})$$

$$= (m_k+\alpha_1m_{k-1}+\alpha_2m_{k-2}+\alpha_3m_{k-3}-\beta_1u_{k-1}-\beta_2u_{k-2})$$

where

$$\begin{aligned} \alpha_1 &= a_1+c_1 & \beta_1 &= b_1 \\ \alpha_2 &= a_2+a_1c_1 & \beta_2 &= b_1c_1 \\ \alpha_3 &= a_2c_1 \end{aligned} \tag{2.8}$$

$$\begin{aligned} \text{Then}\quad F = \sum e_k^2 &= (\sum m_k^2)+2\alpha_i(\sum m_km_{k-1})-2\beta_i(\sum m_ku_{k-1}) \\ &+ \alpha_i\alpha_j(\sum m_{k-1}m_{k-j}) - \alpha_i\beta_j(\sum m_{k-i}u_{k-j}) + \beta_i\beta_j(\sum u_{k-i}u_{k-j}) \end{aligned} \tag{2.9}$$

As all the information about the data needed to evaluate F is contained in the matrices ( ), these can be calculated once only during a pass through the data before the optimisation commences.

The optimisation may then be undertaken by varying A,B,C calculating α,β from 2.8 and then F from 2.9; ∇F can be obtained similarly if required. This method is originally due to Soderström (1974).

A further difficulty of estimating parameters in a, b space is that the resulting system may not correspond to a realisable system of time constant and gains. The relationship between these variables is nonlinear, typically (Dixon, 1973) for a second order system

$$a_1 = -\,(e^{-2/T_1} + e^{-2/T_2})$$

$$a_2 = e^{-2/T_1}\, e^{-2/T_2}$$

$$b_1 = K(1 + \frac{2T_1}{T_2-T_1}\, e^{-2/T_1} - \frac{(T_1+T_2)}{(T_2-T_1)}\, e^{-2/T_2})$$

$$b_2 = K(1 - a_1+a_2).$$

and we see at once that some combinations of a,b do not correspond to any values of K, T_1, T_2. (Any negative values of a_2 for instance).

We can avoid difficulties of this sort by using the K, T_1, T_2 variables for optimisation, and for any values of K, T_1, T_2 calculating the appropriate values of (a,b) and then α and hence F via the precondensed data. This, of course, entails a cost and may well be four times as expensive as minimisation in (a,b) space.

3. On-line estimation/constant parameters

Let us assume now that we have available K_1 sets of measurements and wish to estimate the parameters. It would seem logical to obtain this by minimising

$$F_{K_1} = \frac{1}{K_1} \sum_{k}^{K_1} \xi_k^2 \tag{3.1}$$

then as more data becomes available we may hope to improve the estimate by minimising

$$F_{K_2} = \frac{1}{K_2} \sum_{k}^{K_2} \xi_k^2 \qquad\qquad K_2>K_1$$

If we are using the precondensation technique it is possible to update the data matrices 2.9 whenever new data becomes available. As this could be done as a parallel operation to the minimisation, on a separate microprocessor, there is no longer any need to limit the estimation step to one that can be completed between the arrival of data sets.

Soderström et al.(1974),Ljung (1975) have shown that many recursive techniques that take one step per data item are assymptotically convergent as $K\to\infty$.

From an optimisation viewpoint we may make a general statement. Let us assume the problem statistics are such that the limits

$$F_\infty = \lim_{K\to\infty} \frac{1}{k} \sum_{k}^{K} \xi_k^2 \tag{3.2}$$

and

$$x_\infty = \lim_{k\to\infty} (\arg\min F_K) \tag{3.3}$$

exist, and that constants ρ, μ, k_o exist such that

$$P(\|x_k - x_\infty\| > \frac{1}{k}\rho) < \mu \tag{3.4}$$

all $K > k_o$. Also let us assume the functions F_k are well behaved in the sense of Dixon (1974), (this is true of the likelihood functions discussed in this paper), then any optimisation algorithm that satisfies conditions I, II and III of Wolfes theorem at each iteration will converge to a point where $\nabla F_\infty = 0$ if applied for one or more iterations on a regular subset of F_K.

However in many practical situations we are not so concerned with the behaviour as $K\to\infty$, but rather with the ability to obtain rapid convergence to the region of the minimum in a small number of iterations. If $x^{(K)}$ is the Kth estimate of the solution, it is therefore desirable if $\|x^{(K)} - x_k\|$ decreases rapidly as well as $\|x_k - x_\infty\|$. Obviously

$$\|x^{(K)} - x_\infty\| \leq \|x^{(K)} - x_k\| + \|x_k - x_\infty\|$$

but as the two terms are due to different causes it seems advisable to treat them separately.

As $\|x_k - x_\infty\|$ is independent of the numerical algorithm, the choice of algorithm influences the convergence properties via the term $\|x^{(K)} - x_k\|$. Frequently there will be three stages in an estimation, an initial period in which very little data is available so that x_k is not a good approximation to x_∞ and in this region it does not really matter if $x^{(k)} \sim x_k$. This will be followed by a region where $\|x_k - x_\infty\| \ll \|x^{(k)} - x_k\|$ and finally with a good algorithm this will be followed by a region where $\|x^{(k)} - x_k\| < \|x_k - x_\infty\|$ and the estimate follows the optimal estimate to the solution.

The usefulness of the estimates when limited data is available is largely dependent on how rapidly the algorithm can reduce $\|x^{(k)} - x_k\|$

in the intermediate period. As x_k will hopefully be constant relative to the changes in $x^{(k)}$ in this period, algorithms with a slow rate of convergence such as steepest descent and stochastic approximation will perform much worse than superlinearly convergent processes such as safeguarded Newton-Raphson or Variable Metric algorithms.

The algorithm REVAR developped by James (1980) was constructed to be assymptotically convergent and to give initial estimates for the SISO problem. It can be summarised as

(1) Accept k_i items of data and build the precondensation matrices.

(2) Perform $L \geq 1$ iterations of a variable metric algorithm to F_{k_i} and stop if there is no more data

(3) Accept $k_{i+1}-k_i$ items of data, updating both the condensation matrices and the variable metric matrix with the added information.

(4) Return to step 2. i=i+1.

As this algorithm constructs a variable metric matrix that is an approximation to the Hessian of F_k, the convergence of $x^{(k)}$ to x_k is superlinear. The results quoted in Dixon and James (1978) indicate that systems exist for which the rate of convergence of $x^{(k)}$ to x_∞ is considerably faster, than for the standard recursive generalised least squares method (Hastings James & Sage, 1969) RGLS, though of course for other systems this is perfectly adequate.

4. On line estimation of time varying parameters

If the parameters in a system vary with time then the use of the function F_∞ introduced in section 3 is inappropriate, as the data obtained at any early stage in the process may have been produced by very different parameter values than those existing at the end.

One way (Hastings James & Sage, 1969) of attempting to determine how the parameters vary with time is to introduce a parameter $\rho > 1$ and to define

$$F_k = \frac{1}{k} \sum_{k}^{K} \rho^{K-k} \xi_k^2 \qquad (4.1)$$

This parameter ρ is introduced into the RLS or RGLS algorithms and can be also introduced into the REVAR algorithm. In both the RLS and REVAR algorithms the aim would then be to accept the values of the parameters that minimise F_K (approximately) as the time history of the parameters.

An alternative approach would be to divide the data set into blocks of length L and to minimize

$$F_k = \frac{1}{K-L} \sum_{k=K-L}^{K} \xi_k^2$$

to obtain the estimates of the parameters.

If either of these techniques were employed on the constant parameter problem then they would lead to biassed results, as we could not expect

$$\| x_K - x_\infty \| \to 0 \quad \text{as } K \to \infty.$$

The two techniques are very similar as the choice of L and ρ determine the amount of data that is effectively being considered in each estimate. The choice of these factors can vitally effect the performance. Let us illustrate this with an extreme example:

The constant speed, two point interception problem gives rise to the following system description:

$$\dot{\underline{x}} = \begin{bmatrix} 0 & 1 & 0 & 0 \\ -w_n^2 & -2\xi w_n & 0 & 0 \\ 0 & 0 & -\frac{1}{T} & \frac{1}{T}. \\ -\overline{R} & 0 & 0 & -\frac{2\dot{R}}{R} \end{bmatrix} \underline{x} + \begin{bmatrix} 0 \\ w_n^2 \\ 0 \\ 0 \end{bmatrix} U$$

where x_1 = lateral acceleration of the interceptor.

$x_2 = \dot{x}_1$

x_3 = measured angular rate of the interceptor-target sight line.

x_4 = actual sight line rate.

w_n, ξ, T - are constants of the principal dynamics of the interceptor.

$\dot{R}$ - rate of change of range

R - separation range.

A commonly used control law in aerospace applications is

$$U = 3V_c \quad \text{(sight line rate)}$$

where V_c = closing velocity = $-\dot{R}$.

Therefore in order to implement this control law we must first solve a nonlinear parameter estimation problem for R.

If we select L=1 and consider the problem with constant $\dot{R}$ and no noise on the measurement or in the system, then an oscillating divergent estimate of $\dot{R}$ results even when we start the estimation with the correct value of $\dot{R}$ (Fig. 1).

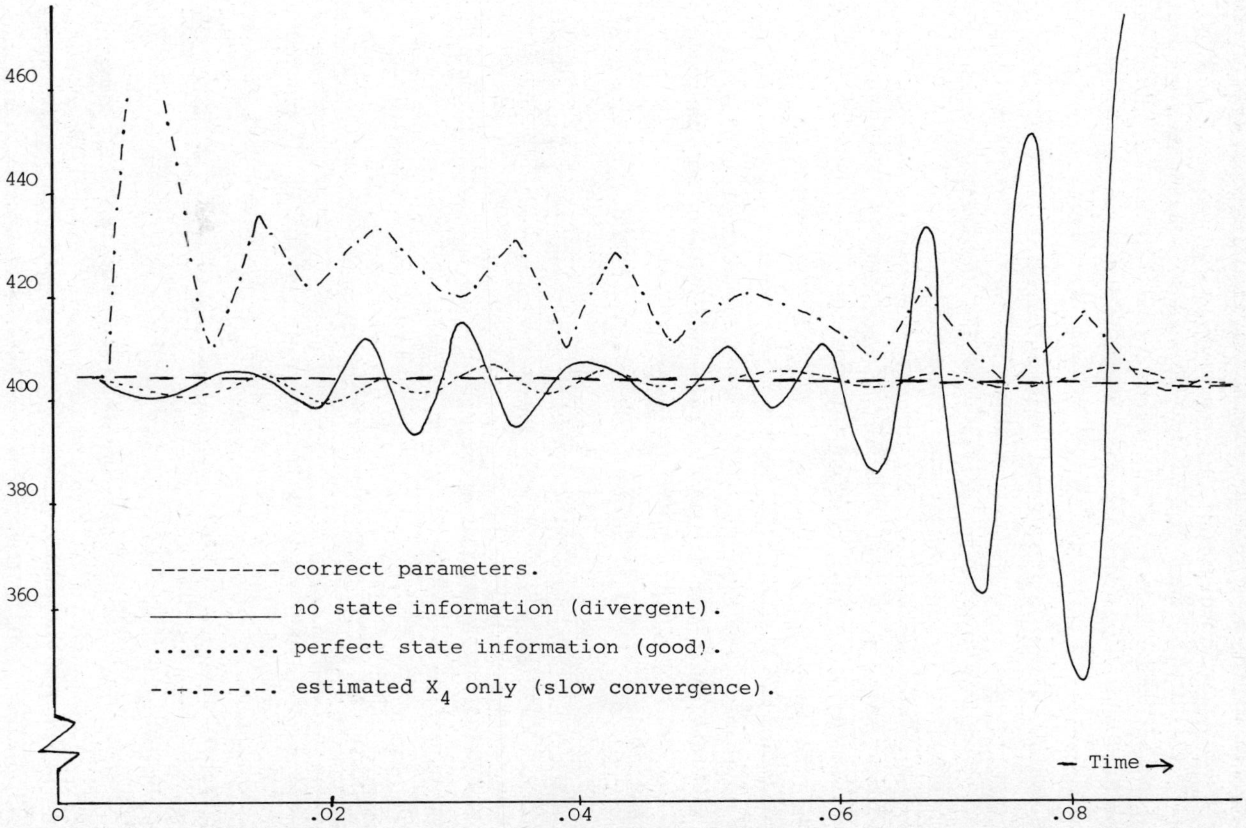

Fig. 1: - Examples of parameter variation.

This is due to the fact that the behaviour of the system over a time interval depends on the state X_k as well as the value of $\dot{R}$. As the estimate of the state depends on the previous estimates of both the state and the parameters this oscillation results. To check this we verified that given the accurate state information $\underline{X}(t)$ the parameter estimates from minimising 4.2 are correct.

Two questions arise: (1) what information contained in $\underline{X}(t)$ needs to be correct for a convergent estimate of the parameters to result? and (2) does this result generalise to other problems?
For this particular problem it was found that provided $x_4(t)$ is estimated independently then the parameter estimate converges (Fig.5). In the next section we will look at the problem of estimating the initial states in such circumstances.

5. The initial state problem

It is obviously true that for any problem the output will depend on both the initial states and on the parameter values. Let us reconsider the two examples we have considered to date, in the SISO problem, where

$$y_k + \sum_j^p a_j y_{k-j} = \sum_i b_i u_{k-i}$$

then the first k_o equations contain unknown information where $k_o = \max(p,q)$. In the equation error method this unknown information is removed if the objective function is modified to

$$F = \frac{1}{K-k_o} \cdot \sum_{k=k_o}^{k} e_k^2 \tag{5.1}$$

Similarly in the generalised least squares approach it is removed if

$$F = \frac{1}{K-k_1} \sum_{k=k_1}^{K} \xi_k^2 \tag{5.2}$$

where $k_1 = k_o + r$, and r is the order of the noise filter C.

In this case the effect of the initial condition is easily removed. One general approach to the problem of removing the effect of the initial states is described in Pearson (1980).An alternative general approach is described below.

For the more general system where we have a state equation
$X_{k+1} = AX_k + \nu_k$

and measurements

$$Y_{k+1} = C\ X_{k+1} + \mu_{k+1},$$

then a standard way of estimating the states is via the Kalman Filter. Knowledge of three statistical distributions is assumed, namely the probability P_1 of errors in X_o, P_2 on the system noise ν_k and P_3 on the measurement noise μ_k. It is of course simplest to assume these are independent and Gaussian with the standard notation $P_1=N(o,P)$, $P_2=N(0,Q)$, $P_3=N(0,R)$.

Given these statistical distributions the likelihood of the sets of measurements being due to a particular set of states X_k, X_{k+1} can be expressed as a simple function whose logarithm is quadratic, giving a simple formula for $\hat{X}_{k+1}$.
However when the matrix A depends on unknown parameters (such as $\dot{R}$ in our example) then the logarithm of the likelihood function is not quadratic but highly nonlinear and a compromise is usually made between the complete minimisation and the time available. The simplest compromise is known as the extended Kalman Filter approach, which is known to diverge on some examples.

Let us now consider a special case. If there is no system or measurement noise i.e. $\nu=\mu=0$ and all the parameters are given, then to determine the n initial values $X(o)$ from scalar measurements Y_k we would need n measurements. If we also wish to determine r parameters, then at least $n+r$ measurements would be required to determine both the initial state and the parameters.

Similarly if we consider the case when $\nu=0$ and there is no prior information available about the initial value of X_o or the values of the parameters, then the log likelihood function reduces to

$$F = \sum_k \mu_k^2$$

and the maximum likelihood values of the states X_o and parameters θ would be obtained by minimising the highly nonlinear function F with respect to both sets of variables.

The assumption that the error in each measurement μ_k is independent is obviously unlikely to be correct and it may well be advisable to introduce a G.L.S. type filter into this problem before minimisation to reduce the biass. This will increase the nonlinearity.

If all the parameters are known then a system is said to be observable if the initial conditions can be recovered from the measurements. This will theoretically be true if the Hessian of the log likelihood function (the information matrix) is nonsingular at the solution. Given that we are now interested in determining both states and parameters, the condition for observability is just that the Hessian matrix with respect to both sets of variables is nonsingular.

However in practise the ability to solve the minimisation problem will depend on the condition number of this Hessian matrix at the solution. Indeed the formula for the magnitude of the error in the solution of a set of linear equations is typically a product involving the condition number of the matrix, (Wilkinson (1977)). A similar result holds for a set of nonlinear equations. It is therefore tempting to introduce the concept of the condition of observability, as the condition number of the Jacobian matrix J of the data at the solution, where

$$J = \left(\frac{\partial \mu_k}{\partial X_i(o)} ; \frac{\partial \mu_k}{\partial \theta_j}\right)$$

Experience has however indicated that this measure is not particularly useful unless we need all the values of the state and parameters. Let us consider the above four dimensional problem with $\dot{R}$ known. Singular value decomposition indicated that the condition number was $.14 \times 10^9$ and the eigen structure

$\lambda_1^2 = 6.9$	$\lambda_2^2 = .13$	$\lambda_3^2 = .1 \times 10^{-9}$	$\lambda_4^2 = .3\ 10^{-15}$
$v_1 = -.16\ 10^{-5}$	$v_2 = -.11\ 10^{-4}$	$v_3 = 1$	$v_4 = 10^{-3}$
$-.15\ 10^{-5}$	$-.13\ 10^{-4}$	10^{-3}	1
$.97$	$-.22$	10^{-6}	10^{-7}
$.22$	$.98$	10^{-4}	10^{-6}

The solution obtained when attempting to solve this problem is highly dependent on the technique used. Some equation solving techniques give large errors in all variables, however when applying a variable metric optimisation routine the variables x_3 and x_4 are determined accurately, whilst the errors are confined to the x_1, x_2 space. This is approximately what would be expected from an optimisation routine on this function. The optimisation routine terminates at a point where

$$||\nabla F|| < e_o^2. \tag{5.3}$$

If Δx is the error in x, it can be expressed in terms of the eigenvectors as

$$\underline{\Delta}x = \sum \xi_i \underline{v}_i \tag{5.4}$$

then if λ are the eigenvalues of $\nabla^2 F$ we have

$$\nabla F = \sum \lambda_i \xi_i \underline{v}_i \tag{5.5}$$

and $\quad ||\nabla F|| = \sum \lambda_i^2 \xi_i^2. \qquad (5.6)$

It follows that if

$$||\nabla F|| = \sum \lambda_i^2 \xi_i^2 < e_o^2$$

then ξ_i is small for eigenvectors with large eigenvalues and can be large for eigenvectors with small eigenvalues
since

$\xi_i^2 < e_o^2 / \lambda_i^2$,

therefore $\quad |\Delta x_j| < \sum_i \frac{e_o}{\lambda_i} v_{ij} \qquad (5.7)$

It will be seen that this bound is quite consistent with the result that x_3 and x_4 can be determined accurately whilst x_1 and x_2 cannot. If $e_o = 10^{-5}$ this implies approx. upper bounds on the errors x_j of 2.3, 10^3, 10^{-4}, 10^{-3} respectively.

For this particular problem it was found that the states $X_1(o)$ and $X_2(o)$ could be estimated accurately if 10 measurements of both X_1 and X_3 were used. In this case the condition number reduces to $.85 \times 10^5$ which is still fairly large but the optimisation routine finds a satisfactory point in 11 iterations.

6. Estimation of initial states parameters

Let us now turn to the problem of estimating the parameters. If this is done in conjunction with the initial states, then although the problem is highly nonlinear we may assume that for small enough ϵ_o, quadratic analysis is sufficient. Considering our example having 4 states and 1 parameter then the Hessian can again be analysed by SVD giving 5 eigenvalues/eigenvectors λ, ξ, and by the analysis given above the error in the parameter x_5 is given by

$$|\Delta x_5| < \sum_i \frac{\epsilon_o}{\lambda_i} v_{i5} \tag{6.1}$$

when it is found by minimising F as a function of the 5 variables. It would be quite possible for this bound to be small when bounds on some initial states are large and vice versa.

The above analysis is not however particularly helpful if the approach being analysed is one in which only certain of the variables are used in the optimisation. If we now split the variables into a set x which is used in the optimisation and a set a determined in some other way, then at the minimum with respect to x we have

$$\frac{\partial f}{\partial x_i}(x,a) = 0 \qquad i=1, \ldots, n \tag{6.2}$$

If a is non optimal by an amount Δa then the resulting change in x is given by

$$\frac{\partial^2 f}{\partial x_i \partial x_j} \Delta x_j + \frac{\partial^2 f}{\partial x_i \partial a_k} \Delta a_k = 0 \tag{6.3}$$

and we see at once that the sensitivity of the optimised variables x to changes in the unoptimised variables depends on the matrices $\frac{\partial^2 f}{\partial x_i \partial x_j}$ and $\frac{\partial^2 f}{\partial x_i \partial a_k}$. In the case considered above of estimating 1 parameter in a 4 state system, if we represent the complete 5x5 Hessian by $H_{i,j}$ then by optimising over an increasing no of variables letting the parameter be x_5, we would be interested in the relative magnitudes of the following matrix structures

$$(1) \quad \begin{bmatrix} H_{55} \end{bmatrix} \qquad \begin{bmatrix} H_{51} & H_{52} & H_{53} & H_{54} \end{bmatrix}$$

$$(2) \quad \begin{bmatrix} H_{55} & H_{51} \\ H_{15} & H_{11} \end{bmatrix} \qquad \begin{bmatrix} H_{52} & H_{53} & H_{54} \\ H_{12} & H_{13} & H_{14} \end{bmatrix} \tag{6.4}$$

$$(3)\quad \begin{bmatrix} H_{55} & H_{51} & H_{52} \\ H_{15} & H_{11} & H_{12} \\ H_{25} & H_{21} & H_{22} \end{bmatrix} \qquad \begin{bmatrix} H_{53} & H_{54} \\ H_{13} & H_{14} \\ H_{23} & H_{24} \end{bmatrix} \quad \text{etc.}$$

For accurate results the choice must be made so that all the large elements are in the first matrix.

If a result is obtained by minimising a function with respect to a subset of variables only, (as in the L.S. approach) then likely error can be computed, if at the solution (x,a) the matrices

$\frac{\partial^2 f}{\partial x_i \partial x_j}$, $\frac{\partial f}{\partial a_k}$, $\frac{\partial^2 f}{\partial a_i \partial a_j}$, $\frac{\partial^2 f}{\partial x_i \partial a_k}$ are computed.

For then the equations

$$\begin{cases} \dfrac{\partial^2 f}{\partial x_i \partial x_j} \Delta x_j + \dfrac{\partial^2 f}{\partial x_i \partial a_k} \Delta a_k = 0 \\[2ex] \dfrac{\partial^2 f}{\partial a_l \partial x_j} \Delta x_j + \dfrac{\partial^2 f}{\partial a_l \partial a_k} \Delta a_k = -\dfrac{\partial f}{\partial a_l} \end{cases} \tag{6.5}$$

can be solved to indicate the magnitude in the errors Δx_j.

This approach can be used to analyse the performance of many algorithms that do not treat all the variables equally or introduce approximations to the likelihood function. It enables counter examples for which such algorithms have poor convergence properties or are very biassed to be easily constructed.

7. Conclusions

The performance of many system estimation algorithms can be understood and improved if they are viewed as optimisation algorithms applied to likelihood functions.

REFERENCES

1 Astrom, K.J. and Bohlin, T., "Numerical Identification of linear dynamic systems from normal operating records". In "Theory of Self-adaptive Control Systems" Editor P.H. Hammond. Plenum Press, 1966.

2 Astrom, K.J. and Eykoff, P., "System Identification - A Survey" Automatica, Vol. 7, pp. 123-162, 1972.

3 Clarke, D.W., "Generalized least squares estimation of parameters of a dynamic system." I.F.A.C. Symposium "Identification and Automatic Control Systems". Prague, 1967.

4 Dixon, L.C.W., "An automatic model adjustment technique". I.F.A.C. Symposium. "Identification and System Parameter Estimation". The Hague, 1973.

5 Dixon, L.C.W., "Numerical optimisation; A survey of the State of the Art". In "Software for Numerical Mathematics" Editor D.J. Evans, Academic press, 1974.

6 Dixon, L.C.W. and James, L., "On Stochastic Variable Metric Methods" presented at the I.M.A. Conference on Stochastic Optimisation, Oxford 1978.

7 Hastings James, R. and Sage, M.W., "Recursive Generalized Least squares procedure for on-line identification of process parameters" Proc. IEE. Vol. 116, 1969.

8 James, L., Private Communication, 1976.

9 James,L., "A recursive Variable Metric Method" in "Numerical Optimization of Dynamic Systems" Editors L.C.W. Dixon and G.P. Szegö, North Holland, 1980.

10 Ljung, L., "Theorems for the asymptotic analysis of recursive stochastic algorithms" Report 7522, Department of Automatic Control Lund,University, 1975.

11 Pearson, A.E., "Equation Error Identification with Model Disturbances Suppressed", this volume.

12 Soderström, T., "Converegence properties of the generalised least square method". Automatica, 1974.

13 Soderström, T., Ljung, L. and Gustavsson, I., "A comparative analysis of recursive identification methods". Report 7427, Department of Automatic Control, Lund University, 1974.

14 Wilkinson, J.H., "Some recent advances in numerical linear algebra". In "The State of the Art in Numerical Analysis" Editor D. Jacobs, Academic Press, 1977.

Numerical Techniques for Stochastic Systems
F. Archetti and M. Cugiani (eds.)
© North-Holland Publishing Company, 1980

EQUATION ERROR IDENTIFICATION WITH MODAL DISTURBANCES SUPPRESSED*

by
A.E. Pearson
Division of Engineering
and
Lefschetz Center for Dynamical Systems
Brown University
Providence, Rhode Island, U.S.A.

With disturbances modeled by arbitrary linear combinations of modal type signals, a formulation is presented of a finite time equation error parameter identification technique in which the unknown disturbance parameters are suppressed resulting in a function for least squares minimization depending solely on the system parameters. Computational aspects of the formulation are discussed for both separable and nonseparable system models, and the underlying computations in each case are effectively streamlined taking into account the assumed order of the disturbances model.

1. INTRODUCTION

As is well known, there are many facets to the application of the classical least squares method to system identification problems. These applications differ in the kinds of models and data sets assumed in the formulation of the problem, as well as differing in the numerical techniques for computing the least squares estimate. Recently, a least squares-equation error parameter identification technique has been introduced (Pearson (1979), Pearson & Chin (1979)) which differs from other known applications of least squares in a number of ways. This approach is characterized by the following attributes: (i) only input-output data is presumed to be given over a fixed finite time interval with no attempt to estimate unknown initial conditions. (ii) The basic equation error model is sufficiently general to include a variety of nonlinear, time varying, differential delay, possibly unstable, multivariable system models. (iii) Unknown disturbance inputs are modeled on the finite observation time interval by arbitrary linear combinations of modal type signals (see Note 1) possessing mode locations which are determined in the esti-

*This research was supported in part by the Air Force Office of Scientific Research under grant AFOSR-75 2793 and in part by the National Science Foundation under grant ENG78-25828.

mation procedure. (iv) The formulation is an exact technique in the sense that minimizing the derived equation error functional over the parameters leads to a unique zero value if the disturbance and systems models are correct and if the finite-time data is sufficiently rich in frequency content to avoid certain degeneracies. (v) In many cases the formulation leads to an explicitly defined function of the parameters which simplifies the computations significantly.

In this paper the formulation of the system identification problem posed in (Pearson (1979)) will be modified in a way that reveals the underlying least squares functional to be minimized over the system parameters unfettered by the parameters characterizing the disturbance modes. At the same time, this formulation will accrue significant benefits in streamlining and simplifying the computations needed to obtain the least squares functional from the observed input-output data. These benefits will be described in Section III following the formulation presented in Section II.

2. FORMULATION

The following forced Van der Pol equation will be used to illustrate the formulation:

$$\ddot{y}(t)-\omega_1[1-y^2(t)]\,\dot{y}(t)+\omega_2 y(t)=\omega_3[u(t-\omega_4)+d(t)] \tag{1}$$

where $[u(t), y(t)]$ is a given input-output pair observed over a finite time interval and $d(t)$ is an unknown forcing function disturbance. In differential operator notation with $D \triangleq \frac{d}{dt}$, this can be rearranged into the form

$$[D^2, D, 1]\begin{bmatrix} y(t) \\ [\frac{1}{3}y^3(t)-y(t)]\omega_1 \\ \omega_2 y(t)-\omega_3 u(t-\omega_4) \end{bmatrix} = \omega_3 d(t) \qquad \text{(see Note 2)} \tag{2}$$

where $\omega = (\omega_1, \omega_2, \omega_3, \omega_4)$ is a vector of system parameters to be determined by the identification procedure. Although the pure time delay parameter, ω_4, is unknown, it is assumed to be bounded above by a given number ω_4^{max}, i.e.

$$0 \leq \omega_4 \leq \omega_4^{max}$$

It is also assumed that the finite observation time interval is sufficiently long to include the unknown time delay, i.e. $[u(t), y(t)]$ is observed over a time interval $-\omega_4^{max} \leq t \leq t_1$ where $t_1 > 0$. The unknown disturbance $d(t)$ is modeled on this time interval by a linear homogeneous differential operator equation:

$$T(D,\delta)d(t) = \sum_{i=0}^{r} \delta_i D^{r-i} d(t) = 0$$
$$\delta_o \triangleq 1 \tag{3}$$

where the order r is assumed to be bounded by a preselected integer r_{max}, i.e. $0 \leq r \leq r_{max}$. Notice that the disturbances model (3) contains (2r) degrees of freedom for a specified integer r corresponding to the r arbitrary initial conditions and the r parameters $\delta = (\delta_1, \ldots \delta_r)$.

Operating on both sides of (2) with the differential operator $T(D,\delta)$ in (3) leads to an equation of the generic form

$$0 = [D^r R(D) \vdots D^{r-1} R(D) \vdots \cdots \vdots R(D)] \begin{bmatrix} \delta_o g(t,\omega) \\ \delta_1 g(t,\omega) \\ \vdots \\ \delta_r g(t,\omega) \end{bmatrix} \tag{4}$$

where in the case of the Van der Pol equation (1) $R(D)$ is the row vector

$$R(D) = \text{Row } (D^2, D, 1) \tag{5}$$

and $g(t,\omega)$ is the column vector function

$$g(t,\omega) = \begin{bmatrix} y(t) \\ [\frac{1}{3}y^3(t) - y(t)]\,\omega_1 \\ \omega_2 y(t) - \omega_3 u(t-\omega_4) \end{bmatrix} \tag{6}$$

In the general formulation $R(D)$ will be a known polynomial matrix in the differential operator $D = \frac{d}{dt}$ and $g(t,\omega)$ will be a given vector valued function of (i) the input-output data on a finite observation time interval, $t_o \leq t \leq t_1$, and (ii) a vector of (unknown) system parameters ω. With respect to smoothness, $g(t,\omega)$ is assumed to be piece-wise continuous in "t" and continuously differentiable in ω for each fixed "t".

An equation error function $z(t,\theta)$ is now defined for (4) by the solution to the linear differential operator equation

$$F(D)z(t,\theta) = [D^r R(D) \mid D^{r-1}R(D) \mid \cdots \mid R(D)] \begin{bmatrix} \delta_o g(t,\omega) \\ \delta_1 g(t,\omega) \\ \vdots \\ \delta_r g(t,\omega) \end{bmatrix} \tag{7}$$

where the parameter vector θ is the pair (δ,ω) and $\delta = (\delta_1 .. \delta_r)$ with $\delta_o = 1$ as indicated in (3). In the above $F(D)$ is a square non singular polynomial matrix chosen by the identifier so that $F^{-1}(s)$ $R(s)s^{r_{max}}$ is a proper transfer function matrix corresponding to the given rectangular polynomial matrix $R(D)$ and a preselected non-negative integer r_{max}. No other restrictions are placed on $F(s)$. However, a specific choice for $F(s)$ is suggested at the close of Section III which has some useful properties pertinent to the formulation at hand.

Equation (7) is the starting point for the particular least squares formulation to be described below. This description will be couched in terms of the differential operator $D=\frac{d}{dt}$, although all the results obtained to date are equally valid for discrete time systems with "D" replaced by "q" - the forward difference operator - and with appropriate interpretation of various differentials and integrals by differences and sums. The earliest version of this formulation for a class of linear systems was reported in Pearson (1976). The basic model (4) and the differential operator equation (7) for the equation error function is a slight modification of the most recent formulation given in Pearson (1979), Pearson & Chin (1979) which has certain computational advantages to be discussed later.

Some digression concerning the equation error model (7) is offered at this point before describing the least squares technique for obtaining an estimate of the pair $\theta = (\delta,\omega)$. Equation (7) pertains to any system model which can be rearranged in the form

$$R(D)g(t,\omega) = S(D,\omega)d(t) \tag{8}$$

where $S(D,\omega)$ is some polynomial matrix in D, possibly depending on the system parameter vector ω but independent of "t," and the unknown disturbances $d(t)$ are modeled by (3) on the finite observation time interval. As in the Van der Pol equation example above, a simple rewriting of the differential or difference equation models utilized by a variety of alternative approaches to parameter identification will result in a model of the form (8). This is true for all linear differential and differential-delay systems, apart from the disturbances model, and has been shown to be the case for a variety of other nonlinear time varying differential systems Pearson (1979), Pearson & Chin (1979) . If the disturbances do not enter linearly in the basic system model, then this approach can be construed as assuming that the equation error for the system can be approximated by a sum of modal type signals as represented by the solution to (3) for some a priori integer $r \leq r_{max}$. Hence, in contrasting this approach with other well known parameter identification techniques, such as Maximum Likelihood [Aström & Bohlin (1966)], a significant difference lies in the model for the disturbances and the assumed data set, in addition to the fact that this approach is "one shot" identification versus methods which are iterative in time. While Maximum Likelihood and other statistical methods of identification [Eykhoff (1974), Mendel (1973)] represent the disturbances by stochastic processes with underlying Markov process representations, whether continuous time or discrete time, the model for unknown disturbance signals in this approach is the deterministic homogeneous differential operator equation (3) where the δ_i's and initial conditions are completely arbitrary. Actually, this model can be regarded as generating a stochastic process if one insists on stochastic models for disturbances, where the δ_i's and initial conditions comprise $2r$ independent random variables with essentially infinite variances. While this model may not be suitable for measurement noise models, which are frequently modeled as "white" noise, the approach assumes only input-output data is available (no state estimation is attempted) so that potentially noisy sensors, such as accelerometers, might be avoided.

On the other hand, this model is thought to be quite suitable for process input noise such as wind gust effects, or bending modes ignored in the basic model for aerospace applications, since the data set is presumed to consist of input-output data observed on a finite observation time interval $0 \le t \le t_1$ (see Note 3). Thus, the shorter the $[0,t_1]$ interval, the more reasonable is the above disturbances model for a modest value of r, although the $[0,t_1]$ interval must be long enough to be able to glean the information about the unknown parameters. As discussed in Pearson (1979), Pearson & Chin (1979), the $[0,t_1]$ interval can be surprisingly short in many cases, i.e. on the order of the dominant system time constant, or less. This is based on simulation studies carried out for the nonlinear Van der Pol equation [Pearson (1979)] and a MIMO helicopter system model [Pearson & Chin (1979)], as well as other simulation studies.

Finally, the least squares nature of the parameter identification technique mitigates the effects of modeling errors in either (3) or (8), such as might be due to "white" measurement noise, provided the energy content of the spurious signals is small compared to the correctly modeled signals (roughly less than ten per cent in an integral squared sense based on simulation studies).

A significant point of departure for this approach to least squares parameter identification and other known approaches is the next step involving the projection of the equation error function $z(t,\theta)$ down into a subspace devoid of all initial condition responses of the solution to (7) on $[0,t_1]$. This is accomplished by means of the "annihilating filter", H, defined by

$$\tilde{z}(t) = H(z) = z(t) - Ce^{At}W^{-1}\int_0^{t_1} e^{A'\tau}C'z(\tau)d\tau \qquad 0 \le t \le t_1 \tag{9}$$

where (A,C) is an observable pair for the differential operator equation $F(D)z(t)=0$ and W^{-1} is the inverse of the observability Gramian

$$W = \int_0^{t_1} e^{A't}C'Ce^{At}dt. \tag{10}$$

The domain of H is the Hilbert space Z of square integrable functions for all solutions to (7) on $[0,t_1]$, while its range is contained in the subspace $(Z-Z_o)$ where $Z_o = \{Ce^{At}x_o,\ 0 \le t \le t_1 : x_o \varepsilon R^{\bar{n}}\}$ and $\bar{n}$ is the dimension of the state space for a minimal realization of $F(D)z(t)=0$. The fact that H in (9) is a projection operator is easily verified by the properties: $H(H)=H$ and $\bar{H}=H$ (the self-adjoint property). Note that $H(z)=0$ for all $z\varepsilon Z_o$.

Since the initial condition response for the solution to (7) is arbitrary and has no physical significance, the solution $z(t,\theta)$ is projected down into the subspace $(Z-Z_o)$ via H, thus annihilating the initial condition response on $[0,t_1]$. Then the inner product norm of $\tilde{z}(t,\theta)=H[z(t,\theta)]$ is formed to define the functional $J_1(\theta)$ for the least squares minimization problem:

$$J_1(\theta) = <\tilde{z}(\theta), \tilde{z}(\theta) = \int_o^{t_1} \tilde{z}'(t,\theta)\tilde{z}(t,\theta)dt. \tag{11}$$

If $\mu(t,\theta)$ denotes the zero state solution to (7) for any particular $\theta=(\delta,\omega)$ i.e. $\mu(t,\theta)$ satisfies

$$F(D)\mu(t,\theta)=[D^r R(D) \vdots \cdots \vdots R(D)] \begin{bmatrix} \delta_o g(t,\omega) \\ \vdots \\ \delta_r g(t,\omega) \end{bmatrix} \begin{cases} \text{zero} \\ \text{Initial} \\ \text{Conditions} \end{cases} \tag{12}$$

then $J_1(\theta)$ can be given the more explicit representation

$$J_1(\theta) = \int_o^{t_1} \mu'(t,\theta)\mu(t,\theta)dt - \nu'(\theta)W^{-1}\nu(\theta) \tag{13}$$

where $\nu(\theta)$ is defined in terms of $\mu(t,\theta)$ via

$$\nu(\theta) = \int_o^{t_1} e^{A't}C'\mu(t,\theta)dt \qquad \text{(see Note 4)} \tag{14}$$

Since J_1 is a positive definite function, minimizing (13) over θ provides a least squares estimate of the vector $\theta=(\delta,\omega)$ in the sense that this procedure minimizes the norm of the projection of the equation error function in the function space $(Z-Z_o)$. If the original models (3) and (8) are correct, then a minimizing value, θ^*, for this problem is such that $J_1(\theta^*)=0$.

An important subclass of the system model (8) is the class for which the vector valued function $g(t,\omega)$ admits to the representation

$$g(t,\omega) = U(t)h(\omega) \tag{15}$$

where $U(t)$ is a matrix valued function of the observed input-output data on $[0,t_1]$ and $h(\cdot)$ is a given vector valued function of the system parameters, ω, which is presumed to be continuously differentiable. In such a case, the system is said to be "separable in the parameters". Separability (defined on p. 74 of Pearson (1979)) is a basic property of any system model proposed for identification in that a given model either is, or is not, separable in the parameters (see Note 5).

For example, the Van der Pol equation model (1)-(2) is separable in the parameters if the time delay ω_4 is a known constant. In this case the vector $g(t,\omega)$ in (6) factors as follows:

$$g(t,\omega) = \begin{bmatrix} y(t) & 0 & 0 & 0 \\ 0 & \frac{1}{3}y^3(t)-y(t) & 0 & 0 \\ 0 & 0 & y(t) & -u(t-\omega_4) \end{bmatrix} \begin{bmatrix} 1 \\ \omega_1 \\ \omega_2 \\ \omega_3 \end{bmatrix} \tag{16}$$

$$= U(t)h(\omega)$$

where $h(\omega)=\mathrm{Col}(1,\omega_1,\omega_2,\omega_3)$. All linear differential systems are separable in the parameters as are a variety of well known nonlinear models, such as the Duffing and Hammerstein equations and a class of bilinear systems. Examples of nonseparable models include the Van der Pol equation (1) above with unknown time lag ω_4 and the time varying Mathieu equation with unknown frequency for the sinusoidal coefficient. (See Pearson (1979) for details and additional examples).

The implication of the separability property is that $J_1(\theta)$ in (13) reduces to an <u>explicitly defined</u> function of the parameters. Specifically, there exists a partitioned symmetric non-negative definite matrix Ω, computable by causal (integral) operations on the given data on $[0,t_1]$, such that $J_1(\theta)$ in (13) has the explicit representation

$$J_2(\theta) = [h'(\omega), \delta' H'(\omega)] \begin{bmatrix} \Omega_{oo} & \Omega_{od} \\ \Omega_{do} & \Omega_{dd} \end{bmatrix} \begin{bmatrix} h(\omega) \\ H(\omega)\delta \end{bmatrix} \tag{17}$$

where $H(\omega)$ is defined by

$$H(\omega) = \underbrace{\begin{bmatrix} h(\omega) & 0 & \cdot & 0 \\ 0 & h(\omega) & & \cdot \\ \cdot & & \ddots & 0 \\ 0 & 0 & \cdot & h(\omega) \end{bmatrix}}_{r \text{ columns}} \tag{18}$$

The matrix Ω is the Gramian:

$$\Omega = \int_0^{t_1} Y'(t)Y(t)dt - N'W^{-1}N \qquad \text{(see Note 6)} \tag{19}$$

where $Y(t)$ is obtained from the zero state solution to the matrix differential equation

$$F(D)Y(t) = [D^r R(D) \,\vdots\, D^{r-1}R(D) \,\vdots\, \cdots \,\vdots\, R(D)]\; U(t) \tag{20}$$

and N is defined by

$$N = \int_0^{t_1} e^{A't} C'Y(t)dt. \tag{21}$$

The derivation of (17)-(21) from (13) follows upon substituting the general solution of (12) into (13) and utilizing the separability property (15). The matrix Ω is effectively a time correlation matrix with some bias removal terms (the matrix $N'W^{-1}N$) which arise from the application of the annihilating filter H. The advantage of working with $J_2(\theta)$ rather than $J_1(\theta)$, even though they are equivalent, is obvious since it is an explicit function of the parameters with the positive definte property $J_2(\theta) \geq 0$. Once the matrix Ω has been computed, any suitable hill climbing technique, such as Newton's method, can be utilized on $J_2(\theta)$ without any further integrations involving the data on $[0, t_1]$. If the system and disturbances models are correct, then a minimizing value θ^* is such that $J_2(\theta^*) = 0$; otherwise, Min $J_2(\theta)$ leads to a least squares estimate of θ as before.

In addition to the separability property, another basic notion which has arisen in this approach is that of "partially decoupled parametrized models" for identification of multivariable systems. To indicate this abstractly, suppose a vector valued input-output relation, $S(u,y;\theta) = 0$, admits to a partitioning $(S_a, S_b, S_c, ..)$ such that $S_a(u,y;\theta_a) = 0$, $S_b(u,y;\theta_a,\theta_b) = 0$, $S_c(u,y;\theta_a,\theta_b,\theta_c,...) = 0$

etc. where (u,y) is the input-output pair and $\theta=(\theta_a,\theta_b,\theta_c,\ldots)$ is a partitioned parameter vector which is to be estimated for identification purposes. Physical models for multivariable systems generally exhibit such a partial decoupling with respect to the parameters characterizing the system, as has been shown in Pearson & Chin (1979) . (The examples delineated in Pearson & Chin (1979) include the MIMO longitudinal dynamics of a helicopter and a STOL aircraft, and a MIMO bilinear system model for anti-tumor drug kinetics). It has been shown in Pearson & Chin (1979) how the above least squares-equation error approach can be used to specify a finite sequence of functional minimizations, $J_\alpha(\theta_\alpha)$, $\alpha=a,b,c,\ldots,$ where J_α depends on the values $\hat{\theta}_\beta$ of the preceding minimizations, $a\leq\beta\leq\alpha-1$. This sequence is used in lieu of a single high dimensional minimization of $J(\theta)$ over $\theta=(\theta_a,\theta_b,\ldots)$ in order to take advantage of the partial decoupling. In practical terms, minimizing the finite sequence of functions results in the possibility of accomodating a higher order disturbances model (3), and hence greater flexibility in the modeling of such disturbances, for a fixed observation time interval $[0,t_1]$. Equivalently, taking advantage of the partial decoupling results in the possibility of achieving satisfactory accuracy in the identified parameters for a shorter time interval $[0,t_1]$ of data relative to that which could be obtained for the overall parameter minimization problem. As shown in [Pearson & Chin (1979)], these possibilities were realized in the case of simulations carried out for the MIMO longitudinal dynamics of a helicopter. As mentioned earlier, a small amount of "white" measurement noise could be superimposed on the data without materially affecting the results due to the least squares nature of the identification technique.

3. LEAST SQUARES FUNCTION FOR THE SYSTEM PARAMETERS

Since the disturbance parameters, $\delta=(\delta_1\ldots\delta_r)$, for a selected order r in (3) enter linearly into the equation error function, as seen by (7) and (12), these parameters enter quadratically in the least squares function of the overall parameter vector $\theta=(\delta,\omega)$. This is immediately evident by inspection of $J_2(\theta)$ in (17) for the separable case, although it is also true for $J_1(\theta)$ in (13). Hence, the necessary conditions for a minimal value of either $J_1(\theta)$ or

$J_2(\theta)$, i.e. the vanishing of the gradients

$$\frac{\partial J}{\partial \delta} = 0 \quad \text{and} \quad \frac{\partial J}{\partial \omega} = 0, \tag{22}$$

can be solved for δ as a function of ω under suitable nondegeneracy conditions.

Considering the separable case first, the first of the necessary conditions in (22) for minimizing $J_2(\theta)$ in (17) yields

$$\hat{\delta} = -[H'(\omega)\Omega_{dd}H(\omega)]^{-1}H'(\omega)\Omega_{do}h(\omega) \tag{23}$$

assuming the indicated inverse exists functionally in ω (see Note 7). Substituting (23) into (17) results in the following function:

$$J_3(\omega) = h'(\omega)[\Omega_{oo} - \Omega_{od}H(\omega)[H'(\omega)\Omega_{dd}H(\omega)]^{-1}H'(\omega)\Omega_{do}]h(\omega). \tag{24}$$

The function $J_3(\omega)$ displays the essential aspects of the least squares minimization problem for obtaining the system parameters ω for separable models when modeling disturbances according to (3). In the case of the Van der Pol example (1) with a known time delay parameter ω_4, the vector function $h(\omega)$ is given by (16) in which the system parameters, $\omega = (\omega_1, \omega_2, \omega_3)$, enter linearly. In general, the resulting $J_3(\omega)$ function, though positive definite, is nonlinear, nonquadratic and not necessarily convex in ω. Although not explicitly present, the effect of the disturbance parameters is manifest in the inverse of the $r \times r$ matrix $H'(\omega)\Omega_{dd}H(\omega)$, an inversion which can be written out functionally only for low orders of the model (3). Suppressing the notational dependence on ω, $J_3(\omega)$ for increasing orders of the disturbance model (3) is as follows:

$$r = 0: \quad J_3 = h'\Omega_{oo}h$$

$$r = 1: \quad J_3 = h'\Omega_{oo}h - \frac{h'\Omega_{od}hh'\Omega_{do}h}{h'\Omega_{dd}h}$$

$$\begin{aligned} r = 2: \quad J_3 = {} & h'\Omega_{oo}h - [h'\Omega_{11}hh'\Omega_{22}h - h'\Omega_{21}hh'\Omega_{12}h]^{-1}[(h'\Omega_{01}hh'\Omega_{22}h \\ & - h'\Omega_{02}hh'\Omega_{21}h)h'\Omega'_{01}h + (h'\Omega_{02}hh'\Omega_{11}h \\ & - h'\Omega_{01}hh'\Omega_{12}h)h'\Omega'_{02}h] \end{aligned}$$

where the subscripted matrices in the latter expression are defined by the following partitions on the matrices Ω_{od} and Ω_{dd} in (17):

$$\Omega_{od} = [\Omega_{01}, \Omega_{02}], \qquad \Omega_{dd} = \begin{bmatrix} \Omega_{11} & \Omega_{12} \\ \Omega_{21} & \Omega_{22} \end{bmatrix}.$$

Consider next the computations underlying the partitions of the matrix Ω in (17). Corresponding to a given order r of the disturbance model (3), let the zero state matrix solution to (20) be partitioned in accordance with the right side, i.e.

$$Y(t) = [\, Y_o(t),\ Y_1(t),\ \ldots\ Y_r(t)] , \qquad 0 \le t \le t_1. \tag{25}$$

Also, partition N in (21) according to

$$N = [N_o, \ldots, N_r] = \int_o^{t_1} e^{A't} C' [Y_o(t), \ldots, Y_r(t)]\, dt. \tag{26}$$

Then the matrix Ω partitioned analogously into $(r+1)^2$ blocks has the block representation:

$$\Omega = [\Omega_{ij}], \qquad \Omega_{ij} = \int_0^{t_1} Y_i'(t) Y_j(t)\, dt - N_i' W^{-1} N_j \tag{27}$$

$$0 \le i \le r, \qquad 0 \le j \le r$$

The leading block in (17), Ω_{oo}, is given directly from (27), while $\Omega_{od} = \Omega'_{do}$ and Ω_{dd} in (17) possess sub-block matrices according to

$$\Omega_{od} = [\, \Omega_{01}, \ \ldots \ \Omega_{or}\,]$$

and

$$\Omega_{dd} = \begin{bmatrix} \Omega_{11} & \cdot\ \cdot & \Omega_{1r} \\ \vdots & & \vdots \\ \Omega_{r1} & \cdot\ \cdot & \Omega_{rr} \end{bmatrix}$$

where Ω_{ij} is given by (27).

Now it is clear from (20) that $Y_{i+1}(t)$ is a pure integration of its predecessor $Y_i(t)$ in the block matrix (25), i.e. $Y_{i+1}=D^{-1}Y_i(t)$, $0\leq t\leq t_1$, $0\leq i\leq r-1$. A similar relation can be determined for the blocks of the matrices comprising the bias removal term $N'W^{-1}N$. Let a matrix function $Z(t_1)$ be defined in terms of N by

$$Z(t_1) = e^{-A't_1}N = e^{-A't_1}\int_0^{t_1} e^{A'\tau}C'Y(\tau)d\tau \tag{28}$$

with a similar partitioning, viz.

$$Z = [Z_o,\ldots, Z_r] = e^{-A't_1}[N_o,\ldots,N_r]$$

Then the bias removal term $N'W^{-1}N$ in (19) is equivalently represented by

$$N'W^{-1}N = Z'(t_1)e^{At_1}W^{-1}e^{A't_1}Z(t_1) = Z'(t_1)\tilde{W}^{-1}Z(t_1) \tag{29}$$

where $\tilde{W}$ is given by (see Eq. (10))

$$\tilde{W} = e^{-A't_1}We^{-At_1} = \int_0^{t_1} e^{-A't}C'Ce^{-At}dt. \quad \text{(see Note 8)} \tag{30}$$

In view of (26) and (28), each block Z_1 is given by

$$Z_i(t_1) = e^{-A't_1}\int_0^{t_1} e^{A'\tau}C'Y_i(\tau)d\tau$$

which satisfies the matrix differential equation

$$\dot{Z}_i(t) = -A'Z_i(t) + C'Y_i(t), \quad Z_i(0) = 0. \tag{31}$$

Since $Y_{i+1}(t)$ is a pure integration of $Y_i(t)$ on $[0,t_1]$ with zero initial conditions, it is easily concluded from (31) that the blocks of the $Z(t)$ matrix function satisfy a similar relation, i.e.

$$Z_{i+1}(t) = D^{-1}Z_i(t), \qquad 0\leq t\leq t_1 \tag{32}$$

Let (A,B,C,E) be a minimal realization for the transfer function matrix $F^{-1}(s)R(s)s^r$, i.e.

$$C(sI-A)^{-1}B + E = F^{-1}(s)R(s)s^r. \tag{33}$$

Then the leading block of the matrix function $Y(t)$ in (20) is generated by the matrix differential equation

$$\begin{aligned} \dot{X}_o(t) &= AX_o(t) + BU(t), \quad X_o(0) = 0 \\ Y_o(t) &= CX_o(t) + EU(t), \quad 0 \leq t \leq t_1 \end{aligned} \tag{34}$$

and each succeeding block in (25) follows from

$$Y_{i+1}(t) = D^{-1}Y_i(t) = \int_o^t Y_i(\tau)d\tau, \qquad 0 \leq t \leq t_1. \tag{35}$$

However, computing $[Y_i(t), Z_i(t)]$, $1 \leq i \leq r$, from (32) and (35) can be quite inefficient because these pure integrations can be transferred to the data matrix $U(t)$ which is generally quite sparse. (For example, see the $U(t)$ matrix in (16)). These considerations together with the results of this section are summarized in the following statement. A related statement concerning the uniqueness of a least squares estimate for $\text{Min } J_3(\omega)$ is contained in Appendix B.

Assertion. Let (A,B,C,E) be a minimal realization for the transfer matrix $F^{-1}(s)R(s)S^r$, i.e. (33) is satisfied with (A,B) and (A,C) controllable and observable pairs respectively, with the further stipulation that $F(s)$ has been chosen by the identifier so that $\det A \neq 0$ (no zero eigenvalues for A). Then a least squares estimate of the system parameter vector ω for the basic differential operator model (8) in the separable case, i.e., with (15) satisfied, and with modal type disturbances $d(t)$ modeled by (3), is obtained by minimizing $J_3(\omega)$ in (24). The matrix pairs $[Y_i(t), Z_i(t_1)]$, $0 \leq i \leq r$, which underlie the blocks of the time correlation matrix (27) with the bias removal terms expressed as in (29) are efficiently determined from the zero state solutions to the following matrix differential equations:

For $i = 0$:

$$\begin{aligned} \dot{X}_o(t) &= AX_o(t) + BU(t) \\ Y_o(t) &= CX_o(t) + EU(t) \\ \dot{Z}_o(t) &= -A'Z_o(t) + C'Y_o(t) \end{aligned} \tag{36}$$

$$\text{For } 1\le i\le r: \quad \begin{aligned} X_i(t) &= A^{-1}[X_{i-1}(t) - BD^{-i}U(t)] \\ Y_i(t) &= CX_i(t) + ED^{-i}U(t) \\ Z_i(t) &= (-A')^{-1}[Z_{i-1}(t) - C'Y_i(t)] \end{aligned} \tag{37}$$

where D^{-i} denotes the operation of i-fold pure integration with zero initial conditions. Moreover, the off diagonal matrix blocks in (27) involving the cross correlation between $Y_i(t)$ and $Y_j(t)$ can be obtained iteratively from the diagonal blocks via

$$\int_0^{t_1} Y_i'(t)Y_{i+j}(t)dt = Y_i'(t_1)Y_{i+j}(t_1) - \int_0^{t_1} Y_i'(t)Y_{i+j-1}(t)dt. \tag{38}$$

Proof. Since the function $J_3(\omega)$ for the least squares minimization problem has already been derived together with the relations (36) for computing the leading blocks in the $[Y(t),Z(t_1)]$ matrix pair, it remains to establish the relations (37) for the succeeding blocks of this matrix pair. The third relation in (37) follows immediately from (31) and (32) under the assumption that $\det A \neq 0$, since (31) implies

$$\begin{aligned} Z_i(t) &= (-A')^{-1}\dot{Z}_i(t) - C'Y_i(t) \\ &= (-A')^{-1}DD^{-1}Z_{i-1}(t) - C'Y_i(t) \\ &= (-A')^{-1}Z_{i-1}(t) - C'Y_i(t). \end{aligned}$$

The second relation in (37) follows if $X_i(t)$ is defined iteratively from $X_{i+1}(t) = D^{-1}X_i(t)$, $0\le i\le r-1$, whereupon the first relation in (37) is seen to be valid from (34). Equation 38 follows from integration by parts using (35).

Relations (37) represent a significant saving in computation not only because the data matrix $U(t)$ is generally sparse, but also because the number of distinct time functions in $U(t)$ is less than the number of nonzero entries. For example, the data matrix $U(t)$ in (16), which applies to the forced Van der Pol equation (1) with a known time delay parameter ω_4, contains as distinct elements to be integrated in (37) the time functions $\{u(t-\omega_4), y(t), y^3(t)\}$. Assuming $F(s)$ is chosen of degree $(r+2)$, which is the lowest degree so that $F^{-1}(s)R(s)s^r$ is proper with $R(s)$ given by (5), a simple counting of integrations reveals that obtaining $[Y_i(t), Z_i(t_1)]$, $1\le i\le r$, from (32) and (35) requires $(12r+4r^2)$ integrations versus $3r$ integrations utilizing the relations (37). This is a significant saving even for low orders of r. Analogous savings stem from (38).

In a similar vein, the disturbance parameters δ can be eliminated from $J_1(\theta)=J_1(\delta,\omega)$ in (13), which applies to nonseparable models. Since $\delta_o=1$ in (12) $\mu(t,\theta)$ can be expressed as

$$\mu(t,\theta) = M_o(t,\omega) + M(t,\omega)\delta$$

where $M_o(t,\omega)$ and $M(t,\omega)$ denote the zero state vector and matrix solutions to the differential operator equations

$$F(D)M_o(t,\omega) = R(D)D^r g(t,\omega) \tag{39}$$

and

$$F(D)M(t,\omega) = [D^{r-1}R(D),..R(D)]\ G(t,\omega) \tag{40}$$

respectively, and where the r column matrix $G(t,\omega)$ is defined by

$$G(t,\omega) = \begin{bmatrix} g(t,\omega) & 0 & .. & 0 \\ 0 & g(t,\omega) & & 0 \\ . & . \quad . & & . \\ . & . \quad\quad . & & . \\ 0 & 0 & .. & g(t,\omega) \end{bmatrix} \tag{41}$$

Solving the linear algebraic equations, $\frac{\partial J_1}{\partial \delta}=0$, for $\hat{\delta}$ and substituting this relation back into (13) leads to the expression

$$\begin{aligned} J_4(\omega) = & \int_o^{t_1} M_o'(t,\omega)M_o(t,\omega)dt - N_o'(\omega)W^{-1}N(\omega) \\ & - [\int_o^{t_1} M_o'(t,\omega)M(t,\omega)dt - N_o'(\omega)W^{-1}N(\omega)]\ [\int_o^{t_1} M'(t,\omega)M(t,\omega)dt \\ & - N'(\omega)W^{-1}N(\omega)]^{-1}\ [\int_o^{t_1} M'(t,\omega)M_o(t,\omega)dt - N'(\omega)W^{-1}N_o(\omega)] \end{aligned} \tag{42}$$

where the quantities $(N_o(\omega), N(\omega))$ in the bias removal terms are given by

$$N_o(\omega) = \int_o^{t_1} e^{A'\tau}C'M_o(\tau,\omega)d\tau\ , \qquad N(\omega) = \int_o^{t_1} e^{A'\tau}C'M(\tau,\omega)d\tau. \tag{43}$$

As expected, the final expression for the nonseparable case, $J_4(\omega)$, is more complex and only defined implicitly for each particular value of ω. Nevertheless, simplifying relations for taking advantage of the pure integrations between successive blocks in $M(t,\omega)$, as implied by (39) and (40), can be derived analogous to those for the separable case, (36) and (37). Based on the previous derivations, these relations will take the following forms in connection with the partitionings

$$M(t) = [M_1(t),..(M_r(t)] \quad \text{and} \quad N(\omega) = [N_1(\omega),..N_r(\omega)],$$

i.e. consistent with the partitioning on the right side of (40) and (43). Assuming the same realization as in (33):

$$\begin{aligned} i = 0: \quad & \dot{X}_o(t,\omega) = AX_o(t,\omega) + Bg(t,\omega) \\ & M_o(t,\omega) = CX_o(t,\omega) + Eg(t,\omega) \\ & \dot{Z}_o(t,\omega) = -A'Z_o(t,\omega) + C'M_o(t,\omega) \end{aligned} \tag{44}$$

$$\begin{aligned} 1 \le i \le r: \quad & X_i(t,\omega) = A^{-1}[X_{i-1}(t,\omega) - BD^{-i}g(t,\omega)] \\ & M_i(t,\omega) = CX_i(t,\omega) + ED^{-i}g(t,\omega) \\ & Z_i(t,\omega) = (-A')^{-1}[Z_{i-1}(t,\omega) - C'M_i(t,\omega)] \ . \end{aligned} \tag{45}$$

In utilizing the above relations, the bias removal terms in (42) are modified analogous to the previous, viz.

$$\begin{aligned} & N_o'(\omega)W^{-1}N(\omega) = Z_o'(t_1,\omega)\tilde{W}^{-1}Z(t_1,\omega) \\ & N'(\omega)W^{-1}N(\omega) = Z'(t_1,\omega)\tilde{W}^{-1}Z(t_1,\omega) \end{aligned} \tag{46}$$

where $\tilde{W}$ is the observability Gramian for the pair (-A,C) (see Eq. (30)). Note that the quantities $\{X_i, M_i, Z_i\}$ are vector valued functions in (44) and (45) as opposed to the matrix valued functions in (36) and (37); also, the differential equations and pure integrations in (44) and (45) must be integrated anew (from the zero state) for each iterate in the sequence $\{\omega(n)\}$, n=1,2, ..., constructed by an appropriate hill climbing technique for minimizing $J_4(\omega)$.

As a final computational consideration, the following choice in the polynomial matrix $F(s)$ has a number of useful properties:

$$F(s) = \prod_{k=1}^{m} (s^2+k^2\omega_o^2)I, \qquad \omega_o \triangleq \frac{2\pi}{t_1} \tag{47}$$

where the integer m is chosen sufficiently large that $F^{-1}(s)R(s)s^r$ is proper. The fundamental solution to the homogeneous equation $F(D)z(t)=0$ involves the functions $\{\sin k\omega_o t, \cos k\omega_o t\}$, $k=1,2,..m$, which are orthogonal over the observation time interval $[0,t_1]$. Hence the W matrix in (10) is diagonal, as is the $\tilde{W}$ matrix in (30), i.e.

$$W = \tilde{W} = \frac{t_1}{2} I. \tag{48}$$

Notice that the resonance frequencies of the filter $F^{-1}(s)$ in (47) coincide with the null frequencies of the annihilating filter H in (9) so that the composite filter $HF^{-1}(D)R(D)D^r$ tends to preserve the useful information in the data at all frequencies.

As a further consideration for separable models, if the observation time interval is not truly fixed but is nonetheless finite, and if $[0,t_1]$ is the smallest time interval over which the parameters could be reasonably expected to be determined from the given input-output data, then a useful procedure to follow would be to check the definiteness of the leading block Ω_{oo} in the least squares $J_3(\omega)$ in (24) at integral multiples of t_1, i.e. at $\bar{t}_1=kt_1$, $k=1,2, \ldots$. The matrices W and $\tilde{W}$ will remain diagonal over such time intervals with the values

$$W = \tilde{W} = \frac{kt_1}{2}, \qquad k = 1,2,...$$

Hence the inversion, $\tilde{W}^{-1}$, required for the bias removal term (29) is trivial for any $k\geq 1$. If such a choice in the length of the observation time interval is flexible, then this procedure enhances the possibility that $J_3(\omega)$ will not be ill conditioned due to insufficient data. The definiteness of Ω_{oo} might be checked indirectly simply by the magnitude of its diagonal elements. As to the choice in t_1, some fraction of the longest expected system time constant is suggested, such as one half or one quarter. This is based on the previous simulation studies reported in Pearson (1979), Pearson & Chin (1979).

4. CONCLUSIONS

Modifications in an equation error formulation for parameter identification with modal type disturbances has resulted in a function for least squares minimization which is solely a function of the system parameters ω. The functions $J_3(\omega)$ and $J_4(\omega)$ for separable and nonseparable models can be used as a basis for any suitable hill climbing technique, although $J_4(\omega)$ in (42) for the nonseparable case is defined only implicitly through the integration of the vector differential and pure integration equations (44) and (45) for each value of the parameter vector ω. In the case of models which are separable in the parameters, the matrix differential and pure integration equations of (36) and (37) need to be integrated only once given the observed input-output data on a finite time interval, after which the function $J_3(\omega)$ in (24) is explicitly defined. In either case, the relations (36) and (37), or (44) and (45), provide a significant saving in the underlying computations over that which would be required by the previous formulation of Pearson (Jan. 1979) for a given order r of the disturbances model (3).

The question as to which type of hill climbing technique might be best for a particular class of system models, e.g. when $h(\omega)$ is linearly dependent on ω as in (16), has not been addressed here. Also, the question of approximations to the underlying computations has not been broached, nor has the question of a simplifying procedure for considering various orders, r, of the disturbances model (3). Considering the latter question, simulation results in Pearson (1979), Pearson & Chin (1979) indicate that an overordered disturbances model (r too large in (3)) is not detrimental to obtaining accurate estimates of the system parameters ω, though it is wasteful of computation. However, an underordered disturbance model can result in poor estimates of ω if the energy content of the deviated disturbance signals exceeds approximately ten per cent of the energy content of the assumed modal representation.

In a related context, a two stage deterministic filter has recently been devised which can separate two sets of finite time modal signals when presented with their arbitrary mixed sum [Pearson & Mocenigo (1979)]. This filter utilizes the same projection operator as in (9) while estimating the modes present in a noise signal modeled as in (3).

Appendix A: Derivation of Equation (13)

Using operator notation, $J_1(\theta)$ in (11) is given by

$$J_1(\theta) = \langle H(\mu), H(\mu)\rangle$$

since $\tilde{z} = H(\mu)$. As a result of the fact that H is self adjoint and a projection,

$$J_1(\theta) = \langle\mu, H^2(\mu)\rangle$$

$$= \mu, H(\mu)\rangle$$

Now H can be decomposed as

$$H = I - W$$

where I is the identity operator and W is defined by (cf. Eq. (9))

$$W(z) = Ce^{At}W^{-1}\int_o^{t_1} e^{A'\tau}C'z(\tau)d\tau, \qquad 0 \leq t \leq t_1 .$$

Hence

$$J_1(\theta) = \langle\mu,(I-W)(\mu)\rangle = \langle\mu,\mu\rangle - \langle\mu,W(\mu)\rangle.$$

But $\langle\mu,W(\mu)\rangle = \nu' W^{-1}\nu$ where ν is given by (14), so that (13) follows directly from the above equation.

Appendix B: Uniqueness for the Separable Case

Suppose a value of $\theta = (\delta,\omega)$ exists such that $J_2(\theta)$ in (17) vanishes for $\theta = \theta^*$. The the projected equation error function, $\tilde{z}(t,\theta^*)$, is identically zero on $[0,t_1]$ and any other value of $\theta \neq \theta^*$ is such that

$$\tilde{z}(t,\theta) - \tilde{z}(t,\theta^*) = \tilde{Y}_o(t)[h(\omega)-h(\omega^*)] + \sum_{i=1}^{r} \tilde{Y}_i(t)[h(\omega)\delta_i - h(\omega^*)\delta_i^*]. \tag{49}$$

Let the vector valued function $h(\omega)$ in (15) have the representation

$$h(\omega) = \begin{bmatrix} 1 \\ f(\omega) \end{bmatrix}, \tag{50}$$

i.e. unity in the first entry, where (without loss of generality) $f(\omega)$ is assumed to be single valued. That is to say,

$$f(\omega) = f(\omega^*) \quad \text{iff} \quad \omega = \omega^* \tag{51}$$

for all ω and ω^*. Using the representation (50), $\tilde{z}(t,\theta)$ in (49) can be expressed as

$$\begin{aligned}\tilde{z}(t,\theta) = {} & \tilde{Y}_{of}(t)[f(\omega)-f(\omega^*)] + \sum_{i=1}^{r} \tilde{Y}_{i\delta}(t)[\delta_i-\delta_i^*] \\ & + \sum_{i=1}^{r} \tilde{Y}_{i\delta f}(t)[f(\omega)\delta_i-f(\omega^*)\delta_i^*]\end{aligned} \tag{52}$$

where $(\tilde{Y}_{0f};\tilde{Y}_{1\delta}, ..\tilde{Y}_{r\delta}; \tilde{Y}_{1\delta f}, ..Y_{r\delta f})$ denote appropriate column partitions of the matrix function $\tilde{Y}(t)$.

In view of the single valued property of $f(\omega)$, a sufficient condition for the unique vanishing of $\tilde{z}(t,\theta)$ in (52) at $\theta = \theta^*$ is linear independence of the columns of the functions $[\tilde{Y}_{of}(t), \tilde{Y}_{1\delta}(t), .. \tilde{Y}_{r\delta}(t)]$ on $[0,t_1]$. However, these columns contain no remnant, or component, in the subspace Z_o (defined following (10)) as a result of the annihilating filter H. Hence a sufficient condition for the uniqueness of solutions to the least squares estimation problem in the separable case is linear independence of the columns of $[Y_{of}(t), Y_{1\delta}(t), ..Y_{r\delta}(t)]$ on $[0,t_1]$. These conditions can also be expressed in terms of the positive definiteness of the associated Gramian for these columns. In turn, these conditions can be seen to be sufficient for the inversion of the matrix $H'(\omega)\Omega_{dd}H(\omega)$ in (24) for all finite values of the parameter vector ω.

NOTES

1) A modal signal is any function of the exponential form $\phi(t)=t^k e^{\lambda t}$, with the complex number λ and integer $k \geq 0$ characterizing the mode in the complex plane.

2) Note that $y^2(t)\dot{y}(t)=\frac{1}{3}Dy^3(t)$.

3) For simplicity, the initial time t_o will be taken as $t_o=0$.

4) The derivation of (13) from (11) is given in Appendix A.

5) The definition of separability in this paper is slightly different from that given on p. 74 of Pearson (1979) . However, it includes the notion of separability given in Pearson (1979) by suitably defining $h(\omega)$ in terms of the single valued function $f(\theta)$ in Pearson (1979) . The reason for this difference is to be able to eliminate the vector δ as will be discussed below.

6) It is noted that Ω is the Gram matrix for the columns of the projected matrix function $\tilde{Y}(t)=H(Y(t))$ on $[0,t_1]$; thus Ω is symmetric and nonnegative definite.

7) Sufficient conditions for the uniqueness of a solution to the least squares problem for the separable case are discussed in Appendix B. These conditions imply the existence of the inverse in (23).

8) This is the observability Gramian for the pair (-A,C).

REFERENCES

[1] Aström, K.J. and Bohlin, T., (1966): Numerical Identification of Linear Dynamic Systems from Normal Operating Records, in Theory of Self-Adaptive Control Systems, (Ed. P.H. Hammond), Plenum Press, New York.

[2] Eykhoff, P., (1974): System Identification, Wiley-Interscience, New York.

[3] Mendel, J.M., (1973): Discrete Techniques of Parameter Estimation: The Equation Error Formulation, Marcel Dekker, New York.

[4] Pearson, A.E. (1976): Finite Time Interval Linear System Identification Without Initial State Estimation, Automatica,

Vol. 12, pp. 577-587.

[5] Pearson, A.E. (1979): Nonlinear System Identification with Limited Time Data, Automatica, Vol. 15, pp. 73-84.

[6] Pearson, A.E. and Mocenigo, J.M. (1979): A Filter for Separating Finite Time Modal Signals, IEEE Trans. on Auto. Contr., Vol. AC-24, pp. 926-932.

[7] Pearson, A.E. and Chin, Y.K. (Aug. 1979): Identification of MIMO Systems with Partially Decoupled Parameters, IEEE Trans. on Auto. Contr., Vol. AC-24, pp 599-604.

Numerical Techniques for Stochastic Systems
F. Archetti and M. Cugiani (eds.)
© North-Holland Publishing Company, 1980

FAST ALGORITHMS FOR RECURSIVE ESTIMATION AND IDENTIFICATION

Lennart Ljung

Department of Electrical Engineering
Linköping University
S-581 83 Linköping, Sweden

Some algorithms for efficient solution of linear equations are reviewed. The underlying idea goes back to solution of equations with Toeplitz coefficient matrices. Application to fast calculation of gain matrices for recursive estimation and identification algorithms are also discussed, as well as applications to numerical solution of integral equations.

1. INTRODUCTION

In the control and estimation literature, there has been a recent interest in so called fast algorithms. By "fast" we mean that the calculations are carried out with an order of magnitude less operations than the conventional method, by making use of intrinsic symmetry properties in the problem.

The basic idea of these algorithms is contained in Levinson's algorithm for solving a certain estimation problem, Levinson (1947). The interest in these algorithms has been renewed recently in connection with various application, see e.g. Kailath (1973) - Lindquist (1974).

The purpose of this paper is to give a tutorial background for these results. We aim at displaying the underlying idea and the basics of its application to various problems. The treatment will be confined to the discrete time case.

2. Linear equations with Toeplitz coefficient matrices

Consider the problem of solving the linear equation

$$R\,x = f \tag{2.1}$$

where R is a NxN Toeplitz matrix. This means that the elements of R depend only on the difference between row and column index. This Toeplitz property can also be expressed as

$$R_{n+1} = \begin{bmatrix} R_n & r_n \\ r_n^T & a \end{bmatrix} = \begin{bmatrix} a & \tilde{r}_n^T \\ \tilde{r}_n & R_n \end{bmatrix} \tag{2.2}$$

where R_n is the upper right hand $n\ n$ block of $R(1 \leq n \leq N)$. (We assume here that R is symmetric, but this is not essential).

To utilize this property, (2.2), it seems to be a good idea to introduce a sequence of vectors f_n with increasing dimension n, such that

$$f_N = f \tag{2.3}$$

and

$$f_{n+1} = \begin{pmatrix} f_n \\ f^{(n+1)} \end{pmatrix} \tag{2.4}$$

Suppose now that we know the solution x_n to the equation

$$R_n x_n = f_n \tag{2.5}$$

We shall now discuss how the solution x_{n+1} can be constructed. The simplest candidate for x_{n+1} would be

$$\begin{pmatrix} x_n \\ 0 \end{pmatrix}$$

This gives

$$R_{n+1} \begin{pmatrix} x_n \\ 0 \end{pmatrix} = \begin{pmatrix} f_n \\ \alpha_n \end{pmatrix} \tag{2.6}$$

where

$$\alpha_n = r_n^T x_n \tag{2.7}$$

In (2.6) the first equality in (2.2) was used. We see that (2.6) is not "far" from the solution. We need only to replace the bottom element of the right hand side by $f^{(n+1)}$. If we knew a vector z_{n+1}, such that

$$R_{n+1} z_{n+1} = \begin{pmatrix} 0 \\ 0 \\ \vdots \\ 0 \\ 1 \end{pmatrix} \tag{2.8}$$

this would be easy:

$$R_{n+1} \left[\begin{pmatrix} x_n \\ \hline 0 \end{pmatrix} + (f^{(n+1)} - \alpha_n)\, z_{n+1} \right] = \begin{pmatrix} f_n \\ \hline \alpha_n \end{pmatrix} + (f^{(n+1)} - \alpha_n) \begin{pmatrix} 0 \\ 0 \\ \vdots \\ 0 \\ 1 \end{pmatrix} = f_{n+1} \tag{2.9}$$

Therefore

$$x_{n+1} = \begin{pmatrix} x_n \\ \hline 0 \end{pmatrix} + (f^{(n+1)} - \alpha_n)\, z_{n+1} \tag{2.10}$$

We must now turn to the question of how to find z_{n+1}, defined by (2.8). Again, we assume that z_n is given:

$$R_n z_n = \begin{pmatrix} 0 \\ \vdots \\ 0 \\ 1 \end{pmatrix}. \tag{2.11}$$

Using (2.11) and the second equality of (2.2) we find that

$$R_{n+1} \begin{pmatrix} 0 \\ \hline z_n \end{pmatrix} = \begin{pmatrix} \beta_n \\ \hline 0 \\ \cdot \\ \vdots \\ 0 \\ 1 \end{pmatrix} \tag{2.12}$$

where

$$\beta_n = \tilde{r}_n^T z_n \tag{2.13}$$

To remove the first element in the right hand side of (2.12), we assume that a vector y_{n+1} is given, such that

$$R_{n+1}y_{n+1} = \begin{pmatrix} 1 \\ 0 \\ \cdot \\ \cdot \\ \cdot \\ 0 \end{pmatrix} . \tag{2.14}$$

Then

$$R_{n+1} \left[\begin{pmatrix} 0 \\ -- \\ z_n \end{pmatrix} - \beta_n y_{n+1} \right] = \begin{pmatrix} 0 \\ \cdot \\ \cdot \\ 0 \\ 1 \end{pmatrix} \tag{2.15}$$

which means that

$$z_{n+1} = \begin{pmatrix} 0 \\ -- \\ z_n \end{pmatrix} - \beta_n y_{n+1} . \tag{2.16}$$

Finally, we must consider how to determine y_{n+1}. If, y_n is given,

$$R_n y_n = \begin{pmatrix} 1 \\ 0 \\ \cdot \\ \cdot \\ 0 \end{pmatrix} , \tag{2.17}$$

we get from (2.17) and (2.2) that

$$R_{n+1} \begin{pmatrix} y_n \\ -- \\ 0 \end{pmatrix} = \begin{pmatrix} 1 \\ 0 \\ \cdot \\ \cdot \\ \cdot \\ 0 \\ --- \\ \gamma_n \end{pmatrix} \tag{2.18}$$

where

$$\gamma_n = r_n^T y_n \tag{2.19}$$

Now, combining (2.18) with (2.12) we find that

$$R_{n+1} \left[\begin{pmatrix} y_n \\ --- \\ 0 \end{pmatrix} - \gamma_n \begin{pmatrix} 0 \\ --- \\ z_n \end{pmatrix} \right] = \begin{pmatrix} 1 - \gamma_n \beta_n \\ 0 \\ \cdot \\ \cdot \\ 0 \end{pmatrix} \tag{2.20}$$

which means that

$$y_{n+1} = \left[\begin{pmatrix} y_n \\ --- \\ 0 \end{pmatrix} - \gamma_n \begin{pmatrix} 0 \\ --- \\ z_n \end{pmatrix} \right] / (1-\gamma_n \beta_n). \tag{2.21}$$

We have now completed the derivation of the algorithm to solve (2.1), which may be summarized as follows:

$$x_1 = f_1/a \tag{2.22a}$$

$$y_1 = z_1 = 1/a \tag{2.22b}$$

$$\gamma_n = r_n^T y_n; \quad \beta_n = \tilde{r}_n^T z_n; \quad \alpha_n = r_n^T x_n \tag{2.23}$$

$$y_{n+1} = \left[\begin{pmatrix} y_n \\ --- \\ 0 \end{pmatrix} - \gamma_n \begin{pmatrix} 0 \\ --- \\ z_n \end{pmatrix} \right] / (1-\gamma_n \beta_n) \tag{2.24}$$

$$z_{n+1} = \begin{pmatrix} 0 \\ --- \\ z_n \end{pmatrix} - \beta_n y_{n+1} \tag{2.25}$$

$$x_{n+1} = \begin{pmatrix} x_n \\ --- \\ 0 \end{pmatrix} - (f^{(n+1)} - \alpha_n) z_{n+1} \tag{2.26}$$

$$x_N = x$$

This algorithm is essentially the one derived by Levinson (1947).

We may note that each iteration of (2.23) - (2.26) requires proportional to n operations. The whole scheme to find x consequently will take proportional to N^2 operations. This should be compared with a direct solution of (2.1) using e.g. Gauss elimination, which would take $\sim N^3$ operations.

3. Solution of general linear equation

If we now turn to a general system of equations:

$$Ax = f \tag{3.1}$$

where A no longer is assumed to be Toeplitz. This means that the second equality of (2.2) does not hold. We have in general for a sequence of A_n-matrices with increasing dimensions:

$$A_{n+1} = \left[\begin{array}{c|c} A_n & a_n \\ \hline \bar{a}_n & b_{n+1} \end{array}\right] = \left[\begin{array}{c|c} c_{n+1} & s_n \\ \hline \bar{s}_n & A_n \end{array}\right] + B_{n+1}, \tag{3.2}$$

where B_{n+1} is a "correction matrix". Suppose this correction has low rank, that does not depend on n:

$$B_{n+1} = \begin{bmatrix} 0 \\ --- \\ h_n \end{bmatrix} \left[\begin{array}{c|c} 0 & \tilde{h}_n^T \end{array}\right] \tag{3.3}$$

where h_n is a $n|\alpha$-matrix (where α is rank B_{n+1}). Most of what was done in the previous chapter then still holds, with the exception of (2.12) which now instead gives

$$A_{n+1}\begin{pmatrix} 0 \\ --- \\ z_n \end{pmatrix} = \begin{pmatrix} \beta_n \\ --- \\ 0 \\ \vdots \\ 0 \\ 1 \end{pmatrix} + \begin{pmatrix} 0 \\ --- \\ h_n \end{pmatrix} \tilde{h}_n^T z_n \tag{3.4}$$

However, by introducing auxiliary variables ξ_n, such that

$$A_n \xi_n = h_n, \tag{3.5}$$

the effect of the last term of the right hand side of (3.4) can be eliminated, quite analogously to the treatment in chapter 2.

Since the aim of the present paper is to display basic ideas, rather than technical details, we will not elaborate on the resulting algorithm, but refer the reader to Friedlander, Morf, Kailath & L. Ljung (1979) for this.

The correction term in (3.2) will lead to an increase in operations when solving (3.1) of approximatively α times compared to the algorithm for Toeplitz coefficient matrices. The basic use of the resulting algorithm therefore is when the structure of the underlying problem leads to such coefficient matrices that give a small α in (3.2) - (3.3). Some such examples will be given in the following chapters.

4. Applications to the numerical solution of integral equations.

Integral equations of the Fredholm type arise in the context of many problems in physics and technology. The basic equation then is

$$\lambda\, f(t) + \int_0^T K(t,s) f(s) ds = g(t) \tag{4.1}$$

that is to be solved for $f(\cdot)$, with $K(.,.)$ and $g(\cdot)$ given. If $\lambda=0$ the equation is said to be of the first kind, otherwise of the second kind.

Often, symmetry and invariance of the underlying problem lead to so called "stationary" kernels with

$$K(t,s) = K(t-s). \tag{4.2}$$

The usual way of solving (4.1) numerically is to discretize the integral using some quadrature rule. Then (4.1) is transformed to a set of linear equation.

$$R_N f_N = g_N \tag{4.3}$$

Here N denotes the dimension of the corresponding vector; i.e. to the number of nodal points involved in the discretization procedure. Now, even if the kernel is stationary, so that (4.2) holds, R_N will typically not be a Toeplitz matrix. The reason is that the weighting pattern used in the quadrature rule (e.g. (1 2 . . . 2 1) for the trapezoidal rule and (1 4 2 4 2 .... 4 2 4 1) for Simpson's formula) will destroy this property. However, the matrix R_N will still be close to Toeplitz (if (4.2) holds) in the sense that it will satisfy (3.2), (3.3) with $\alpha=2$. Hence, the fast algorithms for solving (4.3) described in Section 3 can be applied. The details of the algorithm are given in S. Ljung & L. Ljung (1978). This reference also contains extensive numerical experiences of this approach to solve integral equations. The numerical properties of the algorithm are found to be satisfactory: no numerical instability problems have been detected, extrapolation can be successfully applied, and the promised gains in computing time are verified.

In case $\lambda=0$ in (4.1), the corresponding set of linear equations

(4.3) will usually be ill-conditioned. This problem is typically overcome by so called regularization. By this procedure, instead of solving (4.3), the following set of equations is solved:

$$P_N f_N = h_N, \tag{4.4}$$

where

$$P_N = (R_N^T R_N + \delta L_N^T L_N) \tag{4.5}$$

$$h_N = R_N^T g_N. \tag{4.6}$$

Here L_N typically is a (Toeplitz) matrix corresponding to numerical differentiation and δ is a small positive number, the regularization parameter.

Now, P_N will not be a Toeplitz matrix, even if R_N and L_N are Toeplitz. However it will satisfy (3.2), (3.3) with $\alpha = 4$ and hence the fast methods can be applied. The details of the resulting algorithm are given in S. Ljung (Jan. 1979), together with implementation aspects and numerical experiences.

Integral equations of the form (4.1) also arise when applying boundary methods to certain partial differential equations. Then (4.2) corresponds to some symmetry properties of the boundary of the area over which the partial differential equation holds. Applications of the techniques described in this paper to such problems are discussed in S. Ljung (Sept. 1979).

5. Applications to recursive estimation

A general linear estimation problem can be posed as follows. Suppose that a signal $z(t)$, $t=0, 1, 2, \ldots$ is observed with additive white noise $e(t)$:

$$y(t) = z(t) + e(t). \qquad T=0,1,2 \ldots \tag{5.1}$$

The problem is to estimate $z(t)$ based on measurements of $y(r)$ $r=0,1, \ldots, T$. Confining ourselves to linear estimates, we seek a filter h(t,k;T) such that

$$\hat{z}(t|T) = \sum_{k=0}^{T} h(t,k;T)y(k), \tag{5.2}$$

gives the best possible estimate of z(t), i.e. minimizes the variance of $z(t)-\hat{z}(t|T)$.

It is easy to show that this filter is obtained as the solution h(t;T) of

$$R_T h(t;T) = g(t;T) \tag{5.3}$$

where R_T is the covariance matrix of (y(0), ...y(T)), h(t;T) is the vector

$$h^T(t;T) = (h(t,0;T)....h(t,T;T)) \tag{5.4}$$

and

$$g^T(t;T) = (g(t,0), ..., g(t,T));$$
$$g(t,k;T) = E\ y(k)z(t). \tag{5.5}$$

Now, if the signal z(t) is a stationary stochastic process, R_T in (5.3) will be Toeplitz. Then the procedure of Section 2 can be applied to its solution. In fact, the Levinson algorithm, Levinson (1947), was devised for exactly this problem.

Often the signal z(t) is described by a time invariant state space model

$$x(t+1) = F\ x(t) + G\ v(t)$$
$$z(t) = H\ x(t). \tag{5.6}$$

Then it does not necessarily have to be stationary, even though it will approach a stationary process as t tends to infinity. In that case, the covariance matrix R_T, will be subject to (3.2), (3.3) with a value of α that depends on the initial state covariance matrix

$$\Pi(0) = E\, x(0)x^T(0).$$

For certain values of $\Pi(0)$ (such as e.g. 0) α will be small, and the techniques of Section 3 can be applied when solving for the optimal filter (5.3), see Friedlander, Kailath, Morf & L. Ljung (1978) for details.

Moreover, when t=T in (5.3) the filter and the estimates $\hat{z}(t|t)$ can be obtained recursively in time (the Kalman filter). Utilizing the near Toeplitz-ness of R_T by a technique related to the one in Section 3, leads to the so called Chandrasekhar algorithms for recursive estimation. These relationships are described in detail in Friedlander, Kailath, Morf & L. Ljung (1978). See also Lindquist (1974).

6. Applications to recursive identification

An important problem in systems and control theory is to estimate unknown parameters in mathematical models of the system. The most common method is no doubt the least squares procedure. In this procedure the parameters of an autoregressive system model are estimated:

$$y(t) + a_1 y(t-1) + \dots + a_n y(t-n) = e(t) \tag{6.1}$$

Here $y(t)$ is the observed output sequence and $e(t)$ is a noise sequence. (The case with also an input sequence affecting the output is quite analogous).

With

$$\Theta^T = (a_1 \dots a_n)$$

and

$$\phi^T(t) = (-y(t-1), \dots, -y(t-n)), \tag{6.2}$$

eqn (6.1) can be written

$$y(t) = \Theta^T \phi(t) + e(t). \tag{6.3}$$

The Θ-parameter vector can now be estimated using the well known recursive least squares procedure, Astrom & Eykhoff (1971):

$$\hat{\Theta}(t)=\hat{\Theta}(t-1)+R^{-1}(t)\,\phi(t)\,(y(t)-\hat{\Theta}^T(t-1)\,\phi(t)) \tag{6.4}$$

$$R(t) = R(t-1) + \phi(t)\phi^T(t). \tag{6.5}$$

In practice, (6.5) is implemented in terms of

$$P(t) \triangleq R^{-1}(t)$$

using the matrix inversion lemma:

$$P(t) = P(t-1) + \frac{\phi(t)P(t-1)\phi^T(t)}{1+\phi^T(t)P(t-1)\phi(t)}. \tag{6.6}$$

Eqn (6.6) requires proportional to n^2 arithmetic operations per time step and memory locations. However, we see from (6.4) that only the n-dimensional column vector

$$K(t) \triangleq P(t)\phi(t) \tag{6.7}$$

is needed in order to find the estimates.

We shall now proceed to indicate how (6.7) can be directly updated using only proportional to n operations (and memory locations) using the ideas of Section 2.

From (6.5) we find that

$$R(t) = \sum_{1}^{t} \phi(k)\phi^T(k). \tag{6.8}$$

Introduce

$$\bar{\phi}(t) = \begin{pmatrix} -y(t) \\ \phi(t) \end{pmatrix} = \begin{pmatrix} \phi(t+1) \\ -y(t-n) \end{pmatrix} \tag{6.9}$$

and

$$\overline{R}(t) = \sum_{1}^{t} \overline{\phi}(k)\,\overline{\phi}^{T}(k). \tag{6.10}$$

Then, in veiw of (6.9)

$$\overline{R}(t) = \left(\begin{array}{c|c} \cdot & r^{T}(t) \\ \hline r(t) & R(t) \end{array}\right) = \left(\begin{array}{c|c} R(t+1) & \rho(t) \\ \hline \rho^{T}(t) & \cdot \end{array}\right). \tag{6.11}$$

Now, the "gain vector" $K(t)$ given by (6.7) can be defined as the solution of

$$R(t)K(t) = \phi(t) \tag{6.12}$$

or, using (6.11),

$$\overline{R}(t) \begin{pmatrix} 0 \\ K(t) \end{pmatrix} = \begin{pmatrix} \alpha(t) \\ \phi(t) \end{pmatrix} \tag{6.13}$$

where

$$\alpha(t) = r^{T}(t)K(t).$$

Similarly, $K(t+1)$, can be defined as the solution of

$$\overline{R}(t) \begin{pmatrix} K(t+1) \\ 0 \end{pmatrix} = \begin{pmatrix} \phi(t+1) \\ \beta(t) \end{pmatrix} \tag{6.14}$$

where

$$\beta(t) = \rho^{T}(t)K(t+1).$$

We shall now proceed to show how we can go from (6.13) to (6.14) and hence update $K(\cdot)$. Introduce two auxiliary matrices $A(t)$ and $B(t)$ such that

$$\overline{R}(t)A(t) = \begin{pmatrix} 1 \\ 0 \\ \cdot \\ \cdot \\ 0 \end{pmatrix} ; \quad \overline{R}(t) \begin{pmatrix} B(t) \\ 1 \end{pmatrix} = \begin{pmatrix} 0 \\ \vdots \\ 0 \\ \gamma(t) \end{pmatrix} \tag{6.15}$$

for some value $\gamma(t)$. Notice the similarity with the auxiliary variables (2.8) and (2.17)!

We first seek a vector $\bar{K}(t)$ such that

$$\bar{R}(t)\bar{K}(t) = \bar{\phi}(t). \tag{6.16}$$

Indeed, using (6.13) and (6.15) we find that

$$\bar{R}(t)\left[\begin{pmatrix} 0 \\ K(t) \end{pmatrix} + (-y(t)-\alpha(t))A(t)\right] = \bar{\phi}(t)$$

so that

$$\bar{K}(t) = \begin{pmatrix} 0 \\ K(t) \end{pmatrix} - (y(t)+\alpha(t))A(t) \tag{6.17}$$

Now let the last element of $\bar{K}(t)$ be $\mu(t)$, so that

$$\bar{K}(t) = \begin{pmatrix} m(t+1) \\ \mu(t) \end{pmatrix}$$

for some $m(t+1)$. Then, using (6.15) and (6.9)

$$\bar{R}(t)\left[\bar{K}(t)-\mu(t)\begin{pmatrix} B(t) \\ 1 \end{pmatrix}\right] =$$

$$= \bar{R}(t)\begin{pmatrix} m(t+1)-\mu(t)B(t) \\ 0 \end{pmatrix} = \bar{\phi}(t) - \begin{pmatrix} 0 \\ \cdot \\ \cdot \\ \cdot \\ 0 \\ \mu(t)\beta(t) \end{pmatrix} = \begin{pmatrix} \phi(t+1) \\ -y(t-n)-\mu(t)\beta(t) \end{pmatrix}$$

Comparing this expression with (6.14) we see that

$$K(t+1) = m(t+1) - \mu(t)B(t), \tag{6.18}$$

and the update of the gain vector is done. To complete the algorithm we need also updating formulas for the auxiliary variables A and B in (6.15). These are obtained in a similar fashion to the procedure in Section 2. The full algorithm is derived in L. Ljung, Morf & Falconer (1978), where a more general situation is considered.

The resulting algorithm uses only proportional to n operations to update K, which may be of great importance in certain applications, see, e.g. Falconer & L. Ljung (1978).

7. CONCLUSIONS

We have in this exposition of so called fast algorithms tried to point out what are the simple underlying ideas in a number of algorithms for recursive estimation, for recursive identification, for solving linear equations, and for numerical solution of integral equations. We have also displayed the common features in these algorithms.

In the paper only the discrete time case has been considered. Results for the continuous time case are analogous, and they are given e.g. in Kailath, L. Ljung & Morf (1978), Lindquist (1975) and Kailath, L. Ljung & Morf (1976).

REFERENCES

1 Aström, K.J. and Eykhoff, P. (1971): System Identification: A Survey, Automatica, Vol. 7, pp 123-162.

2 Falconer, D.D. and Ljung, L. (1978): Application of fast Kalman estimation to adaptive equalization; IEEE Trans Comm, Vol. COM-26, No 10, October, pp 1439-1446.

3 Friedlander, B., Kailath, T., Morf, M. and Ljung, L. (1978): Extended Levinson and Chandrasekhar equations for general discrete-time linear estimation problems; IEEE Trans Autom Contr., Vol. AC-23, No 4, August, pp 653-659.

4 Friedlander, B., Morf, M., Kailath, T. and Ljung, L. (1979): New inversion formulas for matrices classified in terms of their distance from Toeplitz matrices; Linear Algebra and its applications, to appear.

5 Kailath, T. (1973): Some new algorithms for recursive estimation in constant linear system; IEEE Trans. Inform. Theory, Vol. IT-19, pp 750-760, November.

6 Kailath, T., Ljung, L. and Morf, M. (1976): Recursive input-output and state space solutions for continuous time linear estimation problems; Proc. 1976 IEEE Conference on Decision and Control, Clearwater, Florida, December , paper WA 5-2, pp 182A-182G.

7 Kailath, T., Ljung, L. and Morf, M. (1978): Generalized Krein-Levinson equations for efficient calculation of Fredholm resolvents of non-displacement kernels; in "Topics in Functional Analysis, Advances in Mathematics Supplementary Studies Vol. 3", Academic Press, pp 169-184.

8 Levinson, N., (1947): The Wiener rms (root-mean square) error criterion in filter design and prediction; J. Math. Phys., vol. 25, pp 261-278.

9 Lindquist, A., (1974): A new algorithm for optimal filtering of discrete-time stationary processes; SIAM J. Control, vol. 4.

10 Lindquist, A., (1975): On Fredholm Integral Equations, Toeplitz Equations and Kalman-Bucy Filtering; Applied Mathematics and Optimization, Vol. 1, No. 4, pp 355-373.

11 Ljung, L., Morf, M. and Falconer, D., (1978): Fast calculation of gain matrices for recursive estimation schemes; Int. J. Contr., Vol. 27, No. 1, January, pp 1-19.

12 Ljung, S. and Ljung, L., (1978): Fast numerical solution of integral equations with stationary kernels; Report LiTH-ISY-I-0200, February .

13 Ljung, S., (1979): A fast algorithm to solve Fredholm integral equations of the first kind with stationary kernels; Report LiTH-ISY-I-0265, January .

14 Ljung, S., (1979): Fast numerical solution of Laplace's equation and the biharmonic equation for circular boundaries; Report LiTH-ISY-I-0320, September .

Numerical Techniques for Stochastic Systems
F. Archetti and M. Cugiani (eds.)
© North-Holland Publishing Company, 1980

SPECTRAL DECOMPOSITION WITH APPLICATION TO IDENTIFICATION

Torsten Söderström

Department of Automatic Control and Systems Analysis
Institute of Technology, Uppsala University
P.O. Box 534, S-751 21 UPPSALA, Sweden

ABSTRACT

It is studied how a given spectral density of a measured signal can be decomposed into two parts. One part of the spectral density refers to the effect of the undisturbed signal while the other part originates from measurement noise. This problem is closely related to system identification from noise-corrupted data.
In this study of the spectral decomposition problem both the signal and the noise are treated as autoregressive moving average processes. The decomposition problem has no unique solution unless additional assumptions are added. Necessary conditions as well as sufficient ones for a unique solution are given. For the scalar case a more detailed and explicit analysis is performed.
The spectral decomposition results are applied to a system identification problem. It is assumed that the measurements of the input as well as the output are corrupted by noise. Analysis of the identifiability properties is performed.

1. INTRODUCTION

When estimating the spectrum of a signal or the dynamics of a system it is often not possible to avoid measurement noise. The presence of this noise can have a considerable influence on the resulting estimate.

In such situations it is of great importance to be able to estimate both the effect of the noise and the effect of the signal or the system. Basically, the type of information assumed available is the spectral density of the measurements. The problem then becomes one of "spectral decomposition", i.e. how should a given spectral density of the measurements be decomposed into two terms, one showing the influence from the noise and the other one describing the signal (or the system). The purpose of this contribution is to discuss and analyse this problem.

It is quite clear that without additional a priori assumptions a unique solution cannot exist. It would e.g. not be possible to distinguish between signal and noise. We will generally model both the undisturbed signal and the measurement noise as ARMA (autoregressive moving average) processes. Additional structural assumptions can then be given concerning the degrees of the involved polynomials, etc.

The paper is organized as follows. In the next section the problem is formulated in mathematical form. Some key concepts are introduced. Various ways to perform the spectral decomposition in practice are discussed in section 3. The scalar case is then considered in section 4. Necessary respectively sufficient conditions for a unique solution to the spectral decomposition problem will be given. The sufficient conditions are generalized to the multivariable case in section 5. Identification of processes with measurement noise both in the input and the output are treated in section 6. A somewhat more detailed treatment of the topic can be found in the author's report, Söderström (1979a).

2. STATEMENT OF THE PROBLEM

Consider a stochastic signal $s(t)$ which is measured with noise according to

$$y(t)=s(t)+n(t) \tag{2.1}$$

It is assumed that $s(t)$ and $n(t)$ are ARMA processes and uncorrelated. They may be multivariable. The spectral densities will then fulfil

$$\phi_y(\omega)\equiv \phi_s(\omega)+\phi_n(\omega) \tag{2.2}$$

The problem to be studied is how to find $\phi_s(\omega)$ and $\phi_n(\omega)$ when

$\phi_y(\omega)$ is given.

Needless to say, this problem has in general no unique solution unless additional assumptions are given. It will turn out that e.g. appropriate degrees of the involved polynomials will guarantee a unique solution.

To phrase the problem more in detail we write

$$s(t)=H(q^{-1})v(t)=(I+\sum_{i=1}^{\infty} H_i q^{-i})v(t)+\sum_{i=1}^{\infty} H_i v(t-i)$$
$$n(t)=F(q^{-1})e(t)=(I+\sum_{i=1}^{\infty} F_i q^{-i})e(t)=e(t)+\sum_{i=1}^{\infty} F_i e(t-i) \tag{2.3}$$

where the finite order filters $H(q^{-1})$ and $F(q^{-1})$ as well as their inverses are asymptotically stable. Further, q^{-1} denotes the backward shift operator, so that $q^{-1}v(t)=v(t-1)$ etc. The noise sources $v(t)$ and $e(t)$ are white and mutually uncorrelated. The noise covariance matrices are given by

$$Ev(t)v(t)^T=\Lambda_v \qquad Ee(t)e(t)^T=\Lambda_e \tag{2.4}$$

It is then easy to get expressions for the spectral densities.

$$\phi_s(\omega)=H(e^{i\omega})\Lambda_v H(e^{-i\omega})^T \qquad \phi_n(\omega)=F(e^{i\omega})\Lambda_e F(e^{-i\omega})^T \tag{2.5}$$

The situation is depicted in figure 2.1.

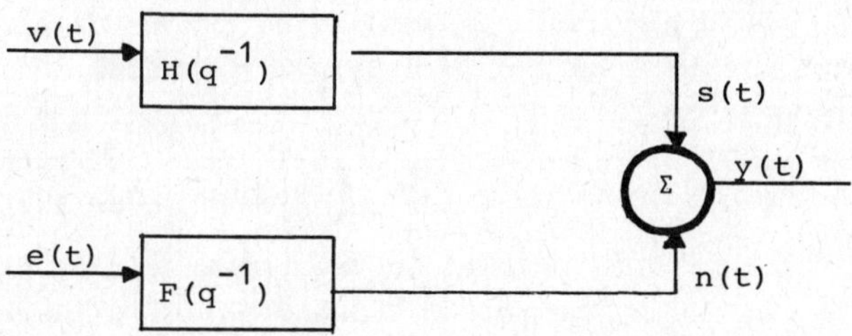

Figure 2.1. Setup for the spectral decomposition problem

The problem will be solved in a parametric way. In practice this means that we perform a parameter estimation to find $\phi_s(\omega)$ and $\phi_n(\omega)$. Introduce then a parameter vector θ of all the unknown para meters of $H(q^{-1})$, $F(q^{-1})$, Λ_v and Λ_e. The information available (or easily estimated) is the output spectral density, which must fulfil

$$\phi_y(\omega) \equiv \hat{H}(e^{i\omega};\theta)\hat{\Lambda}_v(\theta)\hat{H}(e^{-i\omega};\theta)^T + \hat{F}(e^{i\omega};\theta)\hat{\Lambda}_e(\theta)\hat{F}(e^{-i\omega};\theta)^T \tag{2.6}$$

cf (2.2), (2.5). Here $\hat{H}(q^{-1};\theta)$ denotes the estimate of $H(q^{-1})$ etc. The purpose of this paper is to investigate the solutions with respect to θ of (2.6) for some different parameterizations. According to the previous calculating we can rewrite it as (z being an arbitrary complex variable).

$$H(z)\Lambda_v H(z^{-1})^T + F(z)\Lambda_e F(z^{-1})^T \equiv \hat{H}(z;\theta)\hat{\Lambda}_v(\theta)\hat{H}(z^{-1};\theta)^T + \hat{F}(z;\theta)\hat{\Lambda}_e(\theta)\hat{F}(z^{-1};\theta)^T. \tag{2.7}$$

We will in the following refer to (2.7) as <u>the identifiability identity</u>.

Some contributions dealing with the spectral decomposition problem (for <u>white</u> measurement noise) from the identification point of view can be found in Goodrich (1978), Mehra (1971) and Kashyap (1970). Some partial results are also given in Mehra (1970) and Anderson et al (1969). In these two papers the authors work with state space models.

The main restriction of Mehra's paper is that only the noise covariances (roughly corresponding to $\Lambda_v(\theta)$ and $\Lambda_e(\theta)$ in (2.7)) are estimated while the dynamics is assumed either known or already estimated (which can be done under certain assumptions, cf Mehra (1971)). Anderson et al (1969) treats the case when the signal is a multivariable AR(1) process and the noise $n(t)$ is white. This will be a special case of the results here (cf Theorem 5.2).

A similar but not identical problem is treated in signal detection, see e.g. van Trees (1971), Levin and Shinalov (1977). In these and similar contributions the problem is formulated as a statistical test of hypotheses. If only a <u>finite</u> set of models is possible (i.e. if the parameter vector θ can take only a finite number of values) the problem will be to decide which of these that are most appropriate. We can then look at this problem as a multiple hypothesis testing. To be more precise we have to test a set of hypothesis

$\{H_j\}_{j=1}^m$, cf (2.6)

$$H_j: \phi_y(\omega)=\hat{H}(e^{i\omega};\theta_j)\Lambda_v(\theta_j)\hat{H}(e^{-i\omega};\theta_j)^T+\hat{F}(e^{i\omega};\theta_j)\Lambda_e(\theta_j)\hat{F}(e^{-i\omega};\theta_j)^T \tag{2.8}$$

Here $\{\theta_j\}_{j=1}^m$ is the set of possible parameter vectors. However, in the usual case there will be an infinite number of possible θ values. Then the relation to the problem of signal detection is not so clear.

3. SOME PRACTICAL ASPECTS

The spectral decomposition can be treated and solved in some different ways. Some approaches will be outlined here. In practice $\phi_y(\omega)$ may not be known exactly. Merely an estimate based on measurements will be used. If this estimate differ somewhat from the true spectral density $\phi_y(\omega)$ (which, of course, is highly probable) the solvability of (2.6) can be changed.

One approach to treat the problem is to substitute the equation (2.6) with an optimization problem. The parameter vector θ can be chosen as the minimizing argument of the function

$$V(\theta)=\int_{-\pi}^{\pi}||\phi_y(\omega)-\hat{H}(e^{i\omega};\theta)\hat{\Lambda}_v(\theta)\hat{H}(e^{-i\omega};\theta)^T-\hat{F}(e^{i\omega};\theta)\hat{\Lambda}_e(\theta)\hat{F}(e^{-i\omega};\theta)^T||^p d\omega \tag{3.1}$$

(p being a real number ≥ 1).

Another approach is to use the "method of moments". We then use that the measured output $y(t)$ must be a finite order ARMA process. This implies that both sides of (2.6) are rational functions of θ. The identity can therefore be reformulated as a set of non-linear equations. Provided these equations are at least as many as the unknowns a solution might be found. However, using an estimate of $\phi_y(\omega)$ instead of the true value may in some cases imply that no solution exists. This approach is further treated in an indirect way later (Proof of Theorem 4.2).

Still another approach is to apply a prediction error identification method, Ljung (1976). We then assume that a set of measured values $y(1)$ ... $y(N)$ are available. The output $y(t)$ is then written as one ARMA process

$$y(t)=K(q^{-1};\theta)\varepsilon(t) \qquad E\varepsilon(t)\varepsilon(t)^T=Q(\theta) \tag{3.2}$$

where $\{\varepsilon(t)\}$ are the innovations or the prediction errors. The filter $K(q^{-1};\theta)$ and the matrix $Q(\theta)$ are functions of the parameter vector θ. They must fulfil

$$K(z;\theta)Q(\theta)K(z^{-1};\theta)^T \equiv \hat{H}(z;\theta)\hat{\Lambda}_v(\theta)\hat{H}(z^{-1};\theta)^T+\hat{F}(z;\theta)\hat{\Lambda}_e(\theta)\hat{F}(z^{-1};\theta)^T \quad (3.3)$$

K and Q can be computed from $\hat{F}$, $\hat{H}$, $\hat{\Lambda}_e$, $\hat{\Lambda}_v$ through a Ricatti equation using a state space model, cf Anderson and Moore (1979).

The parameter vector θ is then determined as the minimizing element of a suitable function of the prediction errors. Some examples are

$$V_1=\det(\frac{1}{N}\sum_{t=1}^{N}\varepsilon(t)\varepsilon(t)^T) \quad (3.4)$$

$$V_2=\mathrm{tr}(\frac{1}{N}\sum_{t=1}^{N}\varepsilon(t)\varepsilon(t)^T) \quad (3.5)$$

$$V_3=\ln\det\hat{Q}(\theta)+\frac{1}{N}\sum_{t=1}^{N}\varepsilon(t)^T\hat{Q}(\theta)^{-1}\varepsilon(t) \quad (3.6)$$

The identifiability properties of a PEM are given by the following identities, cf Ljung (1976)

$$\hat{K}(z;\theta)\equiv K(z) \quad (3.7)$$

$$\hat{Q}(\theta)=Q \quad (3.8)$$

The identity (3.7) is always valid. Equation (3.8) is valid of either the criterion V_3, (3.6), is used, or if $\hat{K}(q^{-1};\theta)$ and $\hat{Q}(\theta)$ have no common parameters. The latter case is in general not valid for our problem. The quantities $K(z)$ and Q are defined by a similar spectral factorization of the true output spectral density.

4. SCALAR SYSTEMS

We will now investigate the spectral decomposition problem in the scalar case. Necessary conditions as well as sufficient ones for a unique solution will be given.

As the filters $H(q^{-1})$ and $F(q^{-1})$ are of finite order we can write

$$H(z) = \frac{1+\ell_1 z+\ldots+\ell_{n\ell}z^{n\ell}}{1+d_1 z+\ldots+d_{nd}z^{nd}} \triangleq \frac{L(z)}{D(z)} \quad (4.1)$$

$$F(z) = \frac{1+c_1 z+\dots +c_{nc} z^{nc}}{1+a_1 z+\dots +a_{na} z^{na}} \triangleq \frac{C(z)}{A(z)} \tag{4.2}$$

We will assume that $\ell_{n\ell}\neq 0$, $d_{nd}\neq 0$, $c_{nc}\neq 0$, $a_{na}\neq 0$. It will also be generally assumed that the pair $(A(z), C(z))$ as well as $(D(z), L(z))$ is relatively prime. Due to earlier made assumptions all polynomials are restricted to have all zeros outside the unit circle.

The model will now be assumed to take a similar from with all polynomial coefficients as model parameters. To be more precise we have

$$\hat{H}(z;\theta)=\frac{1+\hat{\ell}_1 z+\dots +\hat{\ell}_{\hat{n}\ell} z^{\hat{n}\ell}}{1+\hat{d}_1 z+\dots +\hat{d}_{\hat{n}d} z^{\hat{n}d}} \triangleq \frac{\hat{L}(z)}{\hat{D}(z)} \tag{4.4}$$

$$\hat{F}(z;\theta)=\frac{1+\hat{c}_1 z+\dots +\hat{c}_{\hat{n}c} z^{\hat{n}c}}{1+\hat{a}_1 z+\dots +\hat{a}_{\hat{n}a} z^{\hat{n}a}} \triangleq \frac{\hat{C}(z)}{\hat{A}(z)} \tag{4.4}$$

The model parameter vector θ is given by

$$\theta = [\,\hat{a}_1 \dots \hat{a}_{\hat{n}a}\ \hat{c}_1 \dots \hat{c}_{\hat{n}c}\ \hat{d}_1 \dots \hat{d}_{\hat{n}d}\ \hat{\ell}_1 \dots \hat{\ell}_{\hat{n}\ell}\ \hat{\Lambda}_e\ \hat{\Lambda}_v]^T \tag{4.5}$$

we will generally assume

$$\hat{n}a \geq na \quad \hat{n}c \geq nc \quad \hat{n}d \geq nd \quad \hat{n}\ell \geq n\ell \tag{4.6}$$

which will guarantee existence of at least one solution. In many cases only the case with equalities in (4.6) will be treated explicitly.

Inserting (4.1)-(4.4) into the identifiability identity (2.7) we obtain

$$\frac{L(z)L(z^{-1})}{D(z)D(z^{-1})}\Lambda_v + \frac{C(z)C(z^{-1})}{A(z)A(z^{-1})}\Lambda_e \equiv \frac{\hat{L}(z)\hat{L}(z^{-1})}{\hat{D}(z)\hat{D}(z^{-1})}\hat{\Lambda}_v + \frac{\hat{C}(z)\hat{C}(z^{-1})}{\hat{A}(z)\hat{A}(z^{-1})}\hat{\Lambda}_e \tag{4.7}$$

We then first treat necessary conditions for existence of a unique solution.

Lemma 4.1 Consider the identifiability identity (4.7) where the involved polynomials are given by (4.1)-(4.4). Assume that

i) $\hat{n}a=na \quad \hat{n}c=nc \quad \hat{n}d=nd \quad \hat{n}\ell=n\ell$ (4.8)

ii) The polynomials $A(z)$ and $D(z)$ have precisely p common zeros. Then a <u>necessary</u> condition for existence of a <u>unique</u> solution of (4.7) with respect to the vector θ, (4.5), is that

$$\max(nd-n\ell\ , na-nc) \geq 1+p \tag{4.9}$$

<u>Proof</u> The basic principle used in the proof is that the "effective" number of equations must be at least as large as the number of unknowns.

Denote the common factor to $A(z)$ and $D(z)$ with $\tilde{A}(z)$. Then there are relatively prime polynomials $A^*(z)$ of degree $na-p$ and $D^*(z)$ of degree $nd-p$ such that

$$A(z)=A^*(z)\ \tilde{A}(z) \qquad D(z)=D^*(z)\tilde{A}(z)$$

The left hand side of (4.7) can then be written as

$$\frac{L(z)L(z^{-1})A^*(z)A^*(z^{-1})\Lambda_v+C(z)C(z^{-1})D^*(z)D^*(z^{-1})\Lambda_e}{\tilde{A}(z)\tilde{A}(z^{-1})A^*(z)A^*(z^{-1})D^*(z)D^*(z^{-1})}$$

In this expression the numerator and the denumerator do not contain any common zeros. Since the expression is "symmetric" the "effective" degree of the numerator is

$$n_N=\max(n\ell+na,\ nc+nd)-p$$

while the degree of the denominator is

$$n_D=na+nd-p$$

The number of equations turns out to be n_N+n_D+1+p which must be larger or equal to the number of unknown, i.e. $na+nc+nd+n\ell+2$.

This gives

$$\max(n\ell+na,\ \ nc+nd)-p+na+nd+1 \geq na+nc+nd+n\ell+2$$

which is easily transformed into (4.9).

<u>Remark</u> In the general case when (4.8) is substituted by (4.6) the condition (4.9) becomes

$$\max(\hat{n}d-\hat{n}\ell,\hat{n}a-\hat{n}c) \geq 1+p+\min(\hat{n}a-na+\hat{n}d-nd,\ \max(\hat{n}c+\hat{n}d,\hat{n}a+\hat{n}\ell)-\max(na+n\ell,\ nc+nd)). \tag{4.10}$$

cf Söderström (1979a).

It is shown in Söderström (1979a) that the condition (4.9) is necessary but not generally sufficient. We now turn to some sufficient conditions. There is no real effort made to create as weak conditions as possible.

The first result concerns the sum of an AR process and an MA process.

Theorem 4.1 Consider the identity

$$\frac{\Lambda_v}{D(z)D(z^{-1})} + \Lambda_e C(z)C(z^{-1}) \quad \frac{\hat{\Lambda}_v}{\hat{D}(z)\hat{D}(z^{-1})} + \hat{\Lambda}_e \hat{C}(z)\hat{C}(z^{-1}) \tag{4.11}$$

Assume that $C(z)$, $D(z)$, $\hat{C}(z)$ and $\hat{D}(z)$ have all zeros outside the unit circle and that

$$\hat{n}c \geq nc \geq 0 \tag{4.12}$$

$$\hat{n}d \geq nd \geq 0$$

It then follows that

$$\hat{C}(z) \equiv C(z) \qquad \hat{D}(z) \equiv D(z) \qquad \hat{\Lambda}_e = \Lambda_e \qquad \hat{\Lambda}_v = \Lambda_v \tag{4.13}$$

Proof By comparing the denominators reflecting that both sides of (4.11) must have the same poles, it follows easily that

$$\hat{D}(z) \equiv D(z)$$

A simple rewriting then gives

$$\Lambda_v - \hat{\Lambda}_v \equiv D(z)D(z^{-1})[\hat{\Lambda}_e \hat{C}(z)\hat{C}(z^{-1}) - \Lambda_e C(z)C(z^{-1})]$$

The left hand side is a constant (not a function of z). In the right hand side $D(z)D(z^{-1})$ does vary with z since $nd>0$. Thus it can be concluded that remaining part of the right hand side is zero, or

$$\hat{\Lambda}_e \hat{C}(z)\hat{C}(z^{-1}) \equiv \Lambda_e C(z)C(z^{-1})$$

It then follows from the standard spectral factorization theorem, see Aström (1970), that

$$\hat{C}(z) \equiv C(z) \qquad \hat{\Lambda}_e \equiv \Lambda_e .$$

Insertion of the found results then easily gives $\hat{\Lambda}_v = \Lambda_v$ and the theorem is proved.

Next we consider the case where the signal $s(t)$ is an ARMA process and the noise $n(t)$ is white.

Theorem 4.2 Consider the polynomials

$$D(z)=1+d_1z+\dots+d_{nd}z^{nd} \qquad d_{nd}\neq 0$$
$$L(z)=1+\ell_1z+\dots+\ell_{n\ell}z^{n\ell} \qquad \ell n_{\ell}\neq 0$$
$$\hat{D}(z)=1+\hat{d}_1z+\dots+\hat{d}_{nd}z^{nd}$$
$$\hat{L}(z)=1+\hat{\ell}_1z+\dots+\hat{\ell}_{n\ell}z^{n\ell}$$

which are assumed to have all zeros outside the unit circle. Assume further that the polynomials $D(z)$ and $L(z)$ have no common zeros. Consider the identifiability identity

$$\frac{\hat{L}(z)\hat{L}(z^{-1})}{\hat{D}(z)\hat{D}(z^{-1})}\hat{\Lambda}_v+\hat{\Lambda}_e \quad \frac{L(z)L(z^{-1})}{D(z)D(z^{-1})}\Lambda_v+\Lambda_e \tag{4.14}$$

i) If $n\ell<nd$ there is a unique solution of (4.14). It is given by

$$\hat{L}(z)\equiv L(z) \quad \hat{D}(z)\equiv D(z) \quad \hat{\Lambda}_v=\Lambda_v \quad \hat{\Lambda}_e=\Lambda_e \tag{4.15}$$

ii) If $n\ell\geq nd$ there are infinitely many solutions of (4.14).

Proof No pole-zero cancellation can occur in the right hand side of (4.14). Both sides must have the same poles and zeros. It thus follows that $\hat{D}(z)\equiv D(z)$ and

$$\hat{L}(z)\hat{L}(z^{-1})\hat{\Lambda}_v\equiv L(z)L(z^{-1})\Lambda_v+D(z)D(z^{-1})[\Lambda_e-\hat{\Lambda}_e] \tag{4.16}$$

Consider first the case $n\ell<nd$. By equating the coefficients for z^{nd} we then get $\hat{\Lambda}_e=\Lambda_e$. The result (4.15) then follows from a direct application of the spectral factorization theorem, Aström (1970). Assume then that $n\ell\geq nd$. Let $0\leq\hat{\Lambda}_e<\Lambda_e$ but with $\hat{\Lambda}_e$ else arbitrary. It follows from the spectral factorization theorem that (4.16) has a real valued solution with respect to $\hat{L}(z)$ and $\hat{\Lambda}_v$.

Remark 1 The more general case $\hat{n}d\geq nd$, $\hat{n}\ell\geq n\ell$ is treated in Söderström. The key condition above is $n\ell\leq nd$. It must in the general case be substituted by $\hat{n}\ell<nd$. It has also been shown that solutions can sometimes exist when $\hat{n}d\geq nd$, $\hat{n}\ell<n\ell$.

Remark 2 Part i) of the theorem can be found (more a less explicit) in Goodrich (1978), Mehra (1971) and Kashyap (1970).

Remark 3 Part ii) can in fact be expected from Lemma 4.1.

5. MULTIVARIABLE SYSTEMS

We will now try to extend Theorem 4.2 to the multivariable case. We then start with the following ARMA model of the signal

$$[I+D_1q^{-1}+\dots +D_{nd}q^{-nd}]s(t)=[I+L_1q^{-1}+\dots +L_{n\ell}q^{-n\ell}]v(t) \qquad (5.1)$$

or abbreviated

$$D(q^{-1})s(t)=L(q^{-1})v(t) \qquad (5.2)$$

We will generalize the key assumption in theorem 4.2 ($n\ell<nd$) to

$$n\ell<nd \quad \det D_{nd}\neq 0 \qquad (5.3)$$

The description (5.2) can be transformed. Multiply it from the left with the adjoint of $D(q^{-1})$ [i.e. with $\det D(q^{-1})\,.\,D(q^{-1})^{-1}$].

We then get

$$\delta(q^{-1})s(t)=\tilde{L}(q^{-1})v(t) \qquad (5.4)$$

where $\delta(z)$ is a scalar polynomial and

$$\delta(z)=\det D(z) \qquad (5.5)$$

$$\tilde{L}(z)=\det D(z)\cdot D(z)^{-1}\cdot L(z) \qquad (5.6)$$

The representations (5.2) and (5.4) are equivalent in the sense that they give the same filter, i.e.

$$H(z)\equiv D(z)^{-1}L(z)\equiv\frac{1}{\delta(z)}\tilde{L}(z) \qquad (5.7)$$

Let $s(t)$ and $v(t)$ be p-dimensional. We then have

$$n\delta\triangleq\deg\delta(z)=nd\cdot p \qquad (5.8)$$

$$n\ell\triangleq\deg\tilde{L}(z)\leq\deg\det D(z)\cdot D(z)^{-1}+\deg L(z)\leq nd\cdot(p-1)+n\ell \qquad (5.9)$$

Note that $\det D_{nd}\neq 0$, i.e. (5.3), imply the equality in (5.8).

Let us assume that the parameterized model of the signal has the form

$$H(z;\theta)=\frac{\hat{\tilde{L}}(z)}{\hat{\delta}(z)} \tag{5.10}$$

where $\hat{\delta}(z)$ is scalar polynomial and

$$\hat{\delta}(z)=1+\hat{\delta}_1 z+\dots+\hat{\delta}_{n\delta}z^{n\delta} \tag{5.11}$$

$$\hat{\tilde{L}}(z)=I+\hat{\tilde{L}}_1 z+\dots+\hat{\tilde{L}}_{n\ell}z^{n\ell} \tag{5.12}$$

The identifiability identity (2.7) becomes

$$\frac{\tilde{L}(z)\Lambda_v\tilde{L}(z^{-1})^T}{\delta(z)\delta(z^{-1})}+\Lambda_e\ \frac{\hat{\tilde{L}}(z)\hat{\Lambda}_v\hat{\tilde{L}}(z^{-1})^T}{\hat{\delta}(z)\hat{\delta}(z^{-1})}+\hat{\Lambda}_e \tag{5.13}$$

We then have the following result.

Theorem 5.1 Consider the identity (5.13) where $\hat{\delta}_1\dots\hat{\delta}_{n\delta}$, $\hat{\tilde{L}}_1,\dots \hat{\tilde{L}}_n$, $\hat{\Lambda}_{e\ell},\hat{\Lambda}_v$ are unknowns. Assume that $\delta(z)$ and $\tilde{L}(z)$ are relatively prime and that (5.3) holds. Then (5.13) has a unique solution. It satisfies

$$\hat{\delta}(z)\equiv\delta(z)\qquad \hat{\tilde{L}}(z)\equiv\tilde{L}(z)\qquad \hat{\Lambda}_e=\Lambda_e\qquad \hat{\Lambda}_v=\Lambda_v \tag{5.14}$$

Proof Similar to the proof of Theorem 4.2 we compare the denominators of the identity and conclude that $\hat{\delta}(z)\equiv\delta(z)$. We then get

$$\tilde{L}(z)\Lambda_v\tilde{L}(z^{-1})^T\equiv\hat{\tilde{L}}(z)\hat{\Lambda}_v\hat{\tilde{L}}(z^{-1})^T+(\hat{\Lambda}_e-\Lambda_e)\delta(z)\,\delta(z^{-1})$$

It follows from (5.3), (5.8), (5.9) that $n\delta>n\tilde{\ell}$. By equating the coefficients for $z^{n\delta}$ we get $\hat{\Lambda}_e=\Lambda_e$. The remaining part of the theorem then follows from the spectral factorization theorem for multivariable system, see e.g. Anderson and Moore (1979).

Remark In Söderström (1979a) also the case $\hat{n\delta}\triangleq\deg\hat{\delta}(z)\geq n\delta$, $\hat{n\ell}\triangleq\deg\hat{\tilde{L}}(z)\geq n\tilde{\ell}$ is treated. It is shown that the key condition is $\hat{n\ell}<n\delta$. If multiple solutions exist they all give the same filter $H(z)$ (with some possible common factor) and the same covariance matrix Λ_e and Λ_v.

It is possible to use also other representations of the model. It is well-known that it is not feasible, not even in the noise-free case with $\Lambda_e=0$, to include all the coefficients of $D(z)$ and $L(z)$ as unknowns (unless we have a scalar system with $p=1$). This is related to identification and parameterization of multivariable ARMA processes. However, it follows from the theorem that a sufficient condition for the possibility to use another representation is that its parameter vector is uniquely given from $\delta(z)$ and $\tilde{L}(z)$.

The special case when the system is a pure AR process (i.e. $n\ell=0$ in (5.1)) is rather easy to treat. For that case all parameters of a corresponding $\hat{D}(z)$ polynomial can be used as model parameters.

Theorem 5.2 Consider the identifiability identity

$$D(z)^{-1}\Lambda_v D(z^{-1})^{-T}+\Lambda_e \equiv \hat{D}(z)^{-1}\hat{\Lambda}_v \hat{D}(z^{-1})^{-T}+\hat{\Lambda}_e \qquad (5.19)$$

where

$$D(z)=I+D_1 z+\dots +D_{nd}z^{nd} \qquad nd>0 \qquad \det D_{nd}\neq 0 \qquad (5.20)$$

$$\hat{D}(z)=I+\hat{D}_1 z+\dots +\hat{D}_{\hat{n}d}z^{\hat{n}d} \qquad (5.21)$$

Assume that $\hat{n}d \geq nd$. Regard $\hat{D}_1 \dots \hat{D}_{\hat{n}d}$ $\hat{\Lambda}_e$ as unknowns. It then follows that

$$\hat{D}(z)\equiv D(z) \qquad \hat{\Lambda}_v=\Lambda_v \qquad \hat{\Lambda}_e=\Lambda_e \qquad (5.22)$$

Proof The identity (5.19) can be rearranged as

$$\hat{D}(z)D(z)^{-1}\Lambda_v D(z^{-1})^{-T}\hat{D}(z^{-1})^T \equiv \hat{\Lambda}_v+\hat{D}(z)[\hat{\Lambda}_e-\Lambda_e]\hat{D}(z^{-1})^T$$

Here the right hand side is without poles since $\hat{D}$ is a polynomial. The same must be true for the left hand side. Thus $\hat{D}(z)\equiv F(z)D(z)$ or explicitly

$$[I+\hat{D}_1 z+\dots +\hat{D}_{\hat{n}d}z^{\hat{n}d}]=[I+F_1 z+\dots +F_{nf}z^{nf}][I+D_1 z+\dots +D_{nd}z^{nd}], \quad F_{nf}\neq 0$$

We clearly have $\hat{n}d \leq nf+nd$

If $\hat{n}d < nf+nd$ we get by equating the coefficients for z^{nf+nd}

$$0=F_{nf}D_{nd}$$

and, due to the assumptions, $F_{nf}=0$. This is a contradiction and it can then be concluded that

$$\hat{n}d=nf+nd$$

The identifiability identity can then be written as

$$F(z)\Lambda_v F(z^{-1})^T - \hat{\Lambda}_v \equiv \hat{D}(z)[\hat{\Lambda}_e - \Lambda_e]\hat{D}(z^{-1})^T$$

According to the assumptions it now follows, by equating the coefficients for $z^{\hat{n}d}$, that

$$\hat{\Lambda}_e = \Lambda_e$$

$$\hat{\Lambda}_v \equiv F(z)\Lambda_v F(z^{-1})^T$$

Since the left side is constant and does not vary with z it follows that

$$F(z) \equiv I \qquad \hat{\Lambda}_v = \Lambda_v$$

which makes the end of the proof.

6. APPLICATION TO IDENTIFICATION OF CONTROLLED PROCESSES FROM NOISY DATA

We will now apply the idea of spectral decomposition to the following identification problem. Consider the system in figure 6.1.

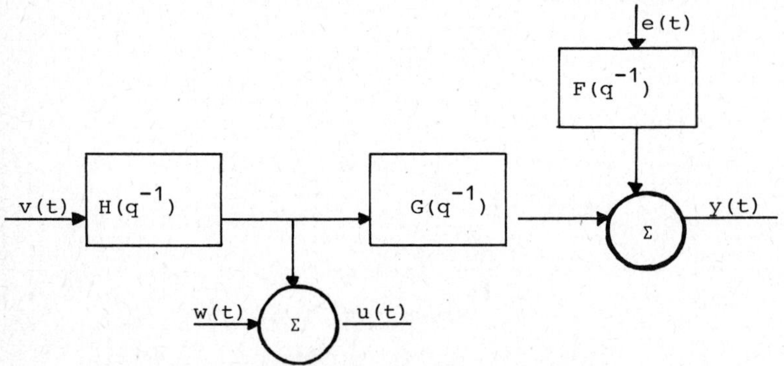

Figure 6.1. Setup for the identification problem

The system dynamics is described with the transfer function G. The input measurements $u(t)$ are corrupted by white measurement noise $w(t)$. The true input signal to the system is assumed to be an ARMA process, $H(q^{-1})v(t)$. Process noise and measurement noise on the output are together modelled as a disturbance $F(q^{-1})e(t)$ at the output. It is assumed that the three noise sources in Figure 6.1

are white and mutually uncorrelated so that

$$E\begin{bmatrix} e(t) \\ v(t) \\ w(t) \end{bmatrix}[e(s)^T \quad v(s)^T \quad w(s)^T] = \begin{bmatrix} \Lambda_e & 0 & 0 \\ 0 & \Lambda_v & 0 \\ 0 & 0 & \Lambda_w \end{bmatrix} \delta_{t,s} \tag{6.1}$$

We also assume that the finite order filters $F(q^{-1})$ and $H(q^{-1})$ have a leading coefficient equal to I, and that these filters and their inverses are asymptotically stable. We are thus modelling the true input $u(t)-w(t)$ and the disturbance at the output as independent (multivariable) ARMA processes.

The identification problem is to estimate the system and noise dynamics from measured data $u(1)$, $y(1), \ldots u(N)$, $y(N)$. This problem has been treated also in Söderström (1979b) where some different approaches are given. A related problem is considered by Maravall (1979). He investigates the <u>local</u> identifiability properties for similar model structures.

Here the problem will be treated using the ideas of spectral decomposition. Then the system is considered to have $u(t)$ and $y(t)$ as outputs. We then have

$$\begin{bmatrix} y(t) \\ \\ u(t) \end{bmatrix} = \begin{bmatrix} F(q^{-1}) & G(q^{-1})H(q^{-1}) & 0 \\ & & \\ 0 & H(q^{-1}) & I \end{bmatrix} \begin{bmatrix} e(t) \\ v(t) \\ w(t) \end{bmatrix} \tag{6.2}$$

We also consider a parametric model of the total process in a similar form:

$$\begin{bmatrix} y(t) \\ \\ u(t) \end{bmatrix} = \begin{bmatrix} \hat{F}(q^{-1};\theta) & \hat{G}(q^{-1};\theta)\hat{H}(q^{-1};\theta) & 0 \\ & & \\ 0 & \hat{H}(q^{-1};\theta) & I \end{bmatrix} \begin{bmatrix} \hat{e}(t) \\ \hat{v}(t) \\ \hat{w}(t) \end{bmatrix} \tag{6.3}$$

$$E\begin{bmatrix} \hat{e}(t) \\ \hat{v}(t) \\ \hat{w}(t) \end{bmatrix}[\hat{e}(s)^T \quad \hat{v}(s)^T \quad \hat{w}(s)^T] = \begin{bmatrix} \hat{\Lambda}_e(\theta) & 0 & 0 \\ 0 & \hat{\Lambda}_v(\theta) & 0 \\ 0 & 0 & \hat{\Lambda}_w(\theta) \end{bmatrix} \delta_{t,s} \tag{6.4}$$

It is then straightforward to obtain the identifiability identity (2.7) for this particular representation. It becomes in partionated form

$$\hat{F}(z;\theta)\hat{\Lambda}_e(\theta)\hat{F}(z^{-1};\theta)^T+\hat{G}(z;\theta)\hat{H}(z;\theta)\hat{\Lambda}_v(\theta)\hat{H}(z^{-1};\theta)^T\hat{G}(z^{-1};\theta)^T$$

$$\equiv F(z)\Lambda_e F(z^{-1})^T+G(z)H(z)\Lambda_v H(z^{-1})^T G(z^{-1})^T \tag{6.5a}$$

$$\hat{G}(z;\theta)\hat{H}(z;\theta)\hat{\Lambda}_v(\theta)\hat{H}(z^{-1};\theta)^T\equiv G(z)H(z)\Lambda_v H(z^{-1})^T \tag{6.5b}$$

$$\hat{H}(z;\theta)\hat{\Lambda}_v(\theta)\hat{H}(z^{-1};\theta)^T+\hat{\Lambda}_w(\theta)\equiv H(z)\Lambda_v H(z^{-1})^T+\Lambda_w \tag{6.5c}$$

Note that (6.5c) is nothing but the identifiability identity for spectral decomposition of the input signal u. Assume now that the parameterization has been chosen such that it is possible to make a unique spectral decomposition of the input signal, i.e. that (6.5c) implies

$$\hat{H}(z;\theta)\equiv H(z) \qquad \hat{\Lambda}_v(\theta)=\Lambda_v \qquad \hat{\Lambda}_w(\theta)=\Lambda_w \tag{6.6}$$

Then it follows directly from (6.5b)

$$\hat{G}(z;\theta)\equiv G(z) \tag{6.7}$$

Finally (6.5a) reduces to

$$\hat{F}(z;\theta)\equiv F(z) \qquad \hat{\Lambda}_e(\theta)=\Lambda_e \tag{6.9}$$

Let us summarize so far: Assume that the spectral density of the measured input can be correctly decomposed into the effects from the measurement noise and from the true unmeasurable input. Then all the filters $F(q^{-1})$, $G(q^{-1})$ and $H(q^{-1})$ are identifiable. It is merely a question of parameterization or canonical representations if we obtain unique parameter estimates or not.

The basic assumption made above was that (6.5c) implies (6.6). This is a sufficient but not necessary condition as illustrated below. Consider now scalar systems. Let the filters be given by

$$F(z)\equiv\frac{C(z)}{A(z)} \quad G(z)\equiv\frac{B(z)}{A(z)} \quad H(z)\equiv\frac{L(z)}{D(z)} \tag{6.10}$$

$$\hat{F}(z;\theta)=\frac{\hat{C}(z)}{\hat{A}(z)} \quad \hat{G}(z;\theta)=\frac{\hat{B}(z)}{\hat{A}(z)} \quad \hat{H}(z;\theta)=\frac{\hat{L}(z)}{\hat{D}(z)} \tag{6.11}$$

where

$$\begin{aligned}
&A(z)=1+a_1z+\dots+a_{na}z^{na} && \hat{A}(z)=1+\hat{a}_1z+\dots+\hat{a}_{na}z^{na}\\
&B(z)=\ \ b_1z+\dots+b_{nb}z^{nb} && \hat{B}(z)=\ \ \hat{b}_1z+\dots+\hat{b}_{nb}z^{nb}\\
&C(z)=1+c_1z+\dots+c_{nc}z^{nc} && \hat{C}(z)=1+\hat{c}_1z+\dots+\hat{c}_{nc}z^{nc}\\
&D(z)=1+d_1z+\dots+d_{nd}z^{nd} && \hat{D}(z)=1+\hat{d}_1z+\dots+\hat{d}_{nd}z^{nd}\\
&L(z)=A+\ell_1z+\dots+\ell_{n\ell}z^{n\ell} && \hat{L}(z)=1+\hat{\ell}_1z+\dots+\hat{\ell}_{n\ell}z^{n\ell}
\end{aligned} \tag{6.12}$$

$a_{na}\neq 0,\ b_{nb}\neq 0,\ c_{nc}\neq 0,\ d_{nd}\neq 0,\ \ell_{n\ell}\neq 0$

The parameter vector is assumed to consist of all the coefficients of the polynomials $\hat{A}$, $\hat{B}$, $\hat{C}$, $\hat{D}$ and $\hat{L}$ as well as the variances $\hat{\Lambda}_e$, $\hat{\Lambda}_v$ and $\hat{\Lambda}_w$. The identifiability results are then as follows.

<u>Theorem 7.1</u> Consider the identifiability identities (6.5a)-(6.5c) and with parameterization of the filters as (6.10)-(6.12). Assume that all the polynomials A, C, D, L, $\hat{A}$, $\hat{C}$, $\hat{D}$ and $\hat{L}$ have all zeros outside the unit circle. Assume further that there is no common zero to A, B and C, nor to D and L. Let $\tilde{B}(z)$ be the unique polynomial with all zeros outside the unit circle and such that $\tilde{B}(z)\tilde{B}(z^{-1})\equiv B(z)B(z^{-1})$. Assume that there is no common zero to AD and BL.

Assume that at least one of the following conditions is satisfied.

i) $nd>0$

ii) $n\ell>0$. Moreover if $B(z_i)=0$ then $B(z_i^{-1})\neq 0$.

Then the identifiability identity has a unique solution given by

$$\hat{A}(z)\equiv A(z)\ \ \hat{B}(z)\equiv B(z)\ \ \hat{C}(z)\equiv C(z)\ \ \hat{D}(z)\equiv D(z)\ \ \hat{L}(z)\equiv L(z)\ \ \hat{\Lambda}_e\equiv\Lambda_e\ \ \hat{\Lambda}_v\equiv\Lambda_v\ \ \hat{\Lambda}_w\equiv\Lambda_w \tag{6.13}$$

<u>Proof</u> Since D and L are relatively prime and both sides of (6.5c) must have the same poles it follows that $\hat{D}(z)\equiv D(z)$. Similarly, using AD and $\tilde{B}L$ relatively prime and invoking (6.5a) it follows that $\hat{A}(z)\equiv A(z)$. The identities can then be rewritten as

$$D(z)D(z^{-1})[C(z)C(z^{-1})\Lambda_e-\hat{C}(z)\hat{C}(z^{-1})\hat{\Lambda}_e]\equiv\hat{B}(z)\hat{B}(z^{-1})\hat{L}(z)\hat{L}(z^{-1})\hat{\Lambda}_v -B(z)B(z^{-1})L(z)L(z^{-1})\Lambda_v \tag{6.14a}$$

$$B(z)L(z)L(z^{-1})\Lambda_v \equiv \hat{B}(z)\hat{L}(z)\hat{L}(z^{-1})\hat{\Lambda}_v \tag{6.14b}$$

$$D(z)D(z^{-1})[\Lambda_w-\hat{\Lambda}_w] \equiv \hat{L}(z)\hat{L}(z^{-1})\hat{\Lambda}_v - L(z)L(z^{-1})\Lambda_v \tag{6.14c}$$

The identity (6.14b) implies that the polynomial $\hat{B}(z)$ fulfils

$$\hat{B}(z) \equiv B(z)\,\frac{L(z)L(z^{-1})}{\hat{L}(z)\hat{L}(z^{-1})}\,\frac{\Lambda_v}{\hat{\Lambda}_v} \tag{6.15}$$

All the poles in (6.15) must be cancelled by zeros. Under condition ii) it then follows that $\hat{L}(z)=L(z)$. We then get $\hat{\Lambda}_v=\Lambda_v$, $\hat{\Lambda}_w=\Lambda_w$ from (6.14c) and thus $\hat{B}(z)\equiv B(z)$. Finally (6.14a) implies $\hat{C}(z)\equiv C(z)$, $\hat{\Lambda}_e=\Lambda_e$ after use of the spectral factorization theorem.

Assume that that condition i) is valid. Insertion of (6.15) into (6.14a) gives after some straightforward calculating

$$C(z)C(z^{-1})\Lambda_e-\hat{C}(z)\hat{C}(z^{-1})\hat{\Lambda}_e \equiv \frac{\tilde{B}(z)\tilde{B}(z^{-1})}{D(z)D(z^{-1})}\,\frac{L(z)L(z^{-1})}{\hat{L}(z)\hat{L}(z^{-1})}\,\frac{\Lambda_v}{\hat{\Lambda}_v}$$

$$[L(z)L(z^{-1})\Lambda_v-\hat{L}(z)\hat{L}(z^{-1})\hat{\Lambda}_v]$$

The poles in the right hand side must be cancelled by corresponding zeros. Since D and BL have no common zero it follows that

$$L(z)L(z^{-1})\Lambda_v \equiv \hat{L}(z)\hat{L}(z^{-1})\hat{\Lambda}_v$$

Again by applying the spectral factorization theorem we conclude that $\hat{L}(z)\equiv L(z)$, $\hat{\Lambda}_v=\Lambda_v$. Then (6.14c) implies $\hat{\Lambda}_w=\Lambda_w$, (6.14b) $\hat{B}(z)\equiv B(z)$ and (6.14a) $\hat{C}(z)\equiv C(z)$, $\hat{\Lambda}_e=\Lambda_e$

Remark The condition $n\ell < nd$, cf Theorem 4.1, necessary to conclude that (6.5c) implies (6.6) is considerably more restrictive than the conditions given in the theorem.

The identification problem posed above has thus very nice identifiability properties. However, the necessary calculations can be cumbersome. If a prediction error method is applied, a numerical optimization routine must be used. In every iteration the corresponding innovations form (3.2) must be computed. This requires the solution of a Ricatti equation or some equivalent nonlinear operations.

An alternative approach using a reparameterization is outlined in Söderström (1979b). The Kalman gain, necessary to form the innovations representation is included in the parameter vector. Then the computational requirements is considerably reduced. However, the identifiability properties will no longer be of the general nature as in Theorem 6.1. In fact identifiability will be lost if $H(q^{-1})$ has smaller order than $G(q^{-1})$.

7. CONCLUSIONS

In various types of spectral analysis it is often not possible to avoid measurement noise. In such cases it is of interest to write a spectral density as a sum of two terms. One term corresponds to the undisturbed signal while the other refers to the measurement noise.

This problem cannot be solved without some a priori structural assumptions. Here both the signal and the measurement noise have been treated as ARMA processes. Note however, that most of the results can also be formulated in continuous time.

For scalar systems an explicit analysis has been done. A simple necessary condition for a unique solution has been established. It involves polynomial degrees. It is based on the fact that the number of independent equations must not be smaller than the number of unknowns. It has also been proved that the decomposition has a unique solution if the signal is an AR process and the noise is an MA process, both of an arbitrary order. Another sufficient condition for existence of a unique solution is to have white measurement noise and that the signal is an ARMA (m, m-k) process with $k>0$.

The results for scalar systems are generalized to the multivariable case where similar results are true, although some additional technical assumptions have been used.

The idea of spectral decomposition has also been applied to an identification problem. It is assumed that non neglectable measurement noise is present both in the input and the output. It is also a priori assumed that the true input is an ARMA process. The technique is to treat the measurements as outputs of a multivariable system. We then have a particular parameterization of the "undisturbed signal". An explicit identifiability analysis is carried out. It is shown that the process can be identified under quite weak conditions.

REFERENCES

[1] Anderson, B.D.O. and Moore, J.B. (1979): Optimal Filtering, Prentice Hall, Englewood Cliffs.

[2] Anderson, W.N. Jr., Kleindorfer, G.B., Kleindorfer, P.R. and Woodroote M.B. (1969): Consistent estimates of the parameters of a linear system. The Annals of Mathematical Statistics, vol. 40, no. 6, pp 2064-2075.

[3] Aström, K.J. (1970): Introduction to Stochastic Control Theory, Academi Press, New York.

[4] Goodrich, R.L. (1978): Stochastic models of human growth. Division of Applied Sciences, Harvard University, Cambridge, Mass, USA.

[5] Kashyap, R.L. (1970): Maximum likehood identification of stochastic linear systems. IEEE Trans. Autom. Control, AC-15, p 25-34.

[6] Levin, B.R. and Shinalov, Yu.S. (1977): Jointly optimum algorithms of signal detection and parameter estimation (a review). Radio Eng. and Electron. Phys., vol. 22, p 1-15.

[7] Ljung, L. (1976): On the consistency of prediction error identification methods. In R.K. Mehra and D.G. Lainiotis, (Eds.): System Identification Advances and Case Studies, Academic Press, New York.

[8] Maravall, A. (1979): Identification in dynamic shock-error models. Springer (Lecture notes in economics and mathematical systems (65)), Berlin.

[9] Mehra, R.K. (1970): On the identification of variances and adaptive Kalman filtering. IEEE Trans. Autom. Control, vol. AC-15, p 175-184.

[10] Mehra, R.K. (1971): On-line identification of linear dynamic systems with applications to Kalman filtering. IEEE Trans. Autom. Control, AC-16, p 12-21.

[11] Söderström,T. (1979a):Spectral decomposition with application to identification. Report UPTEC 79 42R, Inst. of Technology, Uppsala University, Uppsala, Sweden.

[12] Söderström, T. (1979b): Some methods for identification of linear systems with noisy input-output data. Fifth IFAC Symposium on Identification and System Parameter Estimation, Darmstadt.

[13] Van Trees, H.L. (1971): Detection, Estimation and Modulation Theory, Part. III. John Wiley, New York.

Numerical Techniques for Stochastic Systems
F. Archetti and M. Cugiani (eds.)
© North-Holland Publishing Company, 1980

SOME IMPLEMENTATION CONSIDERATIONS OF SELF-TUNING CONTROLLERS

D.W. Clarke

Department of Engineering Science
Parks Road, Oxford, England

The equations of the self-tuning controller of Clarke and Gawthrop are derived as a generalisation of the self-tuning regulator of Aström and Wittenmark. The control objective of minimizing the variance of a generalised system output is shown to be equivalent

(i) to minimizing the conditional expectation of a quadratic function of the system's input, output and set point,

(ii) to a model reference controller with prescribed disturbance control properties,

(iii) to a control for overcoming the system's dead time.

The recursive parameter estimation equations for self-tuning are given, together with routines for the square-root and the UD updating of the covariance matrix involved. These are shown by simulations to have superiour numerical stability when compared with the original Kalman filter formulation. When simulated with short and truncated wordlengths the estimates are seen to become biased. Simulations of the self-tuner using short-word lengths, however, show that its performance is numerically robust as good control is achieved even when individual parameters are away from their optimal values.

Various possibilities for implementing self-tuners on minicomputers and microprocessors are discussed. The equations of a self-tuning commissioning system are derived, for which, unlike previous self-tuners, the loop does not need to be closed when the parameters are being estimated.

1. Introduction

The control of manufacturing processes is difficult in many cases as adequate models of plant and disturbance characteristics are hard to obtain from theoretical considerations, making it impossible to use the synthesis methods which work well for electromechanical systems. In practice, proportional + integral + derivative (PID) controllers are connected to the processes, and the parameters tuned on-line using methods such as that due to Ziegler and Nichols (1942). Although this can give satisfactory control initially, plant non-linearities and drift may lead to poor performance if the parameters are not re-tuned following changes in the operating region or the plant dynamics.

In order to use modern synthesis techniques, the method of system identification has been proposed so that the control strategy can take into account both a model for the disturbances and an input-output process model (e.g. Clarke,Dyer,Hastings-James,Ashton,Emery (1973)). When identification is used in an off-line way, an experiment has to be performed to produce data from which these approximate models can be obtained. This can lead to certain problems:

a) to obtain a good dynamic input-output model, care is often taken to minimize external disturbances and to inject a high-variance test signal; this can often lead to an unrepresentative model for the disturbance, and a regulator designed on this model would be inadequate

b) even if good models of the process and its disturbance are obtained, the plant dynamics may change by the time the controller is implemented

c) most control synthesis methods are 'certainty equivalent' in the sense that they ignore variances of the estimated parameters.

With the advent of process minicomputers and microprocessors, theoretical attention was drawn to recursive parameter estimation and to automated methods of controller synthesis. The optimum procedure, so-called dual control, recognised that the control law had to include probing (to minimise future parameter uncertainties) and cautious (to reflect current uncertainties) components (Jacobs and Patchell (1972)). Unfortunately, dual controllers have been derived for only the simplest of systems, and even these require excessive on-line computations. To simplify matters, therefore, self-tuning algorithms (Astrom,Wittenmark (1973)-Clarke, Gawthrop (1975)-Clarke,

Gawthrop (1979)) have been developed which ignore parameter uncertainties and hence omit cautious and probing control components. Explicit self-tuners identify the system transfer-function directly, using any of the available recursive methods, and at each sample-time perform some algorithmic computation, such as resolving a polynomial identity to place close-loop poles in certain prespecified locations (Wellstead (1978)). Implicit self-tuners, on the other hand, require no intermediate algorithmic computations, as they identify the parameters of the appropriate controller rather than those of the plant transfer-function. An algorithm is called 'self-tuning' if, for constant plant parameters, the controller parameters tend towards those which would have been obtained if the plant parameters were know exactly.

This paper will review certain aspects of the implementation of implicit self-tuners which attempt to minimise particular quadratic performance indices.

A locally-linearised model which includes both disturbances with rational spectra and a transport delay typical of many industrial processes is:

$$y_t+a_1y_{t-1}+\ldots+a_ny_{t-n} = b_0u_{t-k}+b_1u_{t-k+1}+\ldots+b_nu_{t-k-n} +\xi_t+c_1\xi_{t-1}+\ldots+c_n\xi_{t-n}. \tag{1}$$

Here u_t and y_t are deviations from the nominal mean input and output signals $\bar{u}$ and $\bar{y}$, ξ_t is an uncorrelated random sequence and k (which is a positive integer) is the process delay. If u_t and y_t are the actual input and output signals, rather than deviations, a constant d needs to be included in the model of eqn. 1, but it will be omitted here for clarity. By defining q^{-1} to be the backward shift operator (i.e. $q^{-1}u_t=u_{t-1}$) the model becomes:

$$A(q^{-1})y_t = B(q^{-1})u_{t-k} + C(q^{-1})\xi_t \tag{2}$$

where A,B and C are polynomials in q^{-1}; the argument q^{-1} will be later omitted where appropriate.

The above model was used in the self-tuning regulator of Aström and Wittenmark (1973), in which the objective is to choose the control u_t to minimise the variance of the output y_t. To derive the algorithm, eqn. 2 is first rewritten in <u>predictor</u> form:

$$y_{t+k} = \frac{B}{A} u_t + q^k \frac{C}{A} \xi_t.$$

Invoking the well-known polynomial identity:

(3) $\frac{C}{A} = E + q^{-k} \frac{F}{A}$, where $E(q^{-1})$ is a polynomial of order k-1 we have, with ξ_t obtained in terms of u_t and y_t from eqn. 2:

$$y_{t+k} = \frac{B}{A} u_t + E\xi_{t+k} + \frac{F}{A} \left(\frac{Ay_t - Bu_{t-k}}{C}\right)$$

$$= \frac{F}{C} y_t + \frac{B}{AC}(C - q^{-k}F) u_t + E\xi_{t+k} \quad .$$

Then, using (3) again, we obtain:

(4) $$y_{t+k} = \frac{Fy_t + EBu_t}{C} + E\xi_{t+k}$$

or

(5) $$y_{t+k} = y^*_{t+k/t} + \tilde{y}_{t+k/t}.$$

In eqn. 5 y* is the k-step-ahead prediction and $\tilde{y}$ is the prediction error, which is independent of y* as all terms in ξ are of the form ξ_{t+j}, $1 \le j \le k$. Hence $E(y^2_{t+k})$ is minimised by choosing u_t to set y*=0, and the minimum output variance achieved in closed-loop is $E(\tilde{y}^2) = \sigma^2(1+e_1^2+\ldots+e_{k-1}^2)$.

The control using eqn. 4 depends on the parameters of A,B and C being known, and an explicit self-tuner could use a method such as extended least squares (Panuska (1969)) to estimate A, B and C and then use the identity of eqn. 3 to derive E and F parameters. An implicit self-tuner will directly estimate the parameters of the F and G polynomials, where G is defined to be EB, using a regression model of the form:

(6) $$y_t = Fy_{t-k} + Gu_{t-k} + e_t.$$

To see how eqn. 6 is related to eqn. 4, let C^{-1} be expanded as an infinite polynomial in q^{-1} (implying the roots of C are inside the stability region) of the form:

$$C^{-1} = 1+\alpha_1 q^{-1}+\alpha_2 q^{-2}+\ldots+\alpha_r q^{-r}+\ldots, \ |\alpha_r| \to 0, \ r \to \infty.$$

Then eqn. 4 becomes :

$$y_t = Fy_{t-k}+Gu_{t-k}+\ldots+\alpha_r(Fy_{t-k-r}+Gu_{t-k-r})+\ldots+\tilde{y}.$$

Under feedback control, however, $Fy_{t-k-r} + Gu_{t-k-r}$ is set to zero, and so eqn. 6 is an adequate model of the system provided that F and G are exactly known. Note the circularity of the argument here: if the control parameters were known then eqn. 6 is a regression relation allowing the use of linear-least-squares (as e_t is independent of u_{t-k}, y_{t-k}) to obtain a consistent (though not necessarily efficient as e_t is a moving-average process) estimation of F and G. The actual convergence depends on non-linear time-varying equations, but has been verified by many simulations and theoretical discussions (e.g. Gawthrop (1978)) outside the scope of this paper. Suffice is to say that, under certain conditions such as sufficient parameterization and the positive-realness of $C^{-1} - 0.5$, convergence of the self-tuning regulator algorithm has been proved.

2. Objectives of the self-tuning controller

The objective of the self-tuning regulator, in minimising the output variance, is restrictive and applicable to only a few industrial situations such as paper-making where a minimal variance output means that the mean thickness of the paper may be reduced whilst ensuring that only a given percentage of the paper fall below a specified minimum thickness. Such a criterion seems less applicable to process variables such as temperature, flow, pressure etc. Moreover, with the regulator no account is taken of the control effort required to achieve this minimal variance; if this is found to be excessive the only feature that can be altered is the sample interval of the controller. With the self-tuning controller (Clarke, Gawthrop (1975); Clarke, Gawthrop (1979)), greater flexibility is achieved by defining a generalised output ϕ_t of the form:

$$(7) \qquad \phi_{t+k} = Py_{t+k} + Qu_t - w_t .$$

Here $P(q^{-1})$ and $Q(q^{-1})$ are transfer-functions which can be specified by the control designer to achieve one of the variety of objectives. The signal w_t is the set-point (possibly prefiltered to avoid sudden changes) which is available at time t; the use of w_t allows the self-tuner to be further generalised to include tracking as well as regulatory performance. The cost-function which is minimised (given known parameters) is the variance of the generalised output:

(8) $I = E\{\phi^2_{t+k}\}$.

This cost-function in conjunction with eqn. 7 admits several interpretations.

In (5), using direct arguments, I was shown to be equivalent to the cost-function:

$$I' = E\{(Py_{t+k} - w_t)^2 + (Q'u_t)^2 | t\}$$

which is conditioned on data available up to time t, and $q'_o Q' = b_o Q$. For example, if Q is chosen to be a constant, λ, and P is chosen to be 1:

$$I' = E\{(y_{t+k} - w_t)^2 + b_o \lambda u_t^2 | t\}$$

which allows the user to weight deviations of y from the set point w against the magnitude of the control u. Note, however, that the actual weighting unfortunately depends on the parameter b_o and hence will be system dependent. Another possibility is to choose $Q' = \lambda(1-q^{-1})$ leading to a costing depending on changes in the control:

$$I' = E\{(y_{t+k} - w_t)^2 + b_o \lambda (u_t - u_{t-1})^2 | t\}.$$

Other interpretations are due to Gawthrop (1977). If Q=0 then eqns. 5 and 7 show that under a control such that $\phi^*_{t+k|t}=0$,

$$\phi_{t+k} = \tilde{\phi}_{t+k|t} = Py_{t+k} - w_t$$

giving:

$$y_{t+k} = \frac{1}{P}(w_t + \tilde{\phi}_{t+k|t}).$$

The transfer-function P can be chosen to be the inverse M^{-1} of some desired closed-loop model M, such that

(9) $y_{t+k} = Mw_t + M\tilde{\phi}_{t+k|t}$.

This corresponds to a model-reference tracking system in which the disturbance behaviour in closed loop is also prespecified to some extent.

If P=1 and $Q=L^{-1}$, say, where L is a transfer-function such as a PID regulator, the control which minimises I and sets ϕ^* to zero is, from eqn. 7

$$y^*_{t+k|t} + L^{-1}u_t - w_t = 0$$

or

(10) $$u_t = L(w_t - y^*_{t+k|t}).$$

This appears to be in the form of a classical feedback control law, except that the optimal k-step-ahead prediction y^* is fed back rather than the current output y_t. This has the effect of removing the phase lag of the process time delay k and allowing for tighter conventional PID control.

Derivation of the control which minimises the variance of ϕ is straightforward. Substituting eqn. 7 defining ϕ into the system model given by eqn. 2 we obtain

(11) $$A\phi_{t+k} = (PB+QA)u_t + PC\xi_{t+k} - Aw_t.$$

Here the term $-Aw_t$ is known, so except for this term eqn. 11 gives a model

(12.1) $$A'\phi_t = B'u_{t-k} + C'\xi_t$$

where

(12.2) $$\begin{cases} A' = A \\ B' = PB+QA \\ C' = PC \end{cases}$$

For the case where P and Q have both numerator and denominator dynamics, rather than just being polynomials, giving $P=P_n(q^{-1})/P_d(q^{-1})$ and $Q=Q_n(q^{-1})/Q_d(q^{-1})$, a model like eqn. 12.1 emerges, where

(12.3) $$\begin{cases} A'=P_dQ_dA \\ B'=P_nBQ_d+Q_nAP_d \\ C'=P_nCQ_d \end{cases}$$

In each case eqn. 12 is of the same form as eqn. 2, so the derivation of a control for minimising $E\{\phi^2_{t+k}\}$ follows the above derivation for the minimisation of $E\{y^2_{t+k}\}$. From eqn. 11

$$\phi_{t+k} = \frac{(PB+QA)}{A}u_t + \frac{PC}{A}\xi_{t+k} - w_t$$

and the polynomial identity of eqn. 3 becomes

(3.1) $$\frac{P_nC}{P_dA} = E + q^{-k}\frac{F}{P_dA}$$

giving, using the system model, eqn. 2, to relate ξ_t to y_t and u_t

$$\phi_{t+k} = \frac{PB+QA}{A}u_t + E\xi_{t+k} + \frac{F}{P_dA}\{\frac{Ay_t - Bu_{t-k}}{C}\} - w_t$$

$$= \frac{F}{P_dC}y_t + \frac{PB+QA}{A}u_t - \frac{B}{C}(\frac{PC}{A} - E)u_t - w_t + E\xi_{t+k}$$

$$= \frac{Fy_t}{P_dC} + \frac{BE+CQ}{C}u_t - w_t + E\xi_{t+k}$$

(4.1) $$= \frac{FQ_dy_t + P_d(BEQ_d + CQ_n)u_t - P_dQ_dCw_t}{P_dQ_dC} + E\xi_{t+k}$$

or

(5.1) $$\phi_{t+k} = \phi^*_{t+k|t} + \tilde{\phi}_{t+k|t}.$$

Hence the control which minimises $E\{\phi^2_{t+k}\}$ sets $\phi^*_{t+k|t}$ to 0, giving

(13.1) $$u_t = \frac{P_dQ_dCw_T - FQ_dy_t}{P_d(BEQ_d + CQ_n)}$$

(13.2) $$= \frac{Cw_t - Fy_t/P_d}{BE + CQ}.$$

When in a closed-loop, the control given by eqn. 13 gives the following behaviour in terms of the set-point w_t and the disturbance ξ_t:

a) output

(14.1) $$y_{t+k} = \frac{BP_dQ_d}{Q_nAP_d + P_nBQ_d}w_t + \frac{P_dQ_dBE + P_dQ_nC}{Q_nAP_d + P_nBQ_d}\xi_{t+k}$$

b) control

(14.2) $$u_t = \frac{AP_dQ_d}{Q_nAP_d + P_nBQ_d}w_t - \frac{Q_dF}{Q_nAP_d + P_nBQ_d}\xi_t$$

The characteristic equation of the closed-loop system is

$$Q_nAP_d + P_nBQ_d = 0.$$

For the cases of the self-tuning regulator for minimising $E\{y^2\}$ and of the model-reference interpretation of the self-tuning controller there is no control weighting and $Q_n=0$, giving a characteristic equation whose roots are simply those of P_nBQ_d. For stability these roots must lie within the unit-circle, and this implies that the discrete-time model of eqn. 2 must have its zeros within the unit circle.

This is an unreasonable restriction in practice, as systems with fractional time delays, or high-order systems with a short sample interval may well have discrete-time models with zeros outside the stability region. To achieve stability in this context a control weighting where $Q=\lambda$ leads to a characteristic equation

$$\lambda AP_d + P_nB = 0 \ .$$

Hence if the system is open-loop stable (roots of A in the stability region), and if the designer has chosen P_d correctly, the parameter λ may be varied so as to get the closed-loop poles into the stability region. Such a control may be interpreted as "detuned" model-reference control in which control effort is traded off against closeness of model-following. If a parameter such as λ is not available in an implicit self-tuner, stability can only be achieved by either increasing the sample time or the assumed value of the system's delay k.

3. Self-tuning

A naive approach to the use of eqn. 4.1 in self-tuning control would be to follow the argument of the self-tuning regulator and propose a regression model of the form given in eqn. 6:

$$(6.1) \qquad \phi_t = F'y_{t-k} + G'u_{t-k} + H'w_{t-k} + e_t$$

where the polynomials F',G' and H' are, from eqn. 4.1:

$$(15) \qquad \begin{cases} F' = FQ_d \\ G' = P_d(BEQ_d + CQ_n) \\ H' = -P_dQ_dC \end{cases}$$

This is, of course, parametrically inefficient as both P and Q are prespecified by the designer. Moreover, whenever Q (written in the form $\lambda Q'$) is varied to adjust the control variance, the estimation of the G' parameters is affected. Instead, it is convenient to divide eqn. 4.1 through by P_dQ_d and to define y_t^f as the filtered version of y_t by $y_t^f = y_t/P_d$. This gives:

$$(16) \qquad \phi_{t+k} = \frac{Fy_t^f + Gu_t - C(w_t - Qu_t)}{C} + E\xi_{t+k}.$$

Hence a suitable regression model for the self-tuning controller is:

(17) $$\phi_t = Fy^f_{t-k} + Gu_{t-k} + Hw'_{t-k} + e_t$$

where w'_t is the known signal $w_t - Qu_t$. With such a model, on-line changes to Q affect only the data w'_t rather than the parameters of F, G or H. Moreover, defining ϕ^y to be the component in ϕ due to y_t only (i.e. $\phi^y_t = Py_t$) then ϕ^y itself admits to the model

(18) $$\phi^y_{t+k} = \frac{Fy^f_t + Gu_t}{C} + E\xi_{t+k}.$$

Hence a suitable regression model for the F and G parameters alone is

(19) $$\phi^y_t = Fy^f_{t-k} + Gu_{t-k} + e_t.$$

There are several ways to obtain a self-tuning controller from the above equations:

a) using eqn. 17

Use linear-least-squares to estimate $\hat{F}$, $\hat{G}$ and $\hat{H}$ and produce a control signal u_t from

$$\hat{F}y^f_t + \hat{G}u_t + \hat{H}(w_t - Qu_t) = 0.$$

As an arbitrary constant α may be multiplied through this equation, one parameter may be fixed. One useful possibility is to take H_o to be -1, as by assumption $C_o = +1$.

b) using eqn. 19

Use linear-least-squares to estimate $\hat{F}$ and $\hat{G}$ and take C to be 1 to give the control u_t from

$$\hat{F}y^f_t + (\hat{G}+Q)u_t - w_t = 0.$$

This has the advantage of simplicity in that fewest parameters are estimated, although the response to changes in w_t will not be optimal. However, it could be argued that in many practical cases w_t remains constant over long intervals, in which case tracking performance is not important.

c) using extended-least-squares

Defining $\phi^{*y}_{t|t-k}$ to be the k-step-ahead prediction of ϕ^y, eqn. 18 can be written in the form:

$$C\phi^{*y}_{t|t-k} = Fy^f_{t-k} + Gu_{t-k}; \quad \phi^y_t = \phi^{*y}_{t|t-k} + e_t$$

or

(20) $\phi_t^y = Fy_{t-k}^f + Gu_{t-k} - \sum_{i=1}^{n} c_i \phi_{t-i}^{*y} + e_t$

To use extended-least-squares ideas, let ϕ_{t-i}^{*y} be proxied by the calculable approximation $\hat{\phi}$ given by

(21) $\hat{C}\hat{\phi}_{t-i}^y = \hat{F}y_{t-k-i}^f + \hat{G}u_{t-k-i}$

where $\hat{C}$, $\hat{F}$ and $\hat{G}$ are estimates of C, F and G obtained at the previous sample instant. Then a linear-least-squares type of algorithm is used with eqn. 20 to obtain new $\hat{C}$, $\hat{F}$ and $\hat{G}$ estimates and the control is derived from

$$\hat{F}y_t^f + \hat{G}u_t - \hat{C}(w_t - Qu_t) = 0.$$

There is one important practical advantage in the use of this method. The use of linear-least-squares is valid in method (a) above as the feedback ensures that $\phi^*_{t-i|t-k-i}$ is set to zero, so that components analogous to the $\alpha_r q^{-r}$ terms in the self-tuning regulator can be ignored. This is <u>not</u> the case, however, if the self-tuner is in open-loop or if the control signal reaches a limiting value and so cannot ensure $\phi^* = 0$. Method (c), on the other hand, is in the form of a self-tuning predictor and does not assume the self-tuning controller is in closed loop. Any control u_t, such as a pseudo-random binary test signal or resulting from a PID regulator, can be used in the equations 20 and 21 to obtain estimates. For example, during the commissioning phase of a control law various candidate self-tuners -with different assumptions about model order and time delay, or with different choices of P and Q- can be tested by inspecting the control signals that they would produce. It is also possible to use this method off-line with data obtained from a prior identification experiment.

4. <u>Parameter estimation</u>

In all the implementations of self-tuning described here, the equations for parameter estimation may be written in Kalman filter form as:

(22.1) '<u>state dynamics</u>' $\theta_t = \theta_{t-1}$

(22.2) '<u>output</u>' $\phi_t = x_{t-k}^T \theta_t + e_t$

where the vector θ contains the unknown parameters, assumed here to be constant, and the vector x_{t-k} contains past values of data such as y^f_{t-k-i}, u_{t-k-i}, w'_{t-k-i} or $\hat{\phi}_{t-i}$. The appropriate Kalman filter equations for updating estimates $\hat{\theta}$ are well known to be:

(23.1) $$\hat{\theta}_t = \hat{\theta}_{t-1} + K_t(\phi_t - x^T_{t-k}\hat{\theta}_{t-1})$$

(23.2) $$K_t = M^-_t x_{t-k}/(\sigma^2 + x^T_{t-k}M^-_t x_{t-k})$$

(23.3) $$M^+_t = [I - K_t x^T_{t-k}]M^-_t \; ; \; M^-_t = M^+_{t-1}$$

where M^+_t is the posterior covariance of $\hat{\theta}_t-\theta$ and σ^2 is the variance of e_t. As the variance σ^2 is generally unknown, it is conventional to rescale M by defining $M_t=M^+_t/\sigma^2$. Again it is conventional to use self-tuning with nominally (slowly) time-varying systems, so either eqn. 22.1 is modified to include a driving uncorrelated sequence so that θ_t follows a random walk, or more commonly a 'forgetting factor' β (where $\beta=1-\varepsilon$) is used to inflate the prior covariance M^-_t by dividing by β, giving

(24.1) $$K_t = M_{t-1}x_{t-k}/(\beta + x^T_{t-k}M_{t-1}x_{t-k})$$

(24.2) $$M_t = [I - K_t x^T_{t-k}]M_{t-1}/\beta.$$

In view of the inflation of M_t by β^{-1}, $\|M_t\|$ and hence $\|K_t\|$ no longer tends to zero as $t \to \infty$, and the modified estimator is no longer consistent. However, the norm of the asymptotic covariance of the estimates can be controlled by choosing β close to 1 at the expense of speed of adaptation.

Self-tuners are intended to be left unattended over long periods of time; even with the relatively long sample interval of 1 minute typical of many industrial control loops, tens or even hundreds of thousands of samples may be processed without operator intervention. This place particular emphasis on the robustness of the estimation algorithm. Moreover, as microprocessors are likely to be a common means of implementation of self-tuning, the estimators have to be reliable even with relatively short word-lengths and with truncation rather than rounding of floating-point numbers. There are two major difficulties with the use of recursive-least-squares as formulated above:

a) the data may not be 'persistently exciting'. This occurs when the control actuator reaches a limit of its travel and hence charges in demanded u_t no longer affect the process, or when the output measurement is quantized and the self-tuner regulates to within the quantization bound. In such cases the parameters drift so as to produce a control law with increasing gain, until the actuator unsticks or a disturbance arrives which produces a spuriously large control signal

b) the equation for updating M_t, eqn. 24.2, is numerically unsound in that M_t may become negative-definite or even, if updated in its full form, skew-symmetric. The result of this is instability in the parameter estimates which will not be removed until a sufficiently large control signal is sent out to make M_t positive again. In practice this control signal will be sufficient to seriously affect the plant output and lead to poor closed-loop performance.

If the data is insufficiently exciting, e.g. if $x_{T+r}=0$ for $r>0$, then eqn. 24.2 shows that $\|M_t\|$ behaves as $\|M_T\|/\beta^r \to \infty$ as $r \to \infty$. There are several ways to avoid this so-called 'blow-up'; one is to restrict $\|M_t\|$ to some prescribed limit (one possibility is to choose this limit to be $\|M_0\|$), and to stop updating the estimates until $\|M_t\|$ falls below the limit again or to inject a test signal when the limit is reached. These remedies depend on a norm being available, preferably with little extra computing effort. For the case of potential negative definiteness, on the other hand, although a similar solution by imposing a lower bound on $\|M_t\|$ would work, it is preferable to revise the updating equations into numerically more stable forms.

Both solutions to the numerical problem depend on the factorization of M_t into an upper triangular matrix and its transpose, which is possible as M_t is symmetric and positive definite. The algorithms were stimulated by aerospace work into the full Kalman filter equations which suffer from the same numerical difficulties. The first algorithm due to Peterka (1975), factorizes M_t into $S_t S_t^T$ where S is upper triangular, and the updating equation 24.2 is written as:

$$S_tS_t^T = \frac{1}{\beta} S_{t-1} \left\{ 1 - \frac{S_{t-1}^T x_{t-k} x_{t-k}^T S_{t-1}}{\mu^2} \right\} S_{t-1}^T$$

where $\mu^2 = \beta + x_{t-k}^T S_{t-1} S_{t-1}^T x_{t-k}$. By defining the vector $f = S^T x$, and T to be an orthogonal matrix, this may be written as

$$S_tS_t^T = \frac{1}{\sqrt{\beta}} S_{t-1} \left[I \mid \frac{jf}{\mu} \right] TT^T \begin{bmatrix} I \\ \hline j\frac{f^T}{\mu} \end{bmatrix} S_{t-1}^T \frac{1}{\sqrt{\beta}} .$$

Now choose T such that $\left[I \mid \frac{jf}{\mu} \right] T = \left[H \mid \underset{\sim}{0} \right]$ where H is upper trinagular and $\underset{\sim}{0}$ is a vector of zeros, then an upper triangular square-root of M_t can be updated as

$$S_t = \frac{1}{\sqrt{\beta}} S_{t-1} H_t .$$

Clarke, Cope and Gawthrop (1975) give details of how the elements of H are computed, and Table 1 gives a Fortran listing of the square-root algorithm. It is found that the algorithm requires m extractions of a square-root, where m is the number of estimated parameters; this is not a great computational burden, but in small systems the subroutine for extracting the square root could be eliminated if the UD formulation is used instead.

With the UD factorization M_t is written as UDU^T where U is upper triangular with units stored along the diagonal and D is a diagonal matrix corresponding to the variances of the individual parameter estimates. The development of the algorithm, due to Bierman and Thornton for the full Kalman filter case, is given in Bierman and Thornton (1978) and the references therein.

Table 2 shows a Fortran listing of the UD method as applied to parameter estimation; note the similarity in the size of the Fortran routines required. Excluding the loops for calculating PERR and updating THETA, the square-root method involves about $(4m^2+5m)/2$ multiplications plus m square roots per cycle, whereas the UD method involves about $(3m^2+3m)/2$ multiplications per cycle. In addition, in the UD method D is available for diagnostic purpose without the extra computations that would be required in the square-root case, and so the UD method is to be preferred.

In order to test the numerical performance of the algorithms, a fourth-order autoregressive system of the form

(25) $(1 - \alpha q^{-1})^n \; y_t = \xi_t \;;\quad \mathrm{var}\{\xi_t\} = 1$

was simulated, where α was chosen to be 0.93 and n was 4. Roots near to the unit circle were taken to make M nearly singular. A model:

(26) $y_t = \theta_1 y_{t-1} + \ldots + \theta_4 y_{t-4} + e_t$

was chosen for estimation using the Kalman filter (KA), square-root and UD algorithms. In each case M_0 was chosen to be I and β to be 1, and the graphs show the estimates $\bar{\theta}$, whose true value is 3.72 and corresponds to the straight line on the figures. Fig. 1 shows the UD and KA estimates over the first 200 samples. When far from the true value they follow the same trajectory, but when approaching it the numerical features of the algorithms predominate and the estimates drift apart. Fig. 2 shows the same simulation, but now extending the run over 5000 samples. The Kalman filter shows gross numerical instability, whereas the UD estimates have converged. It is this feature which makes the UD method (or the square-root method which produced similar behaviour) mandatory for self-tuning applications. The above simulations were performed on a PDP 11/45 computer in which floating point numbers are stored with 24 bits precision, and where calculations are rounded rather than truncated. To investigate the effect of different word lengths, such as may be used on a microprocessor implementation, the estimation algorithm were rewritten so that results of each numerical operation could be truncated to an arbitrary extent. Figures 3 and 4 show the effect on the estimation of $\hat{\theta}$, using word lengths of 22 and 20 bits: in each case the results of modified UD algorithm, named DU on the graphs, follow the results of the UD estimator initially but drift away later. At no stage is there numerical instability, and this was verified even for very short word-length working. One implication is that near the optimum when the elements of the Kalman gain vector K_t have become small, the truncation is sufficient to nullify the effect of the parameter update equation 23.1. This was verified experimentally when using a forgetting factor β of 0.95; both the UD and DU algorithms produced asymptotically fluctuating estimates of the parameters in which the numerically induced bias in the DU estimates became relativelly small. Even so, the use of short word length working does not seem appropriate for recursive parameter estimation over long periods if the model which is obtained is to be in an explicit self-tuner. The case of the implicit self-tuner, on the other hand, is different in that only the predictive properties of the model are important.

5. Simulation of the self-tuning controller

A system whose continuous transfer function is

$$Y(s) = \frac{1}{s^2+s+1} U(s) + \frac{1}{s+0\cdot 1} \xi(s)$$

when sampled at unit intervals leads to the third order model

$$(1-1.69q^{-1}+1.08q^{-2}-0.33q^{-3})y_t = (0.34-0.07q^{-1}-0.22q^{-2})u_{t-1}+(1-0.79q^{-1}+0.37q^{-2})\xi_t$$

This system was simulated on a PDP 11/45 in conjunction with a self-tuning controller for which P was 1 and control weight Q chosen as $\lambda = 0.1$. As k=1 and n=3, 8 parameters were estimated using the square-root algorithm with a β of 0.995 and with the value of H_0 fixed to its known value of -1. Fig. 5 shows the result of 1000 samples of the self-tuning control in which the initial parameters were chosen as 0, and no limits placed on the control signal amplitude. The graph NSE shows the disturbance which would affect y_t in the absence of control u_t. The dashed lines on the graphs of the estimated parameters correspond to the optimal parameters that would be used if the system parameters were known exactly.
The correlations functions between ϕ_{t+i} and ϕ_t, y_t, u_t and w_t were computed over the last 500 samples; these are used to verify the operation of the self-tuner.
Simulations of self-tuning have also been made in a microcomputer for which numbers are represented using 16 bits and calculations are truncated. Hence these correspond to the case where parameter estimation has been shown to have certain numerical difficulties. The first example corresponds to 1 in Clarke, Gawthrop (1975), which used a PDP 11/45 for simulation, and involves the discrete-time model with an additive constant:

$$(1-1.5q^{-1}+0.7q^{-2})y_t = (1+0.5q^{-1})u_{t-1}+(1-0.5q^{-1})\xi_t+0.4.$$

Here $P=1$, $Q=\lambda=0.5$ and $\beta=0.995$, and fig. 6 corresponds to fig. 2 of Clarke, Gawthrop (1975). The graph CUM is the 'cumulative excess loss' which is $\sum_{i=1}^{t}\{\phi_i^2-\phi_i^{o2}\}$, where ϕ_i^o is the optimum value of ϕ which would be obtained were the parameter known exactly, and would tend to become constant if the self-tuner converges to the exact values. It is interesting to note that, although the parameter trajectories for the 24 bit and 16 bit algorithms differ significantly, the values of cumulative excess loss at t=1000 are remarkably close. The second example corresponds to example 3 of Clarke, Gawthrop (1975), and is the nonminimum phase system

$$(1-0.95q^{-1})y_t = (1+2q^{-1})u_{t-2}+(1-0.7q^{-1})\xi_t$$

where $P=1$ and $Q=\lambda>0.514$ is required for stable control; in the simulation λ was chosen to be 0.7. Fig. 7 shows the result of this simulation; again the performance is similar to that of the 24 bit version given in Clarke, Crawthrop (1975) although the parameter estimates differ significantly.

The above simulations show that self-tuning is effective even when the estimated parameters are well away from the fully converged values. In each case the control performance after 20-30 samples is indistinguishable from that achieved at the end of the run. Hence the implicit self-tuner appears to be considerably more robust than the earlier parameter estimation algorithms. This has been confirmed by simulations of a chemical reactor controller for which respectable results were obtained using only 4 bits of precision. Why is the self-tuner so robust numerically? The most satisfactory explanation is that the objectives of the parameter estimation and of the control law are very closely coupled in implicit self-tuners. The estimator attempts to choose $\hat{\theta}_t$ so that the variance of the prediction error of the model $\phi_t-x_{t-k}^T\hat{\theta}_t$ is as small as possible, whereas the control is attempting to make the variance of the k-step-ahead prediction error $\phi_{t+k}-x_t\hat{\theta}_t$ as small as possible. Such a coupling is not directly obvious with explicit self-tuners, and it is conjectured that these would be more susceptible to the parameter estimation errors resulting from short-word length operation.

6. Possibilities for implementation

When implementing self-tuning two important practical factors need to be considered. Many control actuators are non-linear, having stiction and hysteresis, and a small-signal input-output model of the plant would be poor as it would be dominated by the actuator characteristics. To use self-tuning effectively the actuator needs linearizing using local feedback, with the self-tuner acting in cascade with the loop. For example, if the actuator is a valve, a local flow control loop (e.g. using PI control) should be used with the self-tuner control signal acting as set-point in flow. This has a further advantage in that the self-tuner can be made aware of saturation in control effort and hence freeze its estimation algorithm. The other practical factor is the general requirement to control against constant disturbances such that the average error is zero, as is obtainable with conventional integral control. If there is no integrator in the loop the use of a non-zero value of control weighting Q will produce an offset, even if an additional constant is included in the estimated parameters. Various possibilities for overcoming this problem are outlined in Clarke, Gawthrop (1979); the most promising uses a simple integral control in cascade with the self-tuner. These practical factors mean that the self-tuner is rarely to be used in isolation, and a proper implementation should include allowance for simple controllers in cascade.

A self-tuner can be implemented in a process minicomputer, a microcomputer or even on a single-chip microprocessor. In a minicomputer the self-tuning algorithm can be acting on several loops simultaneously, or it could be used to tune control parameters for individual loops in turn, leaving them with fixed parameter controllers when fully tuned. Using the modified algorithm which does not require the loop to be closed, the operator can commission a new self-tuner by comparing its proposed control signal with that of the currently active control law.

Using an industry-standard 8-bit microprocessor (Intel 8080), a microcomputer was programmed (Clarke, Cope, Gawthrop (1975)) to implement the self-tuning controller. It was found that the approximate storage requirements for components of the algorithm were, in units to Kilo-byte:

a) floating-point arithmetic routines 0.5

b) square-root estimator 0.75
(the UD estimator requires a similar amount of store)

c) self-tuning main program 0.5 - 1.25

d) variables, data, etc. 0.5

With typical industrial problems, requiring the updating of between 6 and 10 parameters, and using a floating-point representation of 8 bits exponent and 16 bits mantissa, the 8080 proved capable of a complete self-tuning calculation in 0.3-1 second. Hence a microcomputer is capable of the self-tuning control of the majority of processes, as typical sample times exceed 1 second.

Typical current single-chip microprocessors have 1 or 2 kilo-byte of program storage, and 64 - 128 bytes of data storage. Hence there is insufficient space for the full self-tuning algorithm, and the effect of simplifications has to be considered. One major component is in the updating and storage of the matrix M_t and the associated gain vector K_t. Replacing K_t by a scalar, such as in stochastic approximation, would drastically reduce computation at the expense of speed and accuracy of adaptation. This may be acceptable for plant with a very low rate of drift, but often self-tuning is proposed for non-linear plant where changes in the process model are due to changes in the operating point, and rapid adaptation is required. The most appropriate policy is to retain the matrix M_t, but in a simplified form, such as a block-Toeplitz matrix in which only the leading elements are explicitly calculated. The relative performance of these methods is a matter of current research.

7. Conclusions

The self-tuning controller has been shown to have a variety of interpretations and to be numerically robust. Reference (Clarke, Gawthrop (1979)) gives a list of current applications, and (Clarke, Gawthrop (1979)) describes some practical plant trials. Simulations outlined in (Atkinson (1977)) show that a self-tuner with proper weighting in control performs at least as well as a fully-tuned PID controller without requiring a detailed plant model. One important development, worthy of further research, is the possibility of using a self-tuner in a commissioning mode without closing the loop. This overcomes one of the practical problems with self-tuning in that the control over the initial stages can be poor when the parameters are being adapted from their initial values.

REFERENCES

1. Astrom, K.J., and Wittenmark,B. 'On self-tuning regulators', Automatica, 1973, Vol. 9, pp. 185-199.

2. Atkinson, R.S. 'Simulation study of self-tuning control laws' OUEL report, 1201/77, Oxford 1977.

3. Clarke, D.W., Dyer, D.A.J., Hastings-James, R., Ashton, R.P. and Emery, J.B. 'Identification and control of a pilot-scale boiling rig' Proceedings, IFAC Symposium on identification and system parameter estimation, The Hagne/Delft 1973, pp. 355-366.

4. Clarke, D.W. and Gawthrop, P.J. 'Self-tuning controller' Proc. I.E.E., 1975, Vol. 122, pp. 929-934.

5. Clarke, D.W. and Gawthrop, P.J. 'Self-tuning control' Proc. I.E.E., 1979, Vol. 126, pp. 633-640.

6. Clarke, D.W. and Gawthrop, P.J. 'Implementation and application of microprocessors-based self-tuners' Proceedings, IFAC Symposium on identification, 1979.

7. Clarke, D.W., Cope, S.N. and Gawthrop, P.J. 'Feasibility study of the application of microprocessors to self-tuning controllers' OUEL report, 1137/75, Oxford 1975.

8. Gawthrop, P.J. 'Some interpretations of the self-tuning controller' Proc. I.E.E., 1977, Vol. 124, pp. 889-894.

9. Gawthrop, P.J. 'On the stability and convergence of self-tuning controllers' Proceedings, IMA Conference on analysis and optimisation of stochastic systems, 1978.

10. Jacobs, O.L.R. and Patchell, J.W. 'Caution and probing in stochastic control' Int. J. of control, Vol. 16,NO1,1972,pp.189-199.

11. Panuska, V. 'An adaptive recursive-least-squares identification algorithm' Proceedings, IEEE Symposium on adaptive processes, 1969.

12. Peterka, V. 'A square-root filter for real-time multivariable regression' Kybernetika, 1975, Vol. 11, pp. 53-67.

13. Thornton, C.L. and Bierman, G.J. 'Filtering and error analysis via the UDU covariance factorization' IEEE Trans., 1978, Vol. AC-23, pp. 901-907.

14. Wellstead, P.E. 'On the self-tuning properties of pole/zero assignment regulators'. Control systems Centre report 402, UMIST, 1978.

15. Ziegler, J.G. and Nichols, N.B. 'Optimum settings for automatic controllers' Transactions ASME, 1942.

```
SUBROUTINE SQRTFL (Y,X,SQRTP,K,THETA,NPAR,PERR,FORGET)
C     PETERKA'S SQUARE-ROOT ALGORITHM
C     Y=MEASURED OUTPUT,X=DATA VECTOR,MOST RECENT IN X(1)
C     SQRTP=SQUARE-ROOT OF COVARIANCE MATRIX, ELEMENTS
            STORED IN A VECTOR (1,1),(1,2),(2,2),(1,3),(2,3),(3,3)...
C     K=KALMAN GAIN VECTOR, THETA=PARAMETER VECTOR
C     NPAR=NUMBER OF PARAMETERS TO BE ESTIMATED-LENGTH OF X,K,THETA
C     PERR=PREDICTION ERROR, FORGET=FORGETTING FACTOR(BETA)
      REAL X(1),SQRTP(1),K(1),THETA(1)
      PERR=Y
      DO 1 I=1,NPAR
1     PERR=PERR-THETA(I)*X(I)
      SIG=FORGET
      SIGSQ=FORGET*FORGET
      IJ=0
      JI=0
      DO 4 J=1,NPAR
      J1=J-1
      FJ=0.0
      DO 2I=1,I
      JI=JI+1
2     FJ=FJ+SQRTP(JI)*X(I)
      A=SIG/FORGET
      B=FJ/SIGSQ
      SIGSQ=SIGSQ+FJ*FJ
      SIG=SQRT(SIGSQ)
      A=A/SIG
      K(J)=SQRTP(JI)*FJ
      SQRTP(JI)=A*SQRTP(JI)
      IF(J1.EQ.0) GO TO 1000
      DO 3I=1,J1
      IJ=IJ+1
      SQP=SQRTP(IJ)
      SQRTP(IJ)=A*(SQP-B*K(I))
      K(I)=K(I)+SQP*FJ
3     CONTINUE
1000  CONTINUE
      IJ=IJ+1
4     CONTINUE
      DO 5 I=1,NPAR
5     THETA(I)=THETA(I)+PERR*K(I)/SIGSQ
      RETURN
      END
```

Table 1   Fortran listing of the square-root filter

```
SUBROUTINE UDFLT(X,Y,D,U,K,THETA,NPAR,PERR,FORGET)
   C    U/D FILTER,AFTER BIERMAN AND THORNTON
   C    Y=MEASURED OUTPUT, X=DATA VECTOR, MOST RECENT IN X(1)
   C    U,D FACTORS OF COVARIANCE MATRIX, U BEING UPPER TRIANGULAR
   C        AND STORED IN A VECTOR(1,2),(1,3),(2,3),(1,4),(3,4)...
   C        D IS A DIAGONAL MATRIX STORED AS A VECTOR
   C    K=KALMAN GAIN VECTOR, THETA=PARAMETER VECTOR
   C    NPAR=NUMBER OF PARAMETERS TO BE ESTIMATED-LENGTH OF X,THETA,K,D
   C    PERR=PREDICTION ERROR, FORGET=FORGETTING FACTOR(BETA)
        REAL X(1),D(1),U(1),K(1),THETA(1)
        PERR=Y
        DO 1 I=1,NPAR
   1    PERR=PERR-THETA(I)*X(I)
        FJ=X(1)
        VJ=D(1)*FJ
        K(1)=VJ
        ALPHAJ =1.0+VJ*FJ
        D(1)=D(1)/ALPHAJ
        IF(NPAR.EQ.1) GO TO 1000
        KF=0
        KU=0
        DO 4 J=2,NPAR
        FJ=X(J)
        DO 2 I=1,J-1
        KF=KF+1
   2    FJ=FJ+X(I)*U(KF)
        VJ=FJ*D(J)
        K(J)=VJ
        AJLAST=ALPHAJ
        ALPHAJ=AJLAST+VJ*FJ
        D(J)=D(J)*AJLAST/(ALPHAJ*FORGET)
        PJ=-FJ/AJLAST
        DO 3 I=1,J-1
        KU=KU+1
        W(KU)=U(KU)+K(I)*PJ
        K(I)=K(I)+U(KU)*VJ
   3    U(KU)=W
   4    CONTINUE
1000    CONTINUE
        DO 5 I=1,NPAR
   5    THETA(I)=THETA(I)+PERR*K(I)/ALPHAJ
        RETURN
        END
```

Table 2   Fortran listing of the U/D filter

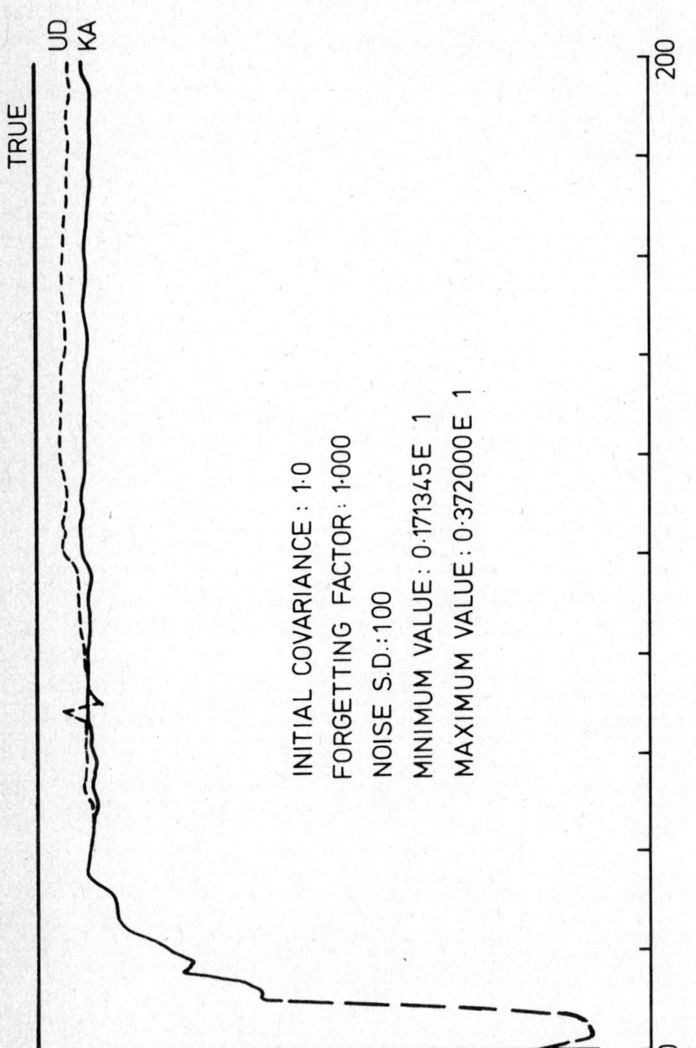

Fig. 1 - UD and Kalman estimates over 200 samples.

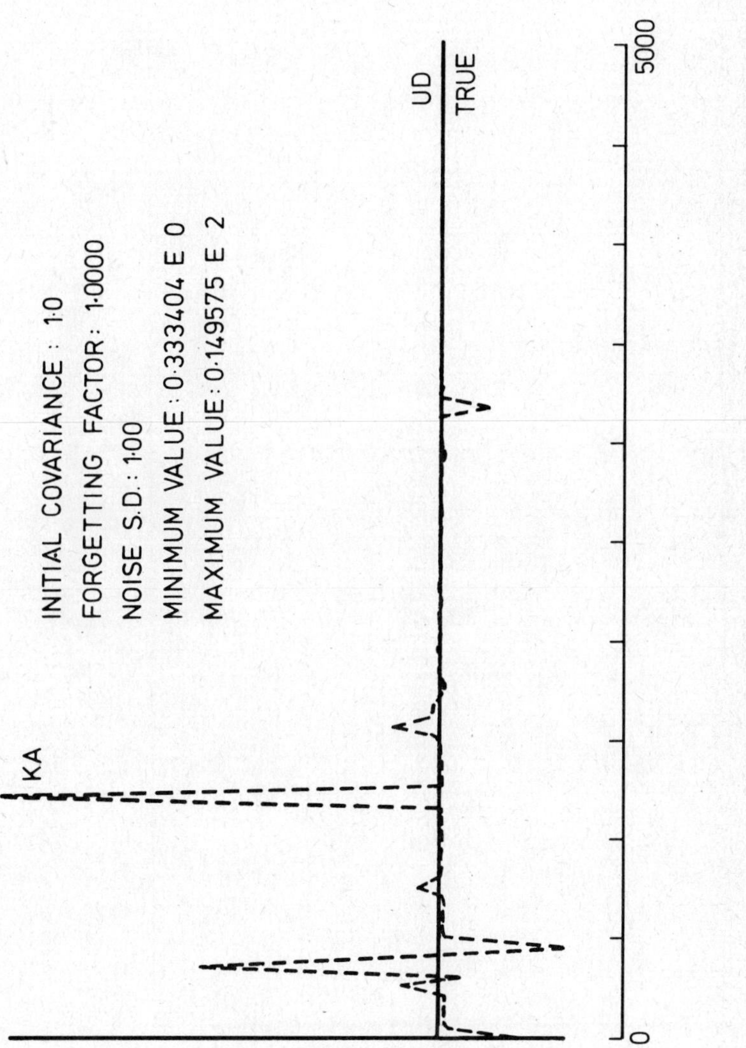

Fig. 2 - UD and KA parameter estimates over 5000 samples.

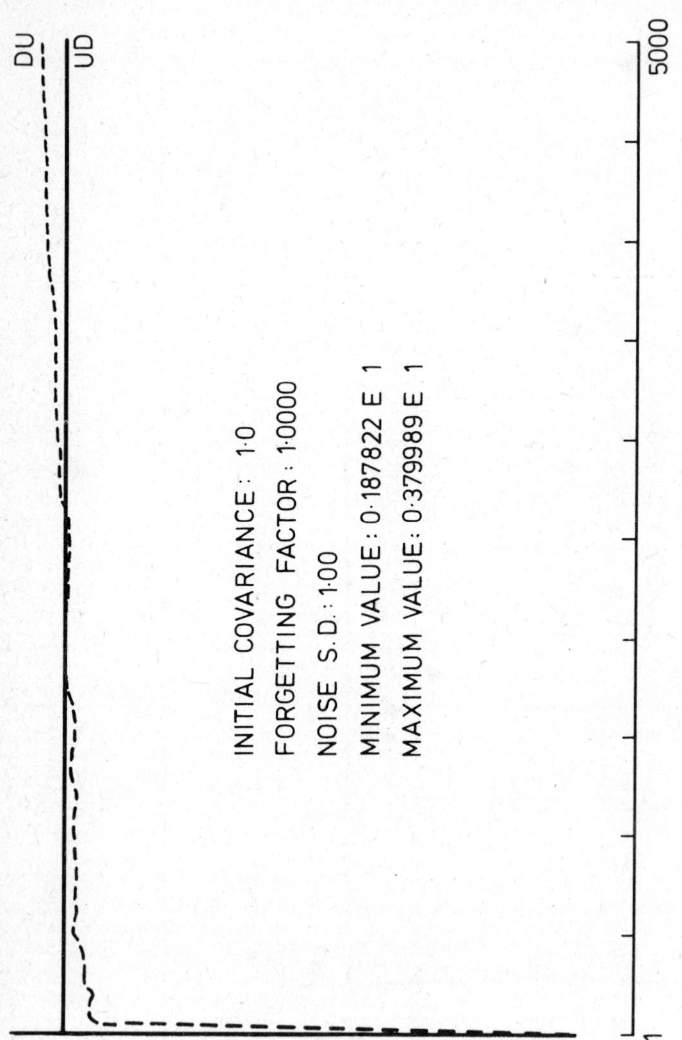

Fig. 3 - UD and 22-bit DU estimates over 5000 samples.

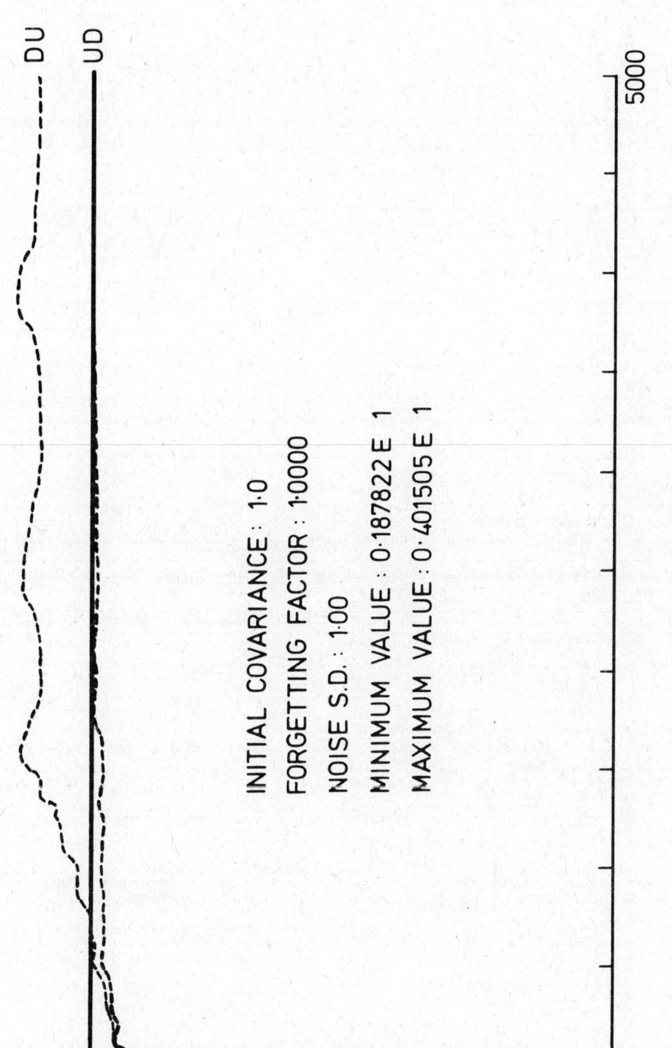

Fig. 4 - UD and 20-bit DU parameter estimates over 5000 samples.

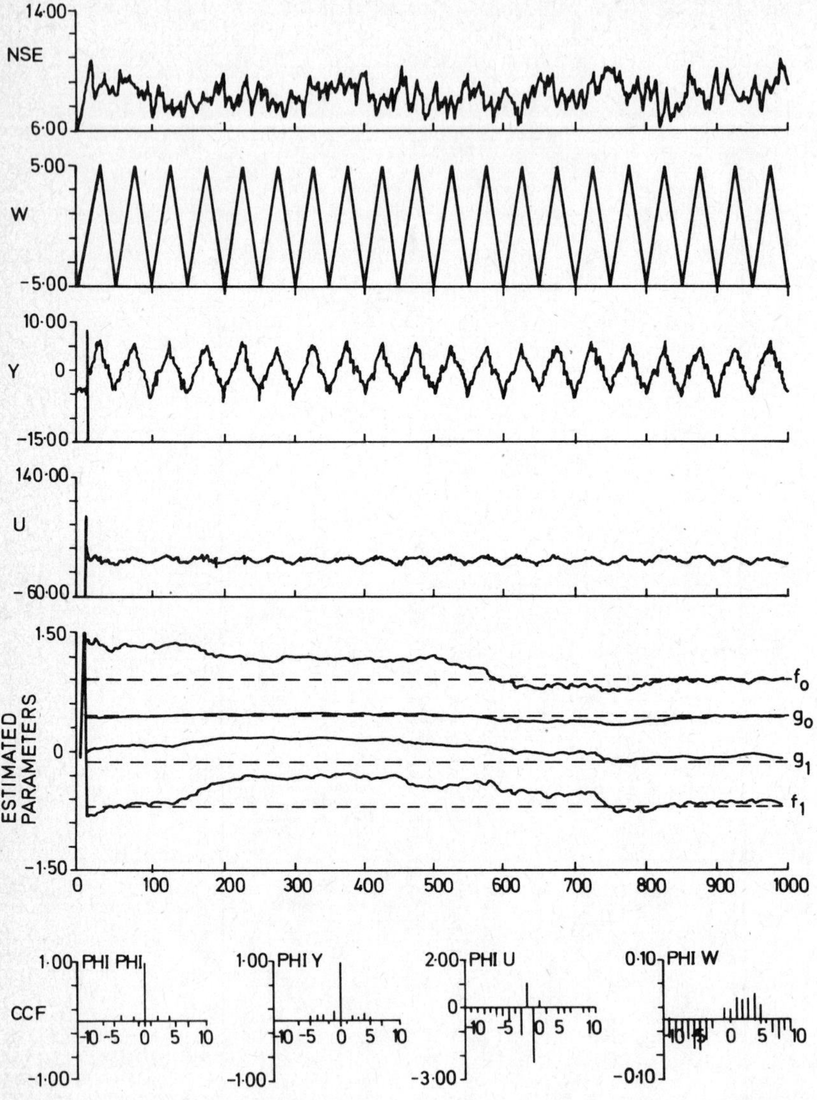

Fig. 5 - Self-tuning control of a simulated second-order system

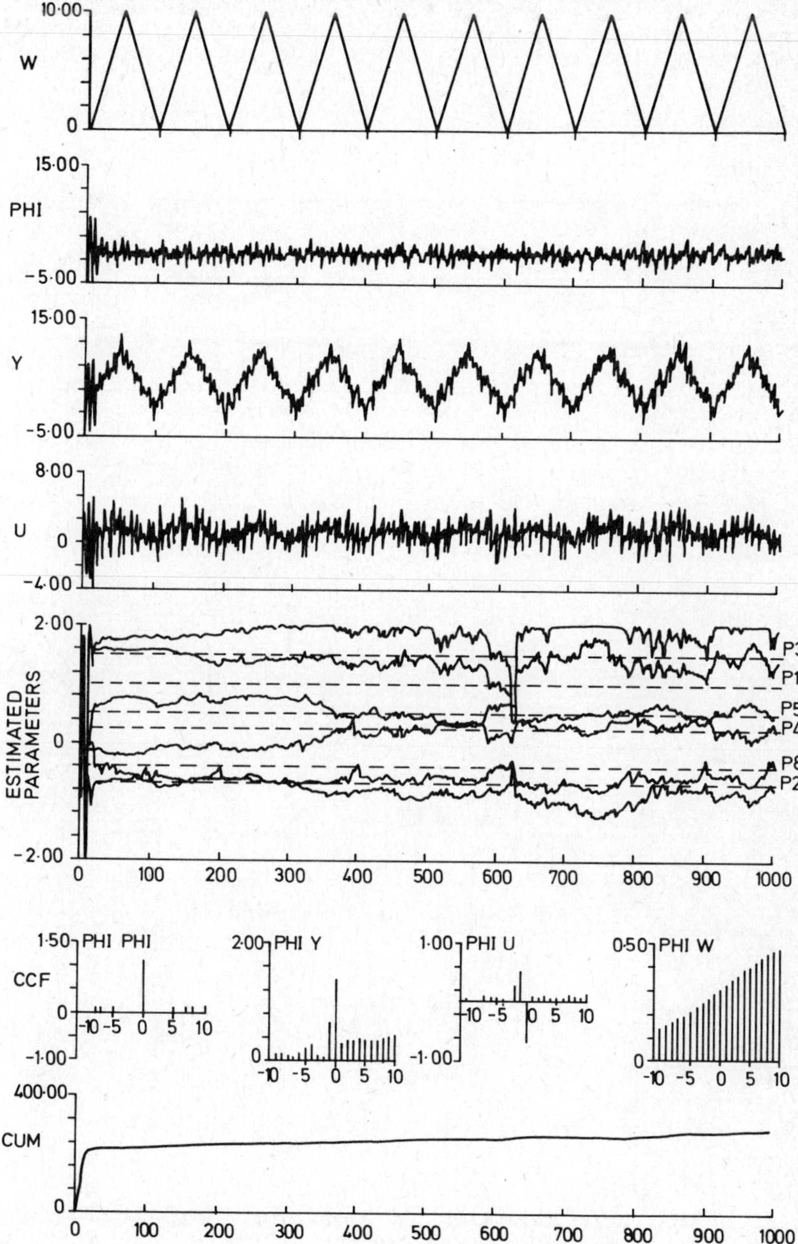

Fig. 6 - Simulation of self-tuning control using a microprocessor

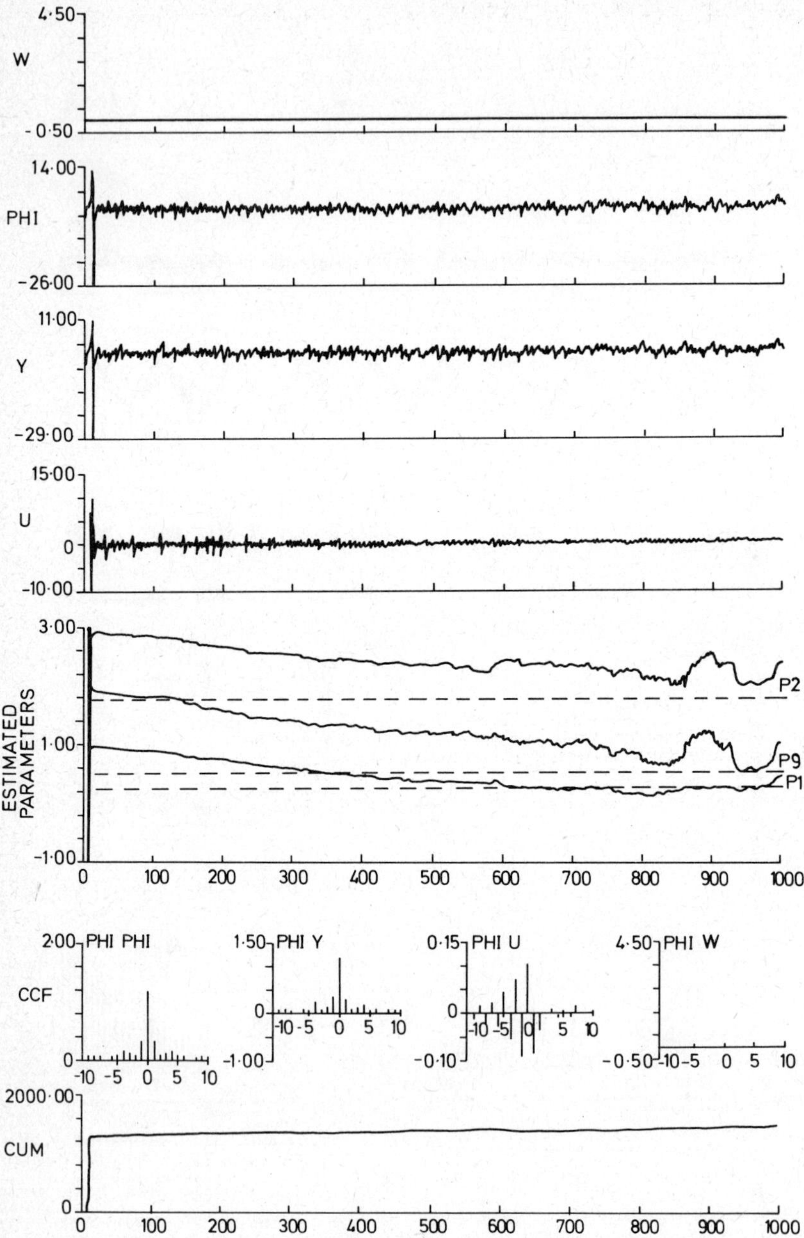

Fig. 7 - Self-tuning control of a non-minimum phase system

Numerical Techniques for Stochastic Systems
F. Archetti and M. Cugiani (eds.)
© North-Holland Publishing Company, 1980

ON A CLASS OF STOCHASTIC DYNAMICAL SYSTEMS: MATHEMATICAL ANALYSIS, SOME OPTIMIZATION PROBLEMS AND APPLICATIONS

Nicola Bellomo
Istituto di Meccanica Razionale - Istituto Matematico
Politecnico - Corso Duca degli Abruzzi 24 -
10129 Torino - Italy -

This work deals with the analysis of a class of stochastic dynamical systems with random constant coefficients and initial conditions. For this class of systems the problem of existence, absolute continuity and evolution of the probability density are studied and some constructive theorems are proposed after the pertinent mathematical analysis. On the basis of these results some direct solution techniques to deal quantitatively with the considered class of systems are firstly proposed, then a class of optimization problems and some applications to mathematical models of systems have been considered and analyzed on the basis of the previously mentioned methods.

INTRODUCTION

It is well understood that an ordinary deterministic differential equation is only a limiting simplified model of a real system. In fact, in mathematical physics, or, more in general, in the mathematical modelling of systems, the identification of some characteristics of the model itself, such as initial conditions, parameters or external actions, can be realized only in some probabilistic sense. This explains the great deal of interest which has very recently (mainly in the last ten years) been given to the mathematical modelling of systems by means of random differential equations; and, consequently, to the mathematical analysis of this class of equations.

Since a similar argument can be applied when a linear differential equation is presented as a mathematical model for nonlinear problem, a good subject of discussion can be based on the choice between a model defined by a deterministic nonlinear differential equation and a model defined by a linear stochastic differential equation. An answer to a question of this kind can be given only after an accurate analysis of the real physical system, which, as always in nature, will be stochastic and nonlinear.

Accordingly, this presentation is mainly related to a particular class of stochastic, ordinary, nonlinear differential equations. As far as bibliography is concerned, besides the important book by Soong (1974), who has supplied a general organization of the subject including existence, uniqueness, constructive solution methodologies and analysis of the stability; other important contributions, which will be afterwards commented on with some more details, are due to Adomian and coworkers (1970-1979) on the search for analytical solutions, and to Boyce and coworkers, refs.(11,23,33,36), on approximate numerical and analytical solutions of linear and nonlinear equations. Studies on the stability of solutions can be found in refs.(28, 30,34,35,47,48). For a general bibliography, see, amongst the others, refs.(9,10,27,41,43) on differential equations and dynamical systems, and refs.(21,22,29,31,32,40,45) on probability theory, ergodic theory and random integral and differential equations.

Of course the bibliography does not cover the extremely wide range of papers published in this area, but is a selected one and is especially pertinent to the presentation of the mathematical methodology proposed by the author and Pistone (1979), which will be the main topic of the next sections.

With regard to the contributions of other authors, it must be pointed out that the most successful attempt to obtain analytical or quasi-analytical solutions of nonlinear stochastic differential equations has been proposed by Adomian and coworkers in refs.(3,4,5) also summarized with some further improvements in ref. (7).

More in detail Adomian studies equations of the type:

$$L\{y\} + N(y, \dot{y}, \ldots) = x(t) \tag{1}$$

where L is a linear stochastic operator and N is a nonlinear stochastic term. The method proposed and developed by Adomian, as a natural extention of a previous work (1) dealing with fully linear operator equations, can be succesfully applied if L can be split into the sum of deterministic operator, the Green's function of which can be determined, and a random operator. Under the above hypothesis, eq.(1) is transformed into a nonlinear stochastic Volterra equation to be solved by a suitable iterative process.

The iterative method he develops and its various later extentions offer significant advantages. Separations of statistical averages occur naturally in this procedure making closure approximations unnecessary and avoiding any "smallness" assumptions or assumptions of white noise. The objective in this work is the direct calculation of first and second order statistics of the solution process directly in terms of input statistics and a "stochastic Green's function". In the quoted papers, moreover, some practical examples also indicate how sufficiently simple analytical solutions can be obtained in the class of problems defined by eq.(1). Recent unpublished work generalizes to linear stochastic partial differential equations and even some nonlinear partial differential equations essentially calculating the multidimensional Green's functions by a constructive method (8).

The papers quoted as refs.(11,23,33,36,37) refer to some-what differents conditions more on the side of the numerical treatment. In fact a first group of the contributions by Boyce and Lax (1976) are devoted to the extention of the Galerkin's methods of moments for systems of linear deterministic ordinary differential equations to systems of stochastic differential equations, defined in a suitable probability space, of the type:

$$\dot{\underset{\sim}{Y}} = [Q(t)]\, \underset{\sim}{Y} + \underset{\sim}{G}(t) \quad , \qquad |t| \leq a \tag{2}$$

where $[Q]$ is an $n \times n$ random matrix and $\underset{\sim}{Y}$ and $\underset{\sim}{G}$ are n-dimensional random column vectors.

The basic point of their theory consists in approximating the stochastic operators and solutions by a suitable sequence of stationary processes and in constructing a sequence of equations such that the solutions converge, in norm, to the original problem. Namely, after eq.(2) and under suitable assumptions about the discretizability of the process, a sequence of m equations is written, such that for each m:

$$\dot{\underset{\sim}{Y}}_m = [\, Q_m]\, \underset{\sim}{Y}_m + \underset{\sim}{G}_m \quad , \quad \underset{\sim}{Y}_m(0) = \underset{\sim}{0} \quad , \quad |t| \le a \tag{3}$$

such that if:

$$\lim_{m\to\infty} \left\{ \max_{\substack{i=1,\dots,n \\ j=1,\dots,n}} \ \underset{\omega\in\Omega}{\text{ess. sup}} \ | \, Q_{ij}(t) - Q_{ij,m}(t) | \right\} = 0 \tag{4}$$

$$\lim_{m\to\infty} \left\{ \max_{\substack{i=1,\dots,n \\ j=1,\dots,n}} \ \underset{\omega\in\Omega}{\text{ess. sup}} \ | \, G_i(t) - G_{i,m}(t) | \right\} = 0 \tag{5}$$

where (ω, Ω) is the probability basis, then:

$$\lim_{m\to\infty} \left\{ \max_{i=1,\dots,n} \ \underset{\omega\in\Omega}{\text{ess. sup}} \ |\hat{Y}_i(t) - \hat{Y}_{i,m}(t)| \right\} = 0 \tag{6}$$

where $\underset{\sim}{\hat{Y}}(t) = \{\underset{\sim}{\hat{Y}}_i(t)\}$ is the solution of the original problem and $\underset{\sim}{\hat{Y}}_m(t) = \{\underset{\sim}{\hat{Y}}_{i,m}(t)\}$ is the solution of the problem defined in eq.(3).

A second group (11,23,33) of the contributions by Boyce, Barry and Koheler is then related to the numerical solutions of, initial value or boundary value, ordinary stochastic differential equations of the type:

$$\dot{u} = K(t, u, r(t, u)) \quad , \quad u(0) = a \tag{7}$$

where K is a deterministic function, r is a given stochastic process and a is a given bounded stochastic variable, see refs.(23,33); or of the type:

$$\ddot{y} = f(t, y, \dot{y}, R_1, \dots, R_n) \quad , \quad t_o \le t \le t_f \tag{8}$$

$$S_1 \cdot y(t_o) + S_2 \cdot \dot{y}(t_o) = S_3 \quad , \quad S_4 \cdot y(t_f) + S_5 \cdot \dot{y}(t_f) = S_6 \tag{9}$$

where R and S are bounded continuous random variables, see ref.(11). The main steps of the proposed integration technique consists in the discretization of the time-variable, in the approximation, for each step, of the stochastic functions by stochastic Taylor expansions and finally, in the numerical integration of the equation itself.

The study of the stochastic stability, see, amongst the others, ref.(42) sec.9 and refs.(28,30,34), follows the same lines of the deterministic stability. The main difference is that, in stochastic differential equations, the stability is considered in terms of boundness and continuity (in some sense to be specified) of the mean expected value or, of some moments, of the dependent variable. On the other hand, in the deterministic case, as known, the afore mentioned arguments are applied directly on the dependent variable itself.

This present work, on the other hand, relates to the class of nonlinear ordinary stochastic differential equations of the type:

$$\dot{\underset{\sim}{x}} = \underset{\sim}{F}(\underset{\sim}{x}, t; \underset{\sim}{\omega}) \quad , \quad \underset{\sim}{x} \in D \subset \mathbb{R}^n \quad , \quad \underset{\sim}{\omega} \in \Omega \subset \mathbb{R}^m \tag{10}$$

where the initial conditions $\underset{\sim}{x}_o$ and the parameters $\underset{\sim}{\omega}$ are given as constant random variables, which are joined to suitable probability densities $P(\underset{\sim}{x}_o)$ and $P(\underset{\sim}{\omega})$ such that:

$$\int P(\underset{\sim}{x}_o)\, d\underset{\sim}{x}_o = \int P(\underset{\sim}{\omega})\, d\underset{\sim}{\omega} = 1 \tag{11}$$

The mathematical theory, relating to the afore-mentioned problem, which has been developed through papers (16,17,18), consists mainly in:

- Transforming eq.(10) into a deterministic equation with random initial conditions on a suitable augmented variable. -
- Deriving, under some regularity and continuity assumptions, an evolution equation on the jacobian of the transformation $\underset{\sim}{x}_o \to \underset{\sim}{x}_t$. -
- Defining the pertinent analytical and numerical methods for obtaining qualitative and quantitative results. -

The methodology is proposed also in view of the mathematical treatment of the class of optimization problems where the probability distribution of the initial data and of the parameters must be optimized in order to obtain an optimum probabilistic output.

It is also worthwhile mentioning that the proposed method has also been used as the necessary basis of the mathematical modelling and analysis of large systems, see refs.(14,15) dealing with the modelling of multi-droplet systems in vapour environment and with the treatment of the derived stochastic equations.

In the second section of this work the mathematical statement of the problem is given including the description of the related optimization problem. In the third section, some mathematical models (in fluid dynamics and in biological sciences) will be outlined in order both to indicate the practical implications of the abstract problem itself and to test the proposed numerical techniques. The said models also allow a preliminary discussion of the fact that the construction of the model itself must consider also the disposable mathematical theory and that every effort must be made to construct models which, at equal reliability, can be placed within the range of applicability of a known and tested mathematical theory. The fourth section is based on the analysis of the problems of existence and absolute continuity of the probabilistic solutions, and in deriving an evolution equation on the jacobian of the transformation from the initial data into the final ones. Finally, in the fifth section some mathematical techniques are proposed in order to obtain approximate solutions in the class of random differential equations which is here considered and tested for the analysis of two particular equations derived from the models outlined in the second section.

As already mentioned, this work can be considered a presentation and a re-organization, with some further extensions and generalizations of the mathematical theory, of the results derived in the quoted papers (16,17,18).

MATHEMATICAL DESCRIPTION OF THE ABSTRACT PROBLEM

The aim of this section is a mathematical description of the abstract problem dealt with in this work and, as already mentioned and indicated in practice in the next section, which includes a large class of physical problems.

Accordingly, let us supply some preliminary definitions. In particular let:
D = a bounded open subset of $\mathbb{R}^n$, $n \geq 1$.

$I = [0,t) \subset \mathbb{R}_+$ an open real interval of the time t.

$\underset{\sim}{x} = \underset{\sim}{x}(t)$: $I \to D$ a point of D with the physical meaning of state variable.

$\underset{\sim}{\omega} \in \Omega \subset R^m$ = set of real valued parameters.

$B = D \cup \Omega \subset \mathbb{R}^p$, $p = n + m$.

σ = Borel σ-algebra on B.

$P(\underset{\sim}{x}, t)$ = probability density of the variable $\underset{\sim}{x}$.

$P(\underset{\sim}{\omega})$ = probability density (constant) of the variable $\underset{\sim}{\omega}$.

Let us also recall that:

$$\int_A P(\underset{\sim}{x};\ t)d\underset{\sim}{x} = \mathbb{P}(\underset{\sim}{x} \in A \subset D;\ t) \tag{12}$$

is the probability of finding an object with the state $\underset{\sim}{x}$ in A Borel subset of D. In the same fashion for the parameters:

$$\int_W P(\underset{\sim}{\omega})\ d\underset{\sim}{\omega} = \mathbb{P}(\underset{\sim}{\omega} \in W \subset \Omega) \tag{13}$$

In this work, the relationship between the probability density and the probability measure μ is assumed to be the following:

$$\mu_t(A) = \int_A P(\underset{\sim}{x}, t)\ d\underset{\sim}{x} \quad , \quad \mu(W) = \int_W P(\underset{\sim}{\omega})\ d\underset{\sim}{\omega} \tag{14}$$

It is also worthwhile mentioning, at least for tutorial aims, that the moments and the correlations of $\underset{\sim}{x}$ are given, by means of P or μ, according to the following equations:

$$< x_i >^p = \xi_i^p = \int_D x_i^p\ P(\underset{\sim}{x}, t)\ d\underset{\sim}{x} = \int_D x_i^p\ d\mu \tag{15}$$

$$< x_{i,j} > = \xi_{i,j} = \int_D x_i\ x_j\ P(\underset{\sim}{x}, t)\ d\underset{\sim}{x} = \int_D x_i\ x_j\ d\mu \tag{16}$$

After these preliminaries, it is possible defining the proposed mathematical system according to the following axioms:

A. 1 - *The deterministic evolution of the variable x is defined by an ordinary vector differential equation of the type:*

$$\underset{\sim}{x}(t) = \underset{\sim}{x}(t{=}0) + \int_0^t \underset{\sim}{F}(\underset{\sim}{x}(s);\ \underset{\sim}{\omega})\ ds \quad , \quad \underset{\sim}{x}(t{=}0) = \underset{\sim}{x}_o \tag{17}$$

which has also the properties of an integrable dynamical system.

A. 2 - $\forall(\underset{\sim}{x}, \underset{\sim}{x}_o) \in D$ *and* $\forall\ \underset{\sim}{\omega} \in \Omega$: $\underset{\sim}{F}$ *is analytical in all its arguments and satisfies a uniform Lipschitz condition:*

$$\forall\, t \in I\,, \quad \forall\, (\underset{\sim}{x}, \underset{\sim}{w}) \in B \;:\; \|\underset{\sim}{F}(\underset{\sim}{x}_1) - \underset{\sim}{F}(\underset{\sim}{x}_2)\| \le L\,\|\underset{\sim}{x}_1 - \underset{\sim}{x}_2\| \tag{18}$$

then a unique flow f exists for the said differential equation such that:

$$\underset{\sim}{x} = \underset{\sim}{f}(t;\, \underset{\sim}{x}_o)\quad , \quad \underset{\sim}{x}_o = \underset{\sim}{x}(t{=}0) \tag{19}$$

A.3 - The probabilistic evolution of the system is defined by eq.(17) with random initial conditions and random constant coefficients:

$$(\underset{\sim}{x}_o,\, \mu_o(\underset{\sim}{x}),\, \mu^{w}) \longrightarrow (\underset{\sim}{x}_t,\, \mu_t(x)) \tag{20}$$

the related stochastic process is here indicated as follows:

$$\dot{\underset{\sim}{X}} = \underset{\sim}{F}(\underset{\sim}{X};\, \underset{\sim}{w},\, \mu_o,\, \mu^{w}) \tag{21}$$

The objectives of this work can be defined, at given $(B,\ \beta,\ \mu_o, \mu^{\omega})$, as consisting in:

a) Studying of the map: $\mu(\underset{\sim}{x},\ t{=}0) \longrightarrow (\underset{\sim}{x},\ t)$*, with an analysis of the problems of existence absolute continuity and stability of the solutions.*

b) Analysis of the qualitative and quantitative methods to construct probabilistic solutions, in analytical forms, of the map defined at point a.

c) Analysis of the optimum problem:

$$opt.\{\mu_o,\, \mu^{w}\} \Rightarrow Min.\{\Gamma(\mu(\underset{\sim}{x},\ t)\} \tag{22}$$

where Γ *is a suitable convex functional on* $\mu(\underset{\sim}{x},\ t)$.

As already mentioned, this analysis will be split into two parts. A first one with the problems of existence, absolute contintuity and evolution of the jacobian of the transformation from the initial conditions into the final ones, and a second one, where analytical-numerical methods are proposed and tested for some particular applications.

SOME EXAMPLES OF STOCHASTIC MATHEMATICAL MODELLING

It is worthwhile, in the author's opinion, to indicate how the abstract mathematical system, described in the preceding section, can be applied as a model of a well defined physical system.

Consider the following two examples:
The first model relates to gas flow with dispersed quasi-spherical particles (13). In particular, let us consider a gas-volume G with known physical conditions $\underset{\sim}{u} = \underset{\sim}{u}(H)$ for every point H of G. Namely:

$$\forall\ t \in I, \quad \forall\ H \in G \subset \mathbb{R}^3 \ :\quad \underset{\sim}{u} = \underset{\sim}{u}(H,\ t) \tag{23}$$

is known; where $\underset{\sim}{u}$ is the set collecting all the physical variables (for instance temperature, density, and vector velocity of the flow) necessary to define the physical state of the system. In G a population of quasi-spherical objects is dispersed. The state of each object is defined by the variable:

$$\underset{\sim}{x} = \{\underset{\sim}{y},\ \underset{\sim}{V},\ T,\ R\} \in D \subset \mathbb{R}^8 \quad , \quad \underset{\sim}{x} = \underset{\sim}{x}(t) \tag{24}$$

that is location $\underset{\sim}{y}$, velocity $\underset{\sim}{V}$, temperature T and radius R of the object.

The differential equation of the dynamics of $\underset{\sim}{x}$ can be deduced by the equations of mass, momentum and energy fluxes to be written as follows:

$$\tfrac{1}{2}\, \rho_\infty V \pi R^2 C_N = d(4 \rho \pi R^3/3)/dt \tag{25}$$

$$\tfrac{1}{2}\, \rho_\infty V \underset{\sim}{V} \pi R^2 C_D = d(4 \rho \pi R^3 \underset{\sim}{V}/3)/dt \tag{26}$$

$$\tfrac{1}{2}\, \rho_\infty V^2 \pi R^2 C_E = d(4 \rho \pi R^3 c\, T/3)/dt \tag{27}$$

$$\rho_\infty = \rho_\infty(H, t) \; ; \quad \rho = \rho(T) \; ; \quad c = c(T) \; ; \quad C_{N,D,E} = C_{N,D,E}(\underset{\sim}{x}, \underset{\sim}{u}; \underset{\sim}{\omega}, \mu^\omega) \tag{28}$$

where ρ is the mass density of the particle, c is its heat capacity, ρ_∞ is the gas mass density and the coefficients C are bounded, analytical functions of $\underset{\sim}{x}$, $\underset{\sim}{u}$ and of a suitable set of identification parameters $\underset{\sim}{\omega}$ which, in general, is known only in some probabilistic sense.

The knowledge of the said coefficients can be realized, by the pertinent analysis with the methods of the kinetic theory of gases, see refs.(19,25,26), of the fluid-dynamical problem of the actions between molecular stream and external body of the quasi-spherical particle.

The scalar components of the vector eq.(26) supply three scalar differential equations, which joined to eqs.(25,27) and to the equations which equal V to the time-derivative of y, yield a vector, nonlinear, ordinary differential equation in the form of eq.(17). For this equation, the identification of the initial conditions and of the parameters is impossible, in practice, in a deterministic sense. Thus a suitable probability measure must be joined to the said parameters and initial conditions. The problem is then the one described in the preceding section and defined by eq.(21). In particular in this problem, the evolution equation is an autonomous dynamical system if the gas volume is in uniform steady thermodynamical conditions and is a nonautonomous system if the conditions are non-uniform or unsteady.

The second model which is here mentioned is a model related to the competition between two species, of a type derived by the well known model proposed by Volterra in 1928. According to this model (24), the rate of variation of the numbers x and y of subjects of the two species in competition is given by:

$$\dot{x} = a_1 x + a_2 xy + a_3 x^2 + a_4(x - \hat{x}) \; ; \quad \dot{y} = b_1 y + b_2 xy + b_3 y^2 + b_4(y - \hat{y}) \tag{29}$$

where the coefficients $\underset{\sim}{a}$ and $\underset{\sim}{b}$ are real-valued identification parameters and $\hat{x}$ and $\hat{y}$ are the number of subjects which is wanted for each specie. The identification of the parameters and of the initial condition can be simplified if the same model is rewritten after the following change of variables:

$$\underset{\sim}{x} = \{x_1, x_2\} \, , \quad x_1 = x/(x + y) \quad , \quad x_2 = (x + y) \tag{30}$$

The mathematical model can be then written in the form of eq.(21) with $\underset{\sim}{\omega} = \{\underset{\sim}{a}, \underset{\sim}{b}\}$.

Since the same arguments on the identification of the parameters and initial conditions applied to the first model hold for this model too, this can also be considered a conceivable application of the abstract mathematical system described in the preceeding section. It must be noted that a large class of models in biology can be written in forms equivalent to the ones of eqs.(29,30), see for instance ref.(46) where some models on carcinogenesis are proposed, and that all the aforementioned arguments on the probabilistic structure of the model itself can be applied.

Let us note that in the first model each object is individually identified in some probabilistic sense. In the second model only global quantities are identified. Therefore, in the first model some objects can get out of the domain of the variable (f.i. when $R \to 0$, or $y \notin G$) and the process can lose, becouse of the variable number of objects, some continuity properties. A similar argument cannot be applied to the second model, where the disappearing of subjects does not implies non-continuity of the process of the variables x and y, or x_1 and x_2.

The problem of the absolute continuity of the process will be discussed in detail in the next section. However, this preliminary discussion has been realized in order to point out that when a mathematical model, probabilistic in the sense of the second section, is constructed, unless required by the inner construction of the model itself, some effort should be done in order to have continuous process.

MATHEMATICAL ANALYSIS

The mathematical analysis, carried out in this section, of the abstract mathematical problem described in the second section is mainly referred to the existence, absolute continuity and evolution of the probability measure on the state variable of the system. Since some of the topics dealt with in this section have been studied in refs.(16-18), the proof of some of the theorems proposed in this section will not be repeated, but some references to the above quoted papers will be given.

As a first preliminary step, let us transform the stochastic differential eq.(21) into a deterministic equation with random initial conditions on a suitable augmented variable. Accordingly, and as suggested by Soong (1974), the following variable is introduced:

$$\underset{\sim}{z} = \{\underset{\sim}{x}, \underset{\sim}{\omega}\} \in B \subset \mathbb{R}^p \tag{31}$$

The time-evolution of $\underset{\sim}{z}$ is defined by the following ordinary differential equation:

$$\dot{\underset{\sim}{z}} = \underset{\sim}{K}(\underset{\sim}{z}) \quad ; \quad 0 < j \leq n \Rightarrow K_j = F_j \quad ; \quad n < j \leq p \Rightarrow K_j = 0 \tag{32}$$

which is deterministic if the initial condition is given in the deterministic sense, namely $\underset{\sim}{z}_o = \underset{\sim}{z}(t=0)$, and is stochastic when the initial condition is random, namely given by the following probability measure or density, respectively:

$$\mu_o = \mu(\underset{\sim}{z};\ t=0) = \mu(\underset{\sim}{x};\ t=0)\cdot\mu(\underset{\sim}{\omega}) \quad ; \quad P_o = P(\underset{\sim}{z};\ t=0) = P(\underset{\sim}{x};\ t=0)\cdot P(\underset{\sim}{\omega}) \tag{33}$$

The exact correlation and equivalence between the problem stated in the second section and the one above described and formally stated by eqs.(32,33) is defined by the following remarks:
Remark I: if eq.(21) contains some deterministic parameters, in addition to the random ones, these parameters can be considered as a limiting case of stochastic parameters and transferred into the initial conditions with delta-function distribution.

Remark II: if, for every $\underset{\sim}{x} \in D$ and $\underset{\sim}{\omega} \in \Omega$, eq.(21) satisfies a uniform Lipschitz condition, also eq.(32) satisfies, for every $\underset{\sim}{z} \in B$, a uniform Lipschitz condition, and the problem of the integration of eq.(32), in the deterministic sense, is well posed. If the solution map of the said equation is indicated as follows:

$$\phi_t(\underset{\sim}{z}_o) = \underset{\sim}{z}(t;\ \underset{\sim}{z}_o) \tag{34}$$

if ϕ_t is an application from B into B, it is a diffeomorphism in B (9,10).

Remark III: If the probabilistic solution of eq.(21) exists and is unique, then such a problem is equivalent to the ones of eqs.(32,33) when this problem has a unique solution. In fact:

$$P(\underset{\sim}{x},\ t) = \int_{\Omega} P(\underset{\sim}{z},\ t)\ d\underset{\sim}{\omega} \tag{35}$$

After eqs.(32,33) and the above Remarks I-III, the problem is then the analysis of the actions of an integrable deterministic dynamical system upon an initial probability density. This kind of process is represented in fig.1:

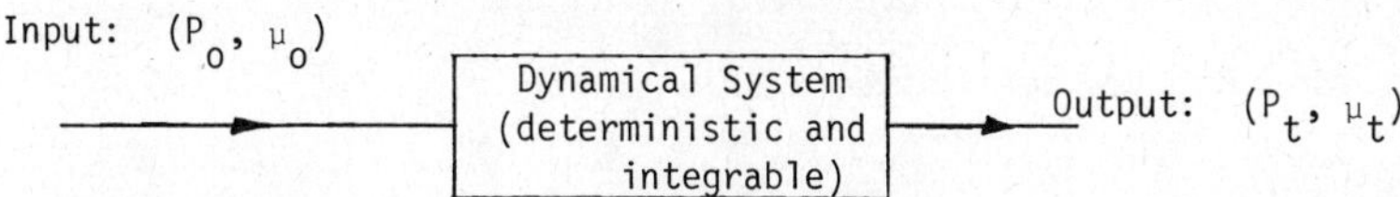

Fig.1 - Flow Chart of the System -

The mathematical theory related to the system represented in fig.1 has been studied in ref.(17).

Let us now consider the problem of the absolute continuity of μ_t with respect to μ_o. This problem is strictly related to the problem of the existence of the probability measure itself. In this sense, the following two definitions can be supplied:

Def. 1 - If μ_t is the probability measure on (B, β), then: $\mu_t = \phi_t(\mu_o) = \mu_o \circ \phi_t^{-1}$

Def. 2 - If μ_t is absolutely continuous with respect to μ_o, $\mu_t \ll \mu_o$, then a density $P_t = d\mu_t/d\mu_o$ exists.

See refs.(16,39).

Accordingly, the line which is here followed, consists in determining some sufficient conditions for the absolute continuity (which implies, by def.2, the existence of a probability density) and in deriving a suitable evolution equation for the probability density (or measure). This is stated by the following Lemmas and Theorems:

Lemma I: *$\mu_t = \phi_t(\mu_o)$ is absolutely continuous with respect to μ_o, $\mu_t \ll \mu_o$, if and only if:*

$$\mu_o(A) = 0 \Longrightarrow \mu_t(A) = \mu_o \circ \phi_t^{-1}(A) = 0 \text{ or } \mu_o(A) > 0 \Longrightarrow \mu_t(A) > 0.$$

Proof:
see ref.(16).

Remark IV: the initial probability measure can be positive, but its immage is not necessarily positive. See fig.2.

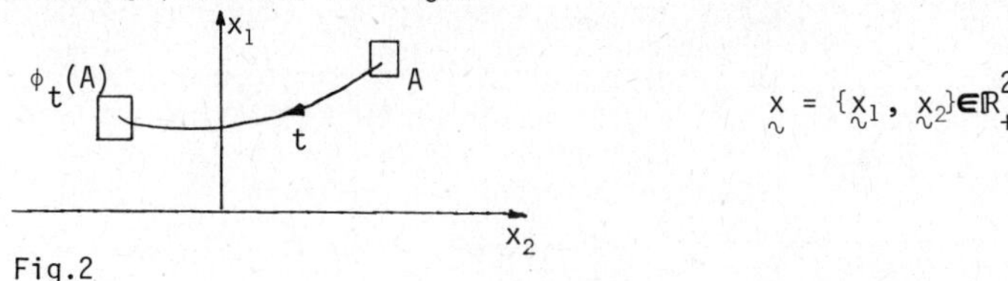

$\underset{\sim}{x} = \{\underset{\sim}{x}_1,\ \underset{\sim}{x}_2\} \in \mathbb{R}_+^2$

Fig.2

In fig.2, $\mu_o(A) > 0$ and $\underset{\sim}{x}$ has physical meaning if x_1 and x_2 are positive quantities. After a time interval t, $x_1 > 0$ and $x_2 < 0$. The process then, is not continuous in the sense of Lemma I.

Let us now denote: $J_t = J(t; \underset{\sim}{z}_o)$ the jacobian of the transformation $\underset{\sim}{z}_o \to \underset{\sim}{z}_t$. According to known definitions:

$$J_t = \det.|\partial \underset{\sim}{z}/\partial \underset{\sim}{z}_o| = \det.|\partial\phi_t/\partial \underset{\sim}{z}_o| \tag{36}$$

and the result of Lemma I can be formulated in terms of existence of the global solution of eq.(32) as follows:

Lemma II: *if the global solution of eq.(32) with given deterministic initial conditions exists and is unique in B, then J_t exists. If the initial condition for such equation is random, then $\mu_t \ll \mu_o$ and:*

$$P(\underset{\sim}{z}, t; \underset{\sim}{z}_o) = J(t; \underset{\sim}{z}_o)\cdot P(\underset{\sim}{z}_o; t=0) \tag{37}$$

Proof:
if the global solution of eq.(32) exists in B and is unique, then ϕ_t is a diffeomorphism and the map $\underset{\sim}{z}_o \to \underset{\sim}{z}_t$ is defined, thus J_t exists. Therefore the implications of Lemma I can be applied to prove the continuity of μ_t with respect to μ_o and then to the existence of a density. Moreover, for every Φ bounded and taking into account the expression of J_t,

$$\int \Phi\, P_t\, d\underset{\sim}{z} = \int (\Phi\circ\phi_t)(<P_o>_t \circ \Phi_t)\, d\underset{\sim}{z}_o = \int \Phi <P_o>_t\, d\underset{\sim}{z} = \int \Phi <P_o>_t\, J_t\, d\underset{\sim}{z}_o$$

which proves the formula of eq.(37). An analogous equation is supplied by Soong, see ref.(42), theorem 6.2.1.

Finally, the following main constructive theorem can be stated:

Theorem I: *if the global solution of eq.(32) exists and is unique in B, then J_t can be calculated by solution of the following augmented differential equation:*

$$\dot{\underset{\sim}{\zeta}} = \underset{\sim}{H}(\underset{\sim}{\zeta}) \quad , \underset{\sim}{\zeta} = \{\underset{\sim}{z}, J\} \quad , \quad \underset{\sim}{\zeta}_o = \{\underset{\sim}{z}_o, 1\} \quad , \quad \underset{\sim}{H} = \{\underset{\sim}{K}, -J(\underset{\sim}{\nabla}\cdot\underset{\sim}{K})\} \tag{38}$$

Proof:
let us consider eq.(32) and apply the equation of the variations:

$\det.((\exp(\underset{\sim}{z})) = \exp(\mathrm{tr}.\underset{\sim}{K}) = \exp(\underset{\sim}{\nabla}\cdot\underset{\sim}{K})$

and considering the definition of J_t:

$$J_t = \exp\left\{-\int_0^t (\underset{\sim}{\nabla}\cdot\underset{\sim}{K}(\underset{\sim}{z}(s)))\, ds\right\}, \text{ namely: } \ln(J_t) = \left\{-\int_0^t (\underset{\sim}{\nabla}\cdot\underset{\sim}{K}(\underset{\sim}{z}(s)))\, ds\right\}$$

and by time-derivation: $\dot{J}_t/J_t = -(\underset{\sim}{\nabla}\cdot\underset{\sim}{K})$, where $t \to 0$ implies that $J \to 1$.

According to Theorem I, eq.(38) can supply solutions in terms of points of the probability density, objective of the considered stochastic process. These points can be obtained by the repeated application of eqs.(37,38) for various initial conditions. Of course the ideal situation is one of the following:

- ϕ_t is analytical in $\underset{\sim}{z}_o$ and t by solution of eq.(32) and J_t is obtained by direct application of its expression $J_t = \det.|\partial\phi_t/\partial\underset{\sim}{z}_o|$. -

- $\underset{\sim}{\zeta}_t$ is analytical in t and in $\underset{\sim}{\zeta}_o$ by solution of eq.(38) and consequently eq.(37) can be directly applied.-

Fig.3 shows the qualitative application of the procedure for a simple one-dimensional problem. It must be noted, however, that since eqs.(32) or (38) can be nonlinear, exact analytical solutions are very unlikely to be obtained, and it becomes fruitful to seek for approximate analytical solutions.

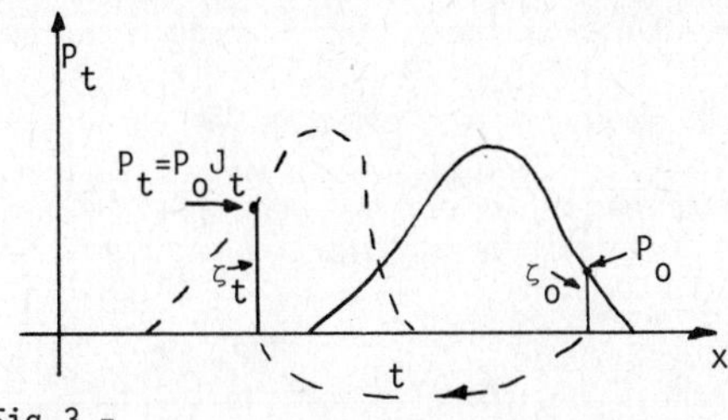

Fig.3 -

A further remark can be formulated as a corollary to Theorem I:

Remark IV: In the above analysis autonomous differential equations have been considered. However, all the arguments applied in deriving eqs.(37,38) have been applied without the need of restrictions to autonomous systems. Consequently, eqs.(37,38) hold also for nonautonomous systems.

The main results of Lemma II and Theorem I can be applied to the analysis of the optimum control problem stated in sec.2. Such a problem, after the above analysis, becomes an inverse application of eqs.(37,38) according to the following Theorem:

Theorem II: suppose there exists, at a given control time t_f, a density P_f such that: $\Gamma(P_f) \to 0$ or $\Gamma(P_f) < \varepsilon$, then the optimum initial density is given by backward integration of eq.(38) and by backward application of eq.(37).

Proof:

the proof is implied by Lemma II and Theorem I themselves. See also ref.(17) on some results on the convergence, in probability measures, to the solution of the considered optimum problem.

In order to complete this section, the problem of the boundness and of the stability of the solutions must be considered. As already mentioned in the introduction, this topic is generally studied as a natural extension of the deterministic stability towards the stochastic stability. In fact the main effort consists in applying to the mean expected value, or to some moments, the same arguments which, in the deterministic stability, are applied to the dependent variable.

General definitions of stability and the basic theorems are given by Soong(1974), ref.(42), sec.9, as well as in refs.(34,35). Let us now extend the afore-results to the class of problems which is here considered. The starting point is then, the following stochastic differential equation derived by eq.(32):

$$\dot{\underset{\sim}{Z}} = \underset{\sim}{K}(\underset{\sim}{Z}, t) \quad , \quad \text{initial conditions } (\underset{\sim}{Z}_o, P_o) \tag{39}$$

which is written as a nonautonomous equation according to Remark IV. Let $\underset{\sim}{Z} = \underset{\sim}{Z}(t; P_o)$ the random solution of eq.(39), the following definitions and results can be supplied:

Def. 1 - The system is said to be stable in mean if: $\lim_{t\to\infty} E\{|\underset{\sim}{Z}(t)|\} < \underset{\sim}{c}$.

Def. 2 - The system is asymptotically stable if: $\lim_{t\to\infty} E\{|\underset{\sim}{Z}(t)|\} \to \underset{\sim}{0}$.

Def. 3 - The equilibrium solution $\underset{\sim}{Z}(t) = \underset{\sim}{0}$ is stable in norm, in the sense of Lyapunov, if two positive constants ε and $\delta(\varepsilon)$ exist such that:

$$E\{\|\underset{\sim}{Z}_o\|\} < \varepsilon \Rightarrow E\{\|\underset{\sim}{Z}(t)\|\} < \delta$$

Moreover, according to Soong, ref.(42), pp.274-275, the following result can be stated:

Theorem III: *If there is a Lyapunov function $V(\underset{\sim}{Z}, t)$ defined over the state space and it satisfies the following conditions:*
i) $V(\underset{\sim}{Z}, t)$ is continuous and derivable in all its arguments,
ii) $V(\underset{\sim}{0}, t) = 0$ and $V(\underset{\sim}{Z}, t) \leq a\|\underset{\sim}{Z}\|$ for some a,
iii) $E\{\dot{V}(\underset{\sim}{Z}, t)\} \leq 0$,
then the equilibrium solution is stable in the norm.

Considering that the analysis of the boundness and stability of the solutions is the preliminary step for their analytical approximation, (in this class of problems the approximation is in the probabilistic sense (42)), it is important deriving some stability results related to Theorem I. Accordingly, the following theorem can be stated:

Theorem IV: *if the deterministic equilibrium solution of eq.(38) is stable in the following sense: $\|\underset{\sim}{\zeta}_o\| < \varepsilon \Rightarrow \|\underset{\sim}{\zeta}_t\| < \delta$, then the random variable $\underset{\sim}{Z}$ is stable in the sense of def.3.*

Proof:
let us consider the following norm: $u(t) = \max_{u_i \in \underset{\sim}{u}}\{|u_i|\}$, then:

$$E\{\|\underset{\sim}{Z}(t)\|\} = \int \|\underset{\sim}{z}(t)\| P(\underset{\sim}{z}, t)\, d\underset{\sim}{z} \leq \|\underset{\sim}{z}_t\| \int P(\underset{\sim}{z}, t)\, d\underset{\sim}{z} = \|\underset{\sim}{z}(t)\|$$

then: $\|\underset{\sim}{\zeta}_t\| < \delta \Rightarrow \|\underset{\sim}{z}_t\| < \varepsilon$ and consequently: $E\{\|\underset{\sim}{Z}(t)\|\} < \|\underset{\sim}{z}_t\| \Rightarrow E\{\|\underset{\sim}{Z}(t)\|\} < \delta$
which proves the theorem.

The result of this theorem is clear since it is evident that if ζ is bounded in some sense, then the boundness can be extended, with the correct metric, to the main expected value. However, it is necessary to verify the boundness of the solutions and then the possibility of their approximation, as indicated in the next section.

SOME MATHEMATICAL METHODS, APPLICATIONS AND DISCUSSION

Let us recall that the main result of the analysis realized in the preceding section has been the ordinary, differential equation of eq.(38), which joined to eq.(37), supplies the translation of the probability density on $\underset{\sim}{z}$ and give the probabilistic solutions of the considered class of stochastic processes. The objective is now finding solutions of the afore-mentioned equations, even with some kind of approximations, but analytical in t and in $\underset{\sim}{z}_o$. These solutions are needed also for the treatment of the considered optimization problem.

The analysis of the solution methods, which will be proposed in this section, is based on the main assumptions that the functions H and K are sufficiently smooth to assure uniqueness and continuous dependance of the solutions on the initial data. Moreover, the analysis is limited to autonomous dynamical systems even if the evolution equation on J can be extended to non-autonomous systems. The solution methods, themselves, can be extended, in general, to non-autonomous systems only after suitable assumptions (or control) on the boundness of the solutions.

The first method, which is here considered, is the extention of the method of the small perturbations to eq.(38). This method can be applied if the original state equation on $\underset{\sim}{x}$ can be written in the following form:

$$\dot{\underset{\sim}{x}} = [A]\underset{\sim}{x} + \varepsilon\cdot\underset{\sim}{g}(\underset{\sim}{x}) = \underset{\sim}{F}(\underset{\sim}{x};\ \varepsilon) \quad , \quad [A] = [a_{ij}] \tag{40}$$

where $[A]$ is a constant matrix, $\underset{\sim}{g}$ is a nonlinear vector function of $\underset{\sim}{x}$, and ε is a small parameter. In eq.(40) the trace of $\underset{\sim}{F}$ is given by:

$$\text{Tr. } \underset{\sim}{F} = \sum_{i=1}^{n} a_{ii} + \varepsilon\cdot \sum_{i=1}^{n} (\partial g_i/\partial x_i) = b + \varepsilon\cdot h(\underset{\sim}{x}) \tag{41}$$

with $h = h(\underset{\sim}{x})$, in general, nonlinear function of $\underset{\sim}{x}$. The corresponding equation on $\underset{\sim}{\zeta}$, namely eq.(38) rewritten for this particular problem, is given by:

$$\dot{\underset{\sim}{\zeta}} = [B]\underset{\sim}{\zeta} + \varepsilon\cdot\underset{\sim}{k}(\underset{\sim}{\zeta}) \tag{42}$$

where:

$$[B] = \begin{bmatrix} a_{11} & a_{12} \ldots a_{1n} & 0 & 0 \ldots & 0 \\ a_{21} & \ldots\ldots\ldots a_{2n} & 0 & \ldots\ldots & 0 \\ \ldots & \ldots\ldots\ldots & 0 & \ldots\ldots & 0 \\ a_{1n} & a_{2n} \quad a_{nn} & 0 & \ldots\ldots & 0 \\ 0 & \ldots\ldots\ldots & \ldots & \ldots\ldots & 0 \\ \ldots & \ldots & \ldots & \ldots & \ldots \\ 0 & \ldots\ldots\ldots & \ldots & \ldots\ldots & -b \end{bmatrix} \qquad \underset{\sim}{k} = \begin{bmatrix} g_1 \\ g_2 \\ \ldots \\ g_n \\ 0 \\ 0 \\ -Jh \end{bmatrix} \tag{43}$$

Let us now expand the searched solution and k in terms of a small parameter in the following truncated form:

$$\underset{\sim}{\zeta}(t) \cong \underset{\sim}{\zeta}_o(t) + \varepsilon\,\underset{\sim}{\zeta}_1(t) \quad , \quad \underset{\sim}{k}(\underset{\sim}{\zeta}) \cong \underset{\sim}{k}(\underset{\sim}{\zeta}_o) + \varepsilon\cdot\underset{\sim}{k}_1(\underset{\sim}{\zeta}_o) \tag{44}$$

replacing eq.(44) into eq.(42) yields:

$$\dot{\underset{\sim}{\zeta}}_o + \varepsilon\cdot\dot{\underset{\sim}{\zeta}}_1 = [B]\,\underset{\sim}{\zeta}_o + \varepsilon[B]\,\underset{\sim}{\zeta}_1 + \varepsilon\cdot\underset{\sim}{k}(\underset{\sim}{\zeta}_o) + \varepsilon^2\cdot\underset{\sim}{k}_1(\underset{\sim}{\zeta}_o) \tag{45}$$

therefore, equating the terms with the same power of ε and neclecting the terms with ε^2, the following sequence of two differential equations with constant coefficients can be obtained:

$$\begin{aligned} \dot{\underset{\sim}{\zeta}}_o &= [B]\,\underset{\sim}{\zeta}_o \longrightarrow \underset{\sim}{\zeta}_o(t) \\ &\qquad\qquad \swarrow \\ \dot{\underset{\sim}{\zeta}}_1 &= [B]\,\underset{\sim}{\zeta}_1 + \underset{\sim}{k}(\underset{\sim}{\zeta}_o) \longrightarrow \underset{\sim}{\zeta}_1(t) \end{aligned} \tag{46}$$

The solution of eq.(46) is in general of a standard type. Of course the expansions of eq.(44) can be carried to higher powers of ε and the method can be formally extended in this sense without any theoretical difficulty, even if the Taylor expansion of the vector function $\underset{\sim}{k}$ can, sometimes, involve tedious calculations. If the original state equation can be written in the form of eq.(40) the method is quite powerful. However, since that form is not always obtained, more general techniques have to be studied.

For this purpose let us consider the set Q of linearly independent functions q_k, each with the structure of a probability density:

$$Q = \{q_k\} \quad , \quad q_k = q_k(\underset{\sim}{z}) \quad , \quad \forall\, k : \int_B q_k(\underset{\sim}{z})\, d\underset{\sim}{z} = 1 \tag{47}$$

P(z; t) can be approximated by a linear combination of q_k as follows:

$$P(\underset{\sim}{z};\, t) \cong \left(1/\sum_k a_k(t)\right) \sum_k a_k(t)\, q_k(\underset{\sim}{z}) = \sum_k b_k(t)\, q_k(\underset{\sim}{z}) \tag{48}$$

The coefficients a_k or b_k can be determined by minimization, at fixed t, of a suitable convex functional on the distance between the probability density calculated in a discrete set of points by numerical integration of eq.(38) and application of eq.(37) and the approximating probability density as it is given by eq.(47). See ref.(18) where this technique is proposed on the basis of the approximation techniques in Hilbert spaces and the related numerical process is also indicated in detail.

An alternative is the one analyzed in ref.(17), namely the approximation of $\underset{\sim}{\zeta}$ with the following expansion:

$$\zeta_u(t;\, \underset{\sim}{\zeta}_o) \cong \sum_{i=1}^{r} a_{u,i}(\underset{\sim}{\zeta}_o)\cdot \xi_i(t) \quad , \quad \zeta_u \in \underset{\sim}{\zeta} \quad , \quad a_{u,i} = \sum_{j=1}^{r} b_{u,i,j}\, \Theta_j(\underset{\sim}{\zeta}_o)$$

therefore:

$$\zeta_u(t;\, \underset{\sim}{\zeta}_o) \cong \sum_{i=1}^{r} \left(\sum_{j=1}^{r} b_{u,i,j}\cdot \Theta_j(\underset{\sim}{\zeta}_o) \right) \xi_i(t) \tag{49}$$

the set of $p\cdot r^2$ coefficients can be obtained by convex minimization of the Bubonov-Galerking functional (12). Namely by minimization of the following:

$$\Delta = \max_{1 \le u \le p} \int_0^t \left[\zeta_u(s;\, \underset{\sim}{\zeta}_o) - \sum_{i=1}^{r} \left(\sum_{j=1}^{r} b_{u,i,j}\cdot \Theta_j(\underset{\sim}{\zeta}_o) \right) \xi_i(s) \right]^2 ds \tag{50}$$

This second method has been accurately studied in ref.(17) where an analysis of the error bounds has been also realized. Of course the applicability of the method itself becomes difficult when the dimension of the state variable is large, but this is a general rule.

In order to test the proposed methods, the following two problems, related to the mathematical models outlined in the third section, and defined by a deterministic nonlinear differential equation with random initial conditions, have been considered:

Problem I:

Vaporization of a population of nondistinguishable water droplets into a vapour ambient at a temperature of 500 °K and a pressure of 10330 kg/cm^2 with negligible motion of the droplets.

Problem II:
Time-evolution of two species in competition in terms of evolution of concentration x_1 of the first specie and total number x_2 of the subjects of the two species.

Results for the Problem I:
In the first problem, which is a particular case of the first model outlined in the third section, if the functions C are expressed by the mathematical model recently proposed by Monaco (1979) and for the numerical values assumed in the particular problem which is here considered, the following differential equation is obtained:

$$\underset{\sim}{x} = \{R = \text{radius},\ T = \text{temperature}\} \quad , \quad D = [0,\ 0.1\ \text{cm}]\cdot[270\ °K,\ 500\ °K]$$

$$dR/dt = .008569(-.207\cdot T^{\frac{1}{2}} + .00078\cdot T\cdot T^{\frac{1}{2}})\cdot g_1(T)\cdot g_2(T)$$

$$dT/dt = .0257(T\cdot T^{\frac{1}{2}}/R)(-.207 + .00078\cdot T)(g_1(T)\cdot g_2(T)\cdot g_3(T) + .386\cdot g_4(T))$$

$$g_1(T) = 6.028(.029(T/290)^2 + .024(T/290)^4)$$

$$g_2(T) = (10330(T/500)^{\frac{1}{2}}/(-.298 + .00112\cdot T))(esp(-s^2) + \quad s(1 + erf.s) - 1)$$

$$g_3(T) = (467.31 + .468\cdot T)/(4180\cdot T)$$

$$g_4(T) = (10330(T/500)^{\frac{1}{2}}/(-.298 + .00112\cdot T))(-exp(-s^2) - \sqrt{\pi}\cdot s(1 + erf.s) +$$
$$+ (500/3.5\cdot T)((exp(-s^2) + \sqrt{\pi}s(1 + erf.s))(s^2 + 3.5) + \tfrac{1}{2}\sqrt{\pi}s(1 + erf.s)))$$

$$s = s(T) = \tfrac{1}{2}(1/\sqrt{\pi})((10330)\ /(-.298 + .00112\cdot T)^2 - 1)$$
$$exp(-(10330)^2/(3.62(-.298 + .00112\cdot T)^2)$$

for such a problem let us fix the error as follows:

$$e = \max_{\substack{t \in I \\ x \in D}} |J_t - J_t^a| \qquad J^a = \text{approximate value of } J \qquad I = [0,\ .5\ \text{sec}]$$

If the maximum value of e is fixed, a priori, at $e \leq .0\ 1$, the following approximate expression of J, by means of a quadratic polynomial approximation, is found:

$$J_t^a = 1 + (\alpha_1 + \beta_1)\cdot t + (\alpha_2 + \beta_2)\cdot t^2$$

$$\alpha_{1,2} = \sum_{k=1}^{2}\sum_{h=1}^{2} k\ a_{kh1,2} R_o^{k-1}\ T_o^h \qquad \beta_{1,2} = \sum_{k=1}^{2}\sum_{h=1}^{2} h\ b_{kh1,2} R_o^k\ T_o^{h-1}$$

with the following numerical values for the coefficients a and b of the expansion:

a_{111}= -.1800	a_{121}= .0005	a_{211}= 84.5000	a_{221}= -.2915
a_{112}=-.11.7799	a_{122}=-.0379	a_{212}=-7.5851	a_{222}= .0252
b_{111}= .0000	b_{121}= .6560	b_{211}=156.6800	b_{221}= -.5247
b_{112}=2613.8700	b_{122}= 9.8200	b_{212}=-1712.1500	b_{222}= 5.8900

The visualization of the results for problem I can easily be obtained by the plot of the above numerical expansion. The considered fixed error is within the range of realibility of the adopted mathematical model, which if compared with the other models available in the pertinent literature, can be considered a quite sophisticated one both from the mathematical and physical viewpoint.

Results for the Problem II:

The second problem is a pertinent application of the second class of mathematical models outlined in the third section. In particular, and after the change of variables indicated in the said section and after the characterization of the coefficients, the following system of differential equations is considered:

$$dx_1/dt = .001\left\{x_1(20 - .99\cdot x_2 - .1/x_2) - (20 - .97\cdot x_2)x_1 + .02\, x_2 x_1^3 + .03/x_2\right\}$$

$$dx_2/dt = .001\left\{(-10 + 20\cdot x_1)x_2 + (.01 - .02\cdot x_1 + .02\cdot x_1^2)x_2 + .1\right\}$$

with given deterministic coefficients and random initial conditions given as follows:

$$P_o = P_{10}(x_{10})\cdot \delta(x_2 - x_{20})$$

the approximate expression of the jacobian can be found as follows:

$$J_t^a = \sum_i c_i(x_{10})\, L_i(t) \quad , \quad L_i = (\tfrac{1}{2}^i \cdot i!)d^i(i(t\ -1)^i)/dt$$

where the coefficients c_i of the Legendre polynome are expanded in the same fashion as follows:

$$c_i = \sum_j b_{ij}\, P_j(x_{10}) \quad , \quad L_j = (\tfrac{1}{2}^j \cdot j!)d^j(j(x_{10} - 1)^j/dx_{10}$$

and consequently:

$$J_t^a = \sum_{i=0}^{n} \sum_{j=0}^{n} b_{ij}\, L_j(x_{10})\, L_i(t)$$

after the calculation of the coefficients b, by means of the the minimization method indicated in this section and with the same error bound of Problem I, the following numerical result is obtained:

$$\begin{aligned} J_t^a = {} & 1 + (.93103 - 1.8351\cdot x_{10} - .07685\cdot x_{10}^2)\cdot t + (.68219 - 1.3350\cdot x_{10} + \\ & + 1.12520\cdot x_{10}^2)\cdot t^2 + (-.04037 + .05655\cdot x_{10} - .15019\cdot x_{10}^2)\cdot t^3 + \\ & + (.11469 - .22001\cdot x_{10} + .12301\cdot x_{10}^2)\cdot t^4 \end{aligned}$$

Then, in conclusion, this paper analyse some mathematical methods for dealing with a class of ordinary differential equations with random initial conditions and random constant parameters. With regard to this class of stochastic equations, the fundamental topics of the existence and absolute continuity of the probability measure and of its evolution, analytical approximation and optimal control have been analysed starting from some previous work (16,17,18) on the said topic.

This mathematical analysis is motivated by the fact that mathematical models in physics and engineering systems can be correctly stated only in some probabilistic sense and are almost never (in the author's opinion and according to the preliminaries of the introduction) strictly deterministic. In fact, two simple examples of mathematical models have been considered in this section and some numerical calculations related to the said models have completed this work. These calculations, in spite of their simplicity, can supply a good example of the applicability itself of the proposed methods.

ACKNOWLEDGMENTS

This work has been realized under the auspicies of the Italian Council for the Research (C.N.R., G.N.F.M.).

The author gratefully thanks G. Geymonat for having encouraged this research and G. Adomian, G. Geymonat and G. Pistone for having carefully read this work and for their helpful suggestions.

REFERENCES

1. Adomian, G., Linear Random Operator Equations in Mathematical Physics, J. of Math. Phys., Part I, 3 (1970) 1069-1084, Part II, 9 (1971) 1944-1948, Part III, 9 (1971) 1948-1955.

2. Adomian, G., The Closure Approximation in the Hierarchy Equations, J. of Stat. Phys., 2 (1971) 127-133.

3. Adomian, G., Nonlinear Stochastic Differential Equations, J. Math. Analysis and App., 2 (1976) 441-452.

4. Adomian, G. and Lynch, T., Stochastic Differential Equations with Random Initial Conditions, J. Math. Analysis and App., 1 (1977) 216-226.

5. Adomian, G. and Sibul, L., Stochastic Green's Formula and Application to Stochastic Differential Equations, J. Math. Analysis and App. 3 (1977) 743-746.

6. Adomian, G., On the Existence of Solutions for Linear and Nonlinear Stochastic Operator Equations, J. Math. Analysis and App., 2 (1978) 229-235.

7. Adomian, G., New Results in Stochastic Equations in the Nonlinear Case, in, Nonlinear Equations in Abstract Spaces, Ed. V. Lakshmikantham, (Academic Press, New York, 1978).

8. Adomian, G., On the Modeling and Analysis of Nonlinear Stochastic Systems, Invited Lecture at the 2nd Int. Conf. on Math. Modeling, St. Louis, 1979, (to be published).

9. Arnold,V., Equations Differentielles Ordinaires,(MIR, Moscou, 1974).

10. Arnold,V. and Avez, A., Théorie Ergodique des Systèmes Dynamiques, (Gauthier-Villas, Paris, 1967).

11. Barry, M., and Boyce, W., Numerical Solution of a Class of Random Boundary Value Problems, J. Math. Analysis and App., 1 (1979) 96-119.

12. Bellmann, R., Methods of Nonlinear Analysis, (Academic Press, New York, 1973).

13. Bellomo, N., Molecular Gas Flow with Dispersed Particles: Methodology for a Mathematical Modelling and Applications, 2nd Int. Conf. on Math. Modelling, St. Louis, 1979, (to be published).

14. Bellomo, N., Loiodice R. and Pistone, G., Time-evolution of a Multidroplet System in the Kinetic Theory of Vaporization and Condensation, in Rarefied Gas Dynamics, J. Potter Ed., Progress in Astron. and Aeronautics n.53, (AIAA, New York, Vol.2, 1977).

15. Bellomo, N., and Monaco, R., Mathematical Models for the Time-evolution of Stochastic Systems in Two-phase Flow, Applied Mathematical Modelling, 3 (1979) 41-44.

16. Bellomo, N. and Pistone, G., Dynamical Systems with a Large Number of Degrees of Freedom: Stochastic Mathematical Analysis for a Class of Deterministic Problems, Mechanics Research Comm., 2 (1979) 75-80.

17. Bellomo, N. and Pistone, G., Direct Solution Method for a Class of Integral Equations with Random Distribution of the Inhomogeneous Part, in Volterra Equations, Lecture Notes in Mathematics n.737, Londen S. O. and Staffans O. J. Eds., (Springer, Berlin, New York, 1979).

18. Bellomo, N., and Pistone G., Time-evolution of the Probability Density Under the Action of a Deterministic Dynamical System, Comm. at ICM 78 Helsinki, (to be published, J. of Mathem. Analysis and App., 1979).

19. Bellomo, N. and Monaco R., On the Mathematical Formulation of the Boundary Conditions in Kinetic Theory: Stochastic Models for Time-dependent Conditions, in Rarefied Gas Dynamics, R. Campargue Ed., (C.E.A. Commissariat a l'Energie Atomique, Paris, 1979).

20. Bensoussan A., Lions, J. and Papanicolau, G., Homogenization in Deterministic and Stochastic Problems, in Stochastic Problems in Dynamics, Clarkson B. Ed., (Pitman, London, 1977).

21. Bharucha-Reid, A., Random Integral Equations, (Academic Press, New York, 1972).

22. Bharucha-Reid, A., Fixed Point Theorems in Probabilistic Analysis, Bulletin of the American Mathem. Society, 82 (1976) 641-657.

23. Boyce W., Approximate Solution of Random Ordinary Differential Equations, Adv. App. Prob., 10 (1978) 172-184.

24. Capurso, G., Problemi di Stabilità Stocastica in Modellistica Matematica, Thesis of the Politecnico of Torino, Italy, N. Bellomo and R. Monaco Eds., 1979.

25. Cercignani, C., Theory and Application of the Boltzmann Equation,(Scottish Academic Press, Edinburgh, 1975).

26. Cercignani, C. and Lampis, M., Half-space Models for Gas Surface Interactions, in Rarefied Gas Dynamics, K. Karamcheti ed., (Academic Press, New York, 1974).

27. Clarkson, B., Stochastic Problems in Dynamics,(Pitman, London, 1977).

28. Curtain, R., Stability of Stochastic Dynamical Systems, Springer Lect. Note in Mathematics n.294, (Springer, Berlin, New York, 1972).

29. Engl, H., Random Fixed Point Theorems, in Nonlinear Equations in Abstract Spaces, V. Lakshmikantham ed., (Academic Press, New York, 1978).

30. Friedman, A., Stability and Angular Behaviour of Solutions of Stochastic Differential Equations, in Stability of Stochastic Dynamical Systems, R. Curtain ed., Springer Lecture Note in Mathematics n.294, (Springer, Berlin, New York, 1972).

31. Gallavotti, G., Lectures on the Billiard, in Dynamical Systems Theory and Applications, J. Moser ed., Springer Lecture Notes in Physics n.38, (Springer, Berlin, New York, 1975).

32. Kallianpur, G., and Kolzow, D., Mesure Theory Applications to Stochastic Analysis, Springer Lecture Notes in Mathematics n.695, (Springer, Berlin, New York, 1978).

33. Kohler, W., and Boyce, W., A Numerical Analysis of Some First Order Stochastic Initial Value Problems, SIAM J. of Applied Math., 1(1974) 167-179.

34. Kozin, F., Stability of Linear Stochastic Systems, in Stability of Stochastic Dynamical Systems, Ed. R. Curtain, Springer Lecture Notes in Math. n.294, (Springer, Berlin, New York, 1972).

35. Kushner, H., Stochastic Stability, in Stability of Stochastic Dynamical Systems, R. Curtain ed., Springer Lecture Notes in Math. N.294, (Springer, Berlin, New York, 1972).

36. Lax, M. and Boyce, W., The Method of Moments for Linear Random Data Initial Value Problems, J. Math. Analysis and Applications, 1 (1976) 111-132.

37. Lax, M., The Method of Moments for Linear Random Boundary Value Problems, SIAM J. of App. Math., 1 (1976) 62-83.

38. Mcshane, E., Stochastic Calculus and Stochastic Models, (Academic Press, New York, 1974).

39. Monaco, R., Time-evolution of Spherical Droplets in Vapour Environment: a Physical Mathematical Model, Int. J. of Heat and Mass Transfer, 6 (1979) 805-812.

40. Prohorov, Y. and Rozanov, Y., Probability Theory, (Springer, Berlin, New York, 1969).

41. Sibirsky, K., Introduction to Topological Dynamics, Noordhoff, Leyden, 1975.

42. Soong, T., Random Differential Equations,(Academic Press, New York, 1974).

43. Souriou, J., Structure dês Systémes Dynamiquês,(Dunod, Paris, 1970).

44. Sunahara, Y., Asakura, T. and Morita, Y., On the Asymptotic Behaviour of Nonlinear Stochastic Dynamical Systems Considering the Initial States, in Stochastic Problems in Dynamics, Ed. B. Clarkson, (Pitman, London, 1977).

45. Tsokos, C. and Padgett, W., Random Integral Equations with Applications to Life Sciences and Engineering, (Academic Press, New York, 1974).

46. Whittemore, A. and Keller, J., Quantitative Theories of Carcinogenesis, SIAM J. of App. Math., 1 (1978) 1.

47. Willems, J., Lyapunov Functions and Global Frequency Domain Stability Criteria for a Class of Stochastic Feedback Systems, in Stability of Stochastic Dynamical Systems, R. Curtain ed., Springer Lecture Notes in Mathematics

n. 294, (Springer, Berlin, New York, 1972).

48. Willems, J., Moment Stability of Linear White Noise and Coloured Noise Systems, in Stochastic Problems in Dynamics, B. Clarkson ed., (Pitman, London, 1977).

Numerical Techniques for Stochastic Systems
F. Archetti and M. Cugiani (eds.)
© North-Holland Publishing Company, 1980

CLASSICAL THEORY VS. WHITE NOISE THEORY: SOME COMMENTS

Paolo d'Alessandro

Istituto di Automatica - Centro di Studio dei Sistemi di Controllo e Calcolo Automatici del C.N.R. - Via Eudossiana 18-00184 Roma

Both the classical and the white noise approach to stochastic differential equations are illustrated. It is shown how white noise theory eliminates all the pathologies of Ito theory, that prevent its practical application for engineering purposes.

1. INTRODUCTION

A vital problem in physics and engineering is to model random phenomena, in particular, in the theory of dynamical systems, stochastic inputs and stochastic observation errors. These, for reasons that include mathematical convenience, are usually considered to be white noise. Under this stipulation the problem becomes that of defining a mathematical object, that might be regarded as reasonable model of white noise, and whereby a rigorous and applicable theory of stochastic differential equations might be constructed.

In the engineering literature white noise has been and is still used without a precise definition. On the other hand after a long history started according to Robert Brown himself [12] in the eighteenth century (accounts of such history, which is profoundly interlaced with the evolution of modern scientific thought can be found e.g. in [10] and in [21]) in the third decade of this century, the mathematics finally succeeded with P. Levy and N. Wiener, in producing a precise formal theory proposing a model for Brownian motion. This is defined to be a stochastic process enjoying suitable properties. Then, and this is the hard step, it is proved that such stochastic process actually exists. Now if for a moment one looks at the mathematically abhorrent engineering idea of what white noise should be, one can euristically conclude that Brownian motion should qualify as pathwise integral of white noise (this argument is illustrated for example in [17]). Unfortunately this idea is misleading because with probability one Brownian motion sample paths are nowhere differentiable. Since this implies that with probability one the sample paths are functions of unbounded variation, Stieltjes integrals cannot be defined by means of Brownian motion either. Here again the mathematics comes to help: sticking with the integrated version of differential equations, Brownian motion is allowed to act in the same way as white noise on a differential equation by means of the technical device of stochastic integrals.

Despite this success, there is a basic weakness in the ensuing theory that can be hardly dismissed. In fact on one hand the input stochastic process, that is Brownian motion, is lacking any physical interpretation because of the above mentioned behaviour of the trajectories, and on the other hand the stochastic differential equations in question unlike any honest differential equation, and in particular unlike the underlying deterministic differential equations, do not provide any smoothing effect, so that the degree of smoothness of the trajectories of the stochastic process, that is solution of the equation, turns out to be (again obviously in probabilistic sense) not better than that of Brownian motion. It follows that the solution is physically meaningless either. Moreover the situation as to the output stochastic process is obviously similar.

These facts are made precise in the next section. Section 3 illustrates the alternative of white noise theory, developed essentially by Balakrishnan (ref's [2] ÷ [9]), and compares this latter with the classical theory. The final section is devoted to some remarks on likelihood functionals.

2. THE CLASSICAL THEORY

Among the various techniques that can be adopted to define Brownian motion the function space approach sketched here follows Parthasarathy [23]. Thus let C be the space of all continuous real valued functions on the interval [0,1] (but any other bounded closed interval would do as well), let $\mathcal{T}$ be the topology of uniform convergence for C, and $\mathcal{B}_C$ be the corresponding Borel σ-algebra of subsets of C (that is the σ-algebra generated by $\mathcal{T}$). Let w be the evaluation map on the product space $[0,1] \times C$, that is the map defined by: for all (t,ω) in $[0,1] \times C$, $w(t,\omega) = \omega(t)$. Recall in the first place that w is a jointly continuous function so that for each t in [0,1], the function $w(t,\cdot)$ is continuous and therefore measurable. It follows that the σ-algebra generated by the family of functions $\{w(t,\cdot): t \in [0,1]\}$ is contained in $\mathcal{B}_C$. That these two σ-algebras coincide is not immediately evident, but this is the case. The proof of this fact, which plays a crucial role in the theory, follows essentially from the separability of $(C,\mathcal{T})$, the Lindelöf theorem, and the separability of [0,1] with the relativized usual topology.

Here comes the main result: there exists a probability measure p_W on $\mathcal{B}_C$ such that the stochastic process $\{w(t,\cdot):t \in [0,1]\}$ on the probability space $(C,\mathcal{B}_C,p_W)$ has the following properties: (i) - the distribution of $w(0,\cdot)$ is the atomic measure concentrated at zero, (ii) - if $0 \leq t_1 < t_2 < \ldots < t_n \leq 1$, where n is any positive integer, the random variables $w(t_2,\cdot)-w(t_1,\cdot),\ldots,w(t_n,\cdot)-w(t_{n-1},\cdot)$ are independent, and finally (iii) - if $0 \leq s < t \leq 1$, the random variable $w(t,\cdot)-w(s,\cdot)$ is gaussian with zero mean and variance t-s.

The stochastic process defined by the function w is what is called Brownian motion. It has been said that it is the most important and fashinating process in probability theory and that the only properties of its path are an absorbing field or study [17]. But there are also most serious drawbacks. In the first place the function space on which the process is sitting. System theory and optimal control theory (see e.g. [15] and [3]), all seem to recommend the use of other function spaces, first and foremost spaces of measurable functions, which easily lead to Hilbert space settings, where a direct definition of white noise may be attempted.

Next comes the erratic behaviour of the trajectories. As is well known the set of all trajectories that are differentiable in at least one point is a subset of a measurable, zero measure, set. This important result is proved e.g. in the book by Breiman [11] via a nice argument due to Dvoretski, Erdös and Kakutani [18]. If it is shown that the linear subspace B_V of C made up of all continuous functions of bounded variation, belongs to $\mathcal{B}_C$, then, since by the Lebesque theorem every function of bounded variation is differentiable almost everywhere (with respect to the Lebesgue measure), it would follow that $p_W(B_V) = 0$.

One simple proof that $B_V \in \mathcal{B}_C$ *runs as follows.* Let D be a countable dense subset of [0,1]. For each finite subset B of [0,1]: $B = \{t_1,\ldots,t_n\}$ with $0 \le t_1 < \ldots < t_n \le 1$ let, for all ω in C, $\sum_B(\omega) = \sum_{i=2}^{n} |\omega(t_i)-\omega(t_{i-1})|$. The continuity of ω is readily seen to imply that for all ω in C:

$$\sup\left\{ \textstyle\sum_B(\omega):B \subset [0,1], B \text{ finite}\right\} = \sup\left\{ \textstyle\sum_B(\omega):B \subset D, B \text{ finite}\right\}$$

(where obviously it is understood that if a subset of R has no upper bound its sup is $+\infty$). Conseguently:

$$B_v = \cup\left\{\cap\{\{\omega:\omega \in C \text{ and } \textstyle\sum_B(\omega) \le K\}: B \subset D, B \text{ finite}\}; K = 1,2,\ldots\right\}$$

But since for each finite subset B of [0,1], $\sum_B(\cdot)$ is a random variable on the probability space $(C, \mathcal{B}_C, p_W)$ and since the family of all finite subset of a countable set is countable, the desired conclusion follows.

Because of these facts it is evident that neither the process in point is a consistent model for the motion of particles observed by Robert Brown, nor there is any chance, besides zero probability events, of ever seeing in practice a realization of the stochastic process Brownian motion, because functions of unbounded variation fall outside the physical experience. But even more important than that is the question as to whether an usable theory of stochastic differential equations can be built on these premises. The remainder of the section is devoted to show that the answer to the question is negative.

First of all, along with the σ-algebra $\mathcal{B}_C$, consider for each t in [0,1], the sub-σ-algebra $\mathcal{B}_t$ of $\mathcal{B}_C$ that is generated by the family of random variables $\{w(s,\cdot):s \in [0,t]\}$, and endow the interval [0,1] with its own Borel σ-algebra $\mathcal{B}$. Actually all these σ-algebras must be substituted by larger ones and the measures on them extended (up to complete measures) in the precise way specified in [14] (all the items thereby obtained are denoted adding a prime to the symbol of the corresponding original item), and then all the topological spaces of measurable functions over the ensuing measure spaces must be substituted by suitable topological quotients. This allows to introduce a conditioning operator, with respect to which Brownian motion is a martingale, and the fundamental notion of Ito integral in a very precise and rigorous manner [14]. However since a streamlined exposition fits better the present purpose and since the thoery given in [13] and in [14] is a secure guide to put it all in the correct way, the error will deliberately made of not insisting on these technical problems, thus handling, for example, L_2 spaces as if they were Hilbert spaces, which is not the case whenever there are nontrivial zero measure measurable sets in the underlying measure space.

It might be useful to introduce in general the definition of strong solution of a stochastic differential equation. Here the main reference is the book by Lipster and Shiryayev [20], with some adaptations to fit the present needs. Suppose that a and b are two functionals on $[0,1] \times C$, measurable with respect to the product σ-algebra $\mathcal{B} \times \mathcal{B}_C$ and adapted to the increaing family of σ-algebras $\{\mathcal{B}_t : t \in [0,1]\}$. Next consider a stochastic process $\{x(t,\cdot):t \in [0,1]\}$ on the probability space $(C,\mathcal{B}'_C,p'_W)$, which is measurable, that is the function x is measurable with respect to the σ-algebra $\mathcal{B}' \times \mathcal{B}'_C$, is continuous, that is it has continuous trajectories almost surely with respect to p'_W

(or more briefly modulo p'_w, in symbols $[p'_w]$) and is adapted to the increasing family of σalgebras $\{\mathcal{B}'_t : t \in [0,1]\}$. For any such process let h_x be the transformation on C into itself defined by: $h_x(\omega)=x(\cdot,\omega)$ for all ω in C for which the function $x(\cdot,\omega)$ is continuous and $h_x(\omega) = 0$ otherwise.

Now *the function* h_x *is claimed to be* $\mathcal{B}'_C - \mathcal{B}_C$ *measurable*. To show this important result begin with denoting by N the subset of C made up of all ω such that $x(\cdot,\omega)$ is not continuous, and note that, because p'_w is the completion of p_w, N belongs to $\mathcal{B}'_C$. Furthermore in view of the aforementioned generating family for $\mathcal{B}_C$ and of a well known result on measurability of transformations (proposition II.1.1 in [22]) the desired conclusion follows after the sole verification that for each t in [0,1], $h_x^{-1}(w(t,\cdot)^{-1}(A)) \in \mathcal{B}'_C$ for all Borel subset A of R. On the other hand this is actually the case since, by the definitions of h_x and w, $h_x^{-1}(w(t,\cdot)^{-1}(A))$ is given by $x(t,\cdot)^{-1}(A) \cap \bar{N}$ if $0 \notin A$ and by

$$(x(t,\cdot)^{-1}(A) \cap \bar{N}) \cup N \text{ if } 0 \in A.$$

A first straingthforward consequence of this result is that if $\tilde{h}_x$ denotes the function on $[0,1] \times C$ into itself defined by $\tilde{h}_x(t,\omega) = (t,h_x(\omega))$ for all $(t,\omega) \in [0,1] \times C$, then $\tilde{h}_x$ is $\mathcal{B}' \times \mathcal{B}'_C - \mathcal{B} \times \mathcal{B}_C$ measurable so that both the functions $a \circ \tilde{h}_x$ and $b \circ \tilde{h}_x$ are measurable with respect to the σ-algebra $\mathcal{B}' \times \mathcal{B}'_C$ (because the composition of two measurable functions is a measurable function).

A stochastic process $\{x(t,\cdot):t \in [0,1]\}$ adapted to $\mathcal{B}'_t:t \in [0,1]\}$ measurable and continuous (in the sense specified above), is said to be a strong solution of the stochastic differential equation, written symbolically as:

$$dx_t = a(t,x)dt + b(t,x)dw_t \; ; \; x_o = \eta$$

where η is a $\mathcal{B}'_o$ measurable random variable, if (the measurability of the bracketed sets below follows from the Fubini's theorem)

$$p'_w(\{\omega:\omega \in C \text{ and } \int |a \circ \tilde{h}_x| d\mu'_L < \infty\}) = 1$$

$$p'_w(\{\omega:\omega \in C \text{ and } \int (b \circ \tilde{h}_x)^2 d\mu'_L < \infty\}) = 1$$

and modulo p'_w, for all $t \in [0,1]$ it is satisfied the equation (giving a precise meaning to the above symbol)

$$x(t,\omega) = \eta(\omega) + \int_{[0,t)} a \circ \tilde{h}_x(s,\omega) d\mu'_L(s) + (\int_{[0,t)} b \circ \tilde{h}_x dw)(\omega)$$

where the second integral is an Ito integral, and the assumption is made that the stochastic process defined by this latter is modified in such a way to be continuous (for the definition of modification of a stochastic process and for a proof of the fact that a continuous modification of such stochastic process exists see [20]).

Before leaving this general context the following important observation is in order. If the strong solution of a stochastic differential equation is unique in a sufficiently strong sense to guarantee that, for a given initial condition, if h_{x_1} and h_{x_2} are the transformations defined as above corresponding to two different strong solutions, then $h_{x_1} = h_{x_2}$ $[p'_W]$, it is possible to associate to the class of strong solutions (for that initial condition) a unique probability measure on $\mathcal{B}_C$, namely $p'_W \circ h^{-1}_{x_1}$, because a quick verification shows that $p'_W \circ h^{-1}_{x_1} = p'_W \circ h^{-1}_{x_2}$. Therefore under these circumstances it is legitimate to regard the stochastic differential equation as transformation of probability measures on $\mathcal{B}_C$ taking the measure p_W onto the measure $p'_W \circ h^{-1}_{x_1}$.

It is convenient for illustration purposes, to look in detail at an important special case, that may provide even better insight into the matter than the general theory. It is assumed that $a = \alpha w$, with $\alpha \in R$, that b is identically equal to one and, for semplicity, that $\eta = 0$. That a is measurable with respect to the σ-algebra $\mathcal{B} \times \mathcal{B}_C$ is a consequence of the fact that so does w (this in turn follows from joint continuity of w, as shown in [14]), whereas all other requirements for a and b are met in a trivial manner (of course the last two after the assumption that a strong solution $\{x(t,\cdot): t \in [0,1]\}$ exists). Thus by a trivial computation for the Ito integral, it can be said that the process $\{x(t,\cdot): t \in [0,1]\}$, adapted to $\{\mathcal{B}'_t: t \in [0,1]\}$ continuous and measurable, is a strong solution for the stochastic differential equation in question (with $\eta = 0$) if and only if it satisfies modulo p'_W the equation

$$x(\cdot,\omega) = \alpha \int_{[0,\cdot)} x(s,\omega)\, d\mu'_L(s) + \omega(\cdot)$$

A pause may be profitable here to appreciate how the form of this integrated equation is intuitively most expected in view of the qualification of Brownian motion as pathwise integral of white noise (if this latter existed; recall also that $\omega(\cdot) = w(\cdot,\omega)$ for all $\omega \in C$).

It is readily seen that the strong solution (henceforth for brevity the qualification strong is dropped) is unique in the sense that if $\{x_1(t,\cdot): t \in [0,1]\}$ and $\{x_2(t,\cdot): t \in [0,1]\}$ are two solutions, then modulo p'_W, $x_1(\cdot,\omega) = x_2(\cdot,\omega)$ [17]. Thus the above condition for associating a transformation of probability measures to the stochastic differential equation is surely satisfied. Usually the solution is given by means of the Wiener integral [17]. *Here it is instead derived in a different form via a simpler purely deterministic method.* Even though the Wiener integral form of the solution has its own advantages, the form obtained below is more convenient for other purposes: for example it provides immediately an explicit expression for the function h_x.

First of all the foregoing equation will be solved for all ω in C. In fact it can be rewritten in the form:

$$(I - \Lambda)x(\cdot,\omega) = \omega(\cdot)$$

where Λ is a continuous kernel Volterra operator on C. The space C is a linear subspace of the space $L_2([0,1], \mathcal{B}', \mu'_L)$ (briefly L_2), which is endowed with the usual inner product and regarded according to a previous stipulation as a Hilbert space (caution: the topology of C is properly stronger than the relativization of the topology of L_2). On the other hand the operator Λ, which is continuous, can be extended to a continuous Volterra operator $\tilde{\Lambda}$ on the whole L_2 in the natural fashion. Because every Volterra operator is quasinilpotent the operator $I-\tilde{\Lambda}$ on L_2 is invertible and actually the inverse operator has the form $I + \tilde{V}$, where $\tilde{V}$ is a Volterra operator on L_2 with continuous kernel. As a consequence of these facts the operator $I-\Lambda$ has a continuous inverse too, which is just $I+\tilde{V}$ restricted to C, and will be denoted by I + V. It might be helpful to remark that when, to make rigorous this argument, L_2 is substituted by the topological quotient $L_2/\{0\}^-$ (where $\{0\}^-$ is the closure of $\{0\}$ in L_2), the proof runs as well thanks to the fact that in each element of $L_2/\{0\}^-$ there is at most one continuous function (reason: any measurable subset of [0,1] with measure 1 is necessarely dense in [0,1]). Summing up this piece of functional analysis shows that if a function x on $[0,1] \times C$ is defined by

$$x(\cdot,\omega) = (I + V)(\omega)$$

for all ω in C, then the integral equation in question is satisfied for all ω in C. It remains to be proved that such a function defines actually the solution of the stochastic differential equation. In the first place note that for each t in $[0,1]$, $x(t,\cdot)=w(t,\cdot) \circ (I+V)$ and recall that $w(t,\cdot)$ is measurable with respect to $\mathcal{B}_C$. Since (I+V) is continuous, it is also measurable with respect to the Borel σ- algebras and since the composition of two measurable functions is a measurable function, $x(t,\cdot)$ is actually measurable with respect to $\mathcal{B}_C$. On the other hand $\mathcal{B}_C \subset \mathcal{B}_C'$, so that $\{x(t,\cdot) : t \in [0,1]\}$ is a bona fide stochastic process on the probability space $(C, \mathcal{B}_C', p_w')$. Note that for such process the function h_x, previously defined, is just the operator I+V and therefore it is $\mathcal{B}_C - \mathcal{B}_C$ measurable. Obviously the process has continuous trajectories for all ω in C and is measurable because $x = w \circ h_x$.

The adaptation of the process, that is causality in stochastic terms, is easily obtained from the fact that $h_x = (I+V)$ *is a causal operator.* Actually the stronger result of adaptation to the increasing family of σ-algebras $\{\mathcal{B}_t : t \in [0,1]\}$ (recall that for all t in $[0,1]$, $\mathcal{B}_t \subset \mathcal{B}_t'$) will be proved.

The causality property of h_x implies that if, for any t in $[0,1]$, f_t is defined to be the linear continuous, and hence $\mathcal{B}_C - \mathcal{B}_C$ measurable, function on C into itself such that for all ω in C, $f_t(\omega)(s) = \omega(s)$ if $0 \leq s \leq t$, and $f_t(\omega)(s) = \omega(t)$ if $s > t$ then:

$$w(t,\cdot) \circ (I+V) = w(t,\cdot) \circ (I+V) \circ f_t$$

Therefore the desired conclusion would follow if the inverse image unber f_t of any element of $\mathcal{B}_C$ is in $\mathcal{B}_t$, or equivalently if for each s in [0,1] the inverse images under f_t of all elements of the σ-algebra $w(s,\cdot)^{-1}(\mathcal{B}_R)$, where $\mathcal{B}_R$ is the Borel σ-algebra of R, are elements of $\mathcal{B}_t$. But that this is the case is seen by a trivial direct verification.

The most interesting property of the solution depends on its sole existence. However to semplify the discussion the present version of the solution is referred to begin with. In view of the integral equation, for all ω in C, $x(\cdot,\omega)$ is the sum of an absolutely continuous function plus ω itself. Thus bearing in mind that any absolutely continuous function has bounded variation and that the subset B_V of C (formed by all functions of bounded variation) is a linear subspace of C, it is clear that for all ω in C $x(\cdot,\omega)$ is a function of bounded variation if and only if ω is a function of bounded variation, that is

$$\{\omega : \omega \in C,\ x(\cdot,\omega) \in B_V\} = B_V$$

whence

$$p'_W(\{\omega : \omega \in C,\ x(\cdot,\omega) \in B_V\}) = p_W(B_V) = 0$$

The same arguments applies immediately to show that, for any other solution, the set of all ω in C for which the corresponding trajectory is a function of bounded variation, is the union of a subset of B_V and a subset of another measurable zero measure set. Therefore the conclusion has been reached that *with probability one the trajectories of any solution of the stochastic differential equation are functions of unbounded variation.* Even worse than that it is obviously possible to define a solution in such a way that none of its trajectories is a function of bounded variation.

Without dwelling on further obvious comments on the lack of any practical meaning of the differential model thereby constructed, consider the observed state process, that is the output process. Usually the noise corrupting the output is assumed to be independent of the noise affecting the state. This is the case covered below, but the final conclusion would not be different in the rather special case where the two noises are equal. Thus consider the product probability space $(C \times C,\ \mathcal{B}'_C \times \mathcal{B}'_C,\ p'_W \times p'_W)$ and let P_1 and P_2 be the projections of $C \times C$ onto its factor spaces. Following the same philosophy as before an integrated output process is now introduced. To avoid tedious and irrelevant technicalities this is simply defined to be the stochastic process $\{y(t,\cdot) : t \in [0,1]\}$ on the probability space $(C \times C,\ \mathcal{B}'_C \times \mathcal{B}'_C,\ p'_W \times p'_W)$ where y is the functional on $[0,1] \times (C\times C)$ such that for all (ω_1, ω_2) in $C \times C$

$$y(\cdot,(\omega_1,\omega_2)) = \int_{[0,1)} x(s,P_1((\omega_1,\omega_2)))\,d\mu'_L(s) + w(\cdot,P_2((\omega_1,\omega_2)))$$

where x is still the functional defined by the foregoing solution. None of the stipulations thereby adopted (e.g. making reference to a certain solution) affects the conclusion reached below, but the simple proof of this claim is omitted. Likewise are omitted the by now straightforward verifications that $\{y(t,\cdot) : t \in [0,1]\}$ is actually a stochastic process on $(C \times C,\ \mathcal{B}'_C \times \mathcal{B}'_C,\ p'_W \times p'_W)$, which is measurable, has all continuous trajectories, and is adapted to the increasing family of σ-algebras generated by the R^2 valued stochastic process $\{n(t,\cdot) : t \in [0,1]\}$ on the same probability space defined by: for each t in $[0,1]$, $n(t,(\omega_1,\omega_2)) = (\omega_1(t),\omega_2(t))$ for all (ω_1,ω_2) in $C \times C$. However the following remark, which is a major clue to such verifications, is given for later use. Let h_y be the function on $C \times C$ into C defined by $h_y((\omega_1,\omega_2)) = (y(\cdot,(\omega_1,\omega_2))$ for all (ω_1,ω_2) in $C \times C$. Obviously $h_y = \hat{\Lambda} \circ (I+V) \circ P_1 + P_2$ (where $\hat{\Lambda}$ is a continuous kernel Volterra operator) so that it is trivially a linear continuous transformation on the linear topological product space $(C \times C,\ \mathcal{T} \times \mathcal{T})$ into the linear topological space $(C,\ \mathcal{T})$. Since $\mathcal{B}_C$ is generated by $\mathcal{T}$, and since the product topology $\mathcal{T} \times \mathcal{T}$ is contained in $\mathcal{B}_C \times \mathcal{B}_C$ (reason: there exists a countable base for the topology $T \times T$ which is contained in $\mathcal{B}_C \times \mathcal{B}_C$, see [14]), it follows that h_y is $\mathcal{B}_C \times \mathcal{B}_C - \mathcal{B}_C$ measurable.

Reasoning in the same way as for the state process, it is clear that, in view of the integral equation defining the output process, it is true that:

$$\{(\omega_1,\omega_2): (\omega_1,\omega_2) \in C \times C,\ y(\cdot,(\omega_1,\omega_2)) \in B_V\} = C \times B_V$$

and therefore

$$p'_w \times p'_w(\{(\omega_1,\omega_2):(\omega_1,\omega_2) \in C \times C, y(\cdot,(\omega_1,\omega_2) \in B_V\}) = p_w \times p_w(C \times B_V) = p_w(B_V) = 0;$$

that can be phrased: *with probability one the trajectories of the output stochastic process are functions of unbounded variation.* Thus, because by the contrary any observed realization of such a process will surely be a function of bounded variation, the model illustrated heretofore, allows only to get experimental information about zero probability events, or else, according to such model, any data obtained by means of measurements is necessarily meaningless.

It might be intresting to retrieve similar results by means of different arguments, that will serve both to introduce Radon-Nikodym derivatives and to show that the illustrated pathologies occur under very general hypotheses, and anyway whenever the theory meets its commitment to make possible to attempt stochastic identification, which is usually just based on Radon-Nikodym derivatives.

With the above notations standing consider the probability measure p_x on $\mathcal{B}_C$ induced by the stochastic process $\{x(t,\cdot):t \in [0,1]\}$. This is defined by:

$$p_x = p'_w \circ h_x^{-1} = p_w \circ h_x^{-1}$$

Since the trajectories of the process $\{x(t,\cdot):\ t \in [0,1]\}$ are continuous the conditions of theorem 7.5 in [20] for the measure p_x to be absolutely continuous with respect to the measure p_w are surely fulfilled. But then, because

$$p_w(\{\omega:\ \omega \in C,\ x(\cdot,\omega) \in B_V\}) = p_w(h_x^{-1}(B_V)) = p_x(B_V)$$

absolute continuity and the fact that $p_w(B_V) = 0$ imply:

$$p_w(\{\omega : \omega \in C,\ x(\cdot,\omega) \in B_V\}) = 0$$

as shown before.

Similarly, let p_y be the probability measure on $\mathcal{B}_C$ induced by the stochastic process $\{y(t,\cdot):\ t \in [0,1]\}$, which is defined by:

$$p_y = (p'_w \times p'_w) \circ h_y^{-1} = (p_w \times p_w) \circ h_y^{-1}$$

As shown e.g. by Balakrishnan in his work on stochastic differential systems [1], the measure p_y is absolutely continuous with respect to the measure p_w. But then because

$$p_w \times p_w(\{(\omega_1,\omega_2):(\omega_1,\omega_2) \in C \times C, y(\cdot,(\omega_1,\omega_2)) \in B_V\}) = p_w \times p_w(h_y^{-1}(B_V)) = p_y(B_V)$$

absolute continuity and the fact that $p_w(B_V) = 0$ imply:

$$p_w \times p_w(\{(\omega_1,\omega_2):(\omega_1,\omega_2) \in C \times C, y(\cdot,(\omega_1,\omega_2)) \in B_V\}) = 0$$

as shown before.

The present subject is left at this point, without any attempt to get involved in infinite dimensional generalizations of the theory, (e.g. covering Hilbert space valued stochastic processes) that would certainly make just much more intricated an already very troublesome situation.

3. WHITE NOISE THEORY

The theory starts in the reverse direction of the foregoing Ito theory, that is in the most natural and direct way: instead of looking for "integrated white noise" a precise mathematical definition of white noise is given in the first place. The definition has infinite dimensionality built in because the space acting as space of equivalence classes of trajectories of white noise is the Hilbert space (topological quotient) derived from a L_2 space over a Hilbert space, which therefore reduces to $L_2/\{0\}^-$ in the scalar case. White noise is a triplet (generalized probability space): the first element is such a Hilbert space, the second is an algebra of subsets of this latter, and more precisely the algebra of cylinder subsets, and the third is a set function on the algebra that differs from a measure by the lack of countable additivity, and of course assigns value one to the whole space. However there is a fair trade between the lost in countable additivity and the gain in algebraic and topological structure obtained entering a Hilbert space setting. In addition, and this is an important technical point, the algebra of cylinder sets is the union of a directed family of σ-algebras, on each of which the set function behaves like a probability measure. The set function will be also referred to as cylinder set measure.

As to stochastic differential equations driven by white noise these are just the same as in the deterministic case, and hence they are naturally defined in the context of the theory of semigroups over Hilbert spaces, the input space being obviously the first element of white noise. Therefore all troubles about stochastic integrals, relationship between deterministic and stochastic differential equations, smoothing effect of a stochastic differential model are over. The output will be again given by the state plus a copy of white noise, independent of noise affecting the state. As explained below the technical setting, whereby independent white noises are defined, requires a somewhat cautious specification. Finally the transformations of cylinder set measures induced by the state and the output are defined in a way that parallels the classical theory.

In order to compare the old theory with the new one, it is necessary to specialize this latter to the finite dimensional case. However this tells only a part of the story, because in white noise theory there are infinite dimensional generalizations of all the relevant results. Such generalizations cannot be discussed here for space reasons. The situation is as follows: if the pathwise integral of white noise is considered, then it defines a "stochastic process" (of course all such terms must be interpreted consistently with the new context i.e. with the understanding of the specification: over a generalized probability space) that has all the same properties of Brownian motion as to finite dimensional distributions, but such that all of its trajectories are absolutely continuous. It is just as if another Brownian motion had been inserted within the zero probability event B_V of the physically meaningful trajectories of the former Brownian motion. Note in this respect that the concept of Brownian motion depends on finite dimensional distributions and not on such technicalities as the σ-additivity of the underlying measure. On the other hand if the appropriate Hilbert space is equipped with the right set function that make its points qualify as equivalence classes of trajectories of white noise, then this function cannot be extended to a probability measure on the σ-algebra generated by its domain (by the way this is just the Borel σ-algebra of the Hilbert space). Here is just a big defeat of measure and probability theory and there is nothing one can do about it.

The trajectories of the state are of course all absolutely continuous functions, and it is also obvious that a trajectory of the output is a continuous function of bounded variation if and only if such is the corresponding noise trajectory. However in the white noise case the only "measurable" subset containing all continuous functions of bounded variation is the whole space. which has measure one.

A brief sketch of the section may be useful: first cylinder set measures and the equivalent concept of weak distributions are introduced, then comes the Gauss standard weak distribution (this part of the exposition follows [16], which can be referred to for proofs and more details). Next the definition of L_2 spaces over Hilbert spaces allows the construction of white noise and the appropriate Cauchy problem that of linear stochastic systems. To conclude this general part it is illustrated how the stochastic model can be viewed as transformation of cylinder set measures. Finally all the above assertions about the comparison of the two theories are proved.

Let W be an infinite dimensional separable Hilbert space over the real field, whose topology will be denoted by S. The finite dimensional linear subspaces of W play an important role in the definition of weak distribution, and the family of all such subspaces of W will be denoted by $\mathcal{F}$. Note that all elements of $\mathcal{F}$ are necessarily closed and that $\mathcal{F}$ is directed by the inclusion relation. To each element F of $\mathcal{F}$ it is associated the corresponding orthogonal projection P_F, which is of course a linear continuous operator of W into itself.

Turning to the domains of set functions, W is endowed in the first place with its Borel σ-algebra to be denoted by $\mathcal{A}$, and then each element F of $\mathcal{F}$ is also endowed with its Borel σ-algebra (that is the σ-algebra of subsets of F, that is generated by the relative topology on F) to be denoted by $\mathcal{A}_F$. Note in passing that, being F closed, $\mathcal{A}_F$ is a sub-σ-ring of $\mathcal{A}$ and that P_F, as a function on W onto F, is continuous and hence $\mathcal{A}$-$\mathcal{A}_F$ measurable. For each F in $\mathcal{F}$, since P_F is continuous and hence measurable, $P_F^{-1}(\mathcal{A})$ is a sub-σ-algebra of $\mathcal{A}$. This is denoted by $\mathcal{A}^F$ and is called the σ-algebra of cylinder subsets of W with base in F. An element of $\mathcal{A}^F$ has the form $P_F^{-1}(A) = A+F^{\perp}$, where $F^{\perp}$ denotes, as usual, the orthogonal complement of F in W.

Next notice that if F_1 and F_2 are in $\mathcal{F}$, then $F_1 \subset F_2$ implies $\mathcal{A}^{F_1} \subset \mathcal{A}^{F_2}$. For let B be any element of $\mathcal{A}^{F_1}$ so that $B = P_{F_1}^{-1}(A)$ for some Borel subset A of F_1. Since obviously $P_{F_1} = P_{F_1}|_{F_2} \circ P_{F_2}$ it follows that $B = P_{F_2}^{-1}(P_{F_1}^{-1}|_{F_2}(A))$. But $P_{F_1}|_{F_2}$ is obviously continuous and hence $\mathcal{A}_{F_2}$ - $\mathcal{A}_{F_1}$ measurable so that $P_{F_1}^{-1}|_{F_2}(A) \in \mathcal{A}_{F_2}$ and hence also $B \in \mathcal{A}^{F_2}$ as required.

This fact has important consequences. First, along with the already mentioned circumstance that $\mathcal{F}$ is directed by inclusion, it immediately implies that the family $\{\mathcal{A}^F;\ F \in \mathcal{F}\}$ is also directed by inclusion. From this it readily follows in turn that $\cup\{\mathcal{A}^F : F \in \mathcal{F}\}$ is a subalgebra of $\mathcal{A}$, namely the algebra of cylinder subsets of W. That this algebra, which will be denoted by A_o, generates A is proved e.g. in [24].

Turning to set functions assume to begin with, that a probability measure p is given for the measurable space (W, A), and from this define a probability measure p_F for each measurable space (F, A_F) letting $p_F = p \circ P_F^{-1}$. The family of probability measures $\{p_F: F \in \mathcal{F}\}$ is called the family of finite dimensional distributions of p. Since whenever F_1 and F_2 are in $\mathcal{F}$ and $F_1 \subset F_2$ then as observed earlier $P_{F_1} = P_{F_1|F_2} \circ P_{F_2}$, it follows that for all such F_1 and F_2

$$p_{F_1} = p \circ P_{F_1}^{-1} = p \circ P_{F_2}^{-1} \circ P_{F_1|F_2}^{-1} = p_{F_2} \circ P_{F_1|F_2}^{-1}$$

This condition is called the compatibility condition and can be expressed in many equivalent ways taking advantage of the fact that for all A in A_{F_1}

$$P_{F_1|F_2}^{-1}(A) = P_{F_1}^{-1}(A) \cap F_2 = A + (F_1^{\perp} \cap F_2)$$

Thus for all such A:

$$p_{F_1}(A) = p_{F_2}(P_{F_1}^{-1}(A) \cap F_2) = p_{F_2}(A + (F_1^{\perp} \cap F_2))$$

Conversely any family of probability measures $\{p_F: F \in \mathcal{F}\}$ each on the measurable space (F, A_F), satisfying the compatibility condition, that is such that for all F_1 and F_2 in $\mathcal{F}$ with $F_1 \subset F_2$:

$$p_{F_1} = p_{F_2} \circ P_{F_1|F_2}^{-1}$$

is said to be a weak distribution for the Hilbert space W.

Of course the family of finite dimensional distributions of a probability measure p on A is a particular instance of weak distribution, but unfortunately it is not true that for any weak distribution $\{p_F: F \in \mathcal{F}\}$ there exists a probability measure p on A such that $p_F = p \circ P_F^{-1}$ for all F in $\mathcal{F}$, a counterexample being just the Gauss standard weak distribution. When conversely this situation occur, the weak distribution $\{p_F: F \in \mathcal{F}\}$ is said to be generated by the probability measure p.

If by generalized probability measure on an algebra of subsets of a space it is meant a real valued nonnegative additive function on the algebra assuming value one at the space, then it can be said that a weak distribution for W is equivalent to a generalized probability measure on A_o whose restriction to each A^F is a probability measure. Any function on A_o enjoying these properties will be called a cylinder set measure. More precisely if a weak distribution $\{p_F: F \in \mathcal{F}\}$ is given then there exists a unique cylinder set measure $\hat{p}$ on A_o such that $p_F = \hat{p} \circ P_F^{-1}$ for all $F \in \mathcal{F}$, and conversely given a cylinder set measure $\hat{p}$ on A_o it determines a unique weak distribution $\{p_F: F \in \mathcal{F}\}$ satisfying the condition $p_F = \hat{p} \circ P_F^{-1}$ for all F in $\mathcal{F}$. More over if the starting weak distribution is the family of finite dimen sional distributions of a probability measure on $\mathcal{A}$, then the corresponding cylinder set measure has a unique extension to a probability on $\mathcal{A}$ which coincides with the original probability measure.

In [24] a number of necessary and sufficient conditions are given for a weak distribution to be the family of finite dimensional distributions of a probability measure. Among these, the celebrated Minlos-Sazonov theorem is distinguished for its primary mathematical importance.

At this point the definition of Gauss standard weak distribution for an arbitrary infinite dimensional separable Hilbert space W over the reals is in order. This is constructed as follows. First for each F in $\mathcal{F}$ with dim F = n > 0 a Lebesque measure m_F is defined exploiting the isomorphism of the Hilbert spaces F and R^n. Let $\{e_1,\ldots,e_n\}$ be an arbitrary orthonormal base for F, and let U be isomorphism of R^n onto F defined by: for all $x = (x_1,\ldots,x_n)$ in R^n, $U(x) = \sum_{i=1}^{n} x_i e_i$. If λ^n is the Lebesque measure for R^n (defined on the product σ-algebra $\mathcal{B}_R^n$) then the Lebesgue measure for $(F, \mathcal{A}_F)$ is given by $m_F = \lambda^n \circ U^{-1}$. It is readily seen that this definition does not depend on the particular orthonormal base considered. Now if for F in $\mathcal{F}$, dim F = 0 then p_F is defined to be the atomic measure concentrated at the origin, otherwise if dim F = n > 0, then p_L is defined to be the measure absolutely continuous with respect to m_L with a density given by $(2\pi)^{-n/2} e^{-\frac{1}{2}\|\cdot\|_L^2}$. The ensuing family of probability measures $\{p_F : F \in \mathcal{F}\}$ is actually a weak distribution [16] and is called Gauss standard weak distribution.

To show that such weak distribution is not generated by a probability measure, and to prove the Minlos-Sazonov theorem, it is necessary to develop a theory of integration with respect to an arbitrary weak distribution $\{p_F : F \in \mathcal{F}\}$, for real valued functions on W (see [16]). One of the possible notions of integral is recalled below. Let f be a real valued functions on W and suppose that, for some F in $\mathcal{F}$, it is measurable with respect to the σ-algebra $\mathcal{A}^F$. Then f is called cylinder function and it is readly seen that it has necessarily the form $f = f_F \circ P_F$ where f_F is a real valued function on F, which is measurable with respect to the σ-algebra $\mathcal{A}_F$. Conversely any function of this form is surely measurable with respect to $\mathcal{A}^F$. If in addition $f_F \in L_1(F, \mathcal{A}_F, p_F)$, then f is called integrable cylinder function and its integral with respect to the weak distribution in question is defined to be given by $\int f_F d\, p_F$. An application of the concept of cylinder function (and of its trivial extension to the case of R^n valued functions) is given at the end of the section.

The specification of W still remains. As before consider the interval [0,1], but again any closed bounded interval would do as well. Let H be any separable Hilbert space over the field of real numbers let ω be a function on [0,1] to H. Such function is said to be weakly measurable if for any v in H, the function $(\omega(\cdot),v)_H$ is Lebesgue measurable. If that is so it is not difficult to show that the function $\|\omega(\cdot)\|_H^2$ is also Lebesgue measurable. Finally W is defined as a set to be the collection of all weakly measurable functions ω such that $\|\omega(\cdot)\|_H^2$ is integrable (with respect to the Lebesgue measure μ_L' of course). The following assertions regarding W are true: W is a linear space over R under the usual pointwise linear operations, a functional $(\cdot,\cdot)_W$ on $W \times W$ can be defined letting for all ω_1 and ω_2 in W

$$(\omega_1, \omega_2)_W = \int (\omega_1(\cdot), \omega_2(\cdot))_H d\mu_L' ;$$

such functional is a nonstrictly positive inner product, and W with the corresponding pseudonorm topology is complete and separable. The space W is named L_2 space over the Hilbert space H. The closure of $\{0\}$ in W with the pseudonorm topology is simply the linear subspace of all ω in W such that $\|\omega(t)\|_H$ is zero for almost every t in [0,1]. The representation of W by means of the topological quotient of W divided by such a subspace must be used in order to define a bona fide Hilbert space, but here, as before for L_2 spaces, the exposition will ignore the problem, leaving to the reader to work out the required technicalities.

The triplet (W, A_o, p_G), where A_o is the algebra of cylinder subsets of W and p_G is the cylinder set measure on A_o corresponding to the Gauss standard weak distribution, is defined to be white noise.

Now the definition of stochastic systems driven by white noise is obtained in the framework of the theory of semigroups over Hilbert spaces, and more specifically by means of a beautiful Cauchy problem after Balakrishnan [3]. This result must be invoked because, obviously, any element of W must be allowed to act as input of the stochastic differential equation and therefore no smoothness assumption is made on the input. Consequently the Riemann integral for continuous Hilbert space valued functions is not adequate to handle the differential equation and resort must be made to the more general Pettis integral.

This latter is defined as follows. If ω is a weakly measurable function on [0,1] to H such that $\|\omega(\cdot)\|_H$ is integrable (which is always the case if $\omega \in W$ because of the Hölder inequality), then for each z in H, the function $(\omega(\cdot), z)_H$ is integrable since view of the Schwartz inequality:

$$|(\omega(t), z)_H| \leq \|\omega(t)\|_H \|z\|_H$$

for all t in [0,1]. Now the functional I on H defined by: $I(z) = \int (\omega(\cdot), z)_H d\mu_L'$ for all z in H, is of course linear, but it is also continuous because

$$|\int (\omega(\cdot),z)_H d\mu_L')| \leq \int |(\omega(\cdot),z)_H| d\mu_L' \leq \int \|\omega(\cdot)\|_H \|z\|_H d\mu_L' = \|z\|_H \int \|\omega(\cdot)\|_H d\mu_L'$$

Thus in view of the Riesz theorem there exists an unique element v in H such that

$$I(\cdot) = \int (\omega, \cdot)_H d\mu_L' = (v, \cdot)_H$$

The element v is called the Pettis integral of ω and is denoted by $\int \omega d\mu_L'$. If A is any measurable subset of [0,1] then surely $\chi_A \omega$ is also weakly measurable and has integrable norm. It is therefore set

$\int_A \omega d\mu'_L = \int \chi_A \omega d\mu'_L$. If ω is continuous then it is weakly measurable and $\|\omega(\cdot)\|_H$ is integrable. In this case the Riemann integral of ω is defined and coincides with its Pettis integral. Space reasons do not allow to treat in detail all properties of the Pettis integral. Such analysis shows that the Pettis integral, with the appropriate handling, turns out to be as an easy instrument as the Lebesgue integral for the scalar case.

Let now $\{T(t):t \in [0,1]\}$ be a strongly continuous semigroup of linear continuous operators on H into itself, let A be the corresponding infinitesimal generator and A^* be the adjoint operator of A. Then for all x_o in H and ω in W there exists one ond only one function x on [0,1] to H, which is weakly continuous, and such that for all y in domain A^*

$$(x(\cdot),y)_H = (x_o,y)_H \chi_{[0,1]}(\cdot) + \int_{[0,\cdot)} (x(s),A^*y)_H d\mu'_L(s) + \int_{[0,\cdot)} (\omega(s),y)_H d\mu'_L(s)$$

and this function, denoting explicitly the dependence on ω, is given by:

$$x(\cdot)(\omega) = T(\cdot)x_o + \int_{[0,\cdot)} T(\cdot - s)\omega(s) d\mu'_L(s)$$

where the integral is a Pettis integral.

Of course this result can be easily generalized to the case where x and ω take values in different Hilbert spaces, but it is not worth dwelling on these details here. Rather it is essential to note that whatever is x_o, since x is weakly continuous, its range is weakly compact, thus weakly bounded and hence bounded. Again weak continuity of x imply that x is weakly measurable, so that from these two conclusions altogether it follows that for all ω in W, x is also an element of W. Applying this to the case where $x_o = 0$, it is seen that the function L on W defined by: for all ω in W, $L(\omega)(\cdot) = \int_{[0,\cdot)} T(\cdot - s)\omega(s) d\mu'_L(s)$, maps W into itself. Such function is obviously linear and the next natural question is as to whether it is continuous. Relying on the properties of the Pettis integral and of the semigroup it is not difficult to prove that L is actually bounded and hence continuous, and more precisely that if M is any constant such that $\|T(t)\| \leq M$ for all $t \in [0,1]$ (the existence of this constant is guaranteed by the strong continuity of the semigroup), then $\|L\| \leq M$.

Here is a crucial point because, making again for semplicity the assumption that $x_o = 0$, L is the transformation that takes each input trajectory onto the corresponding state trajectory and the fact that it is linear and continuous allows to associate a transformation of cylinder set measures on A_o to the differential equation. This is made possible by the following important result, which is stated in somewhat more general terms than needed here, and in particular removing the separability assumption; in this respect it must be premised that the definition of cylinder set for the nonseparable case is identical to that given above.

Let M *and* K *be arbitrary infinite dimensional Hilbert spaces over the field of real numbers, and* Q *be any continuous linear transformation on* M *into* K*. Then the inverse image under* Q *of any cylinder subset of* K *is a cylinder subset of* M.

A simple proof of this result is given below.

Let F be any finite dimensional subspace of K and B any Borel subset of F. It should be verified that $Q^{-1}(P_F^{-1}(B))$ is a cylinder subset of M. To this purpose since $Q^{-1}(P_F^{-1}(B)) = (P_F \circ Q)^{-1}(B)$ it is convenient to denote by T the operator $P_F \circ Q$. Trivially T is a finite rank operator. If N is the kernel of T, since $T_{|N^\perp}$ is a linear isomorphism of $N^\perp$ onto range T, and since this latter is finite dimensional, $N^\perp$ is finite dimensional and $T_{|N^\perp}$ is also a topological isomorphism. At this point it is sufficient to show that

$$T^{-1}(B) = T^{-1}_{|N^\perp}(B) + N$$

because $T^{-1}_{|N^\perp}(B)$ is a Borel subset of the finite dimensional subspace $N^\perp$ of M, and, being T continuous, its kernel is closed so that $N = (N^\perp)^\perp$. But on one hand

$$T(T^{-1}_{|N^\perp}(B) + N) = T(T^{-1}_{|N^\perp}(B)) \subset B$$

and therefore:

$$T^{-1}_{|N^\perp}(B) + N \subset T^{-1}(B)$$

On the other hand since N and $N^\perp$ are complementary subspaces of M, any element z of M can be written in one and only one way as the sum of an element z_N of N and an element $z_{N^\perp}$ of $N^\perp$. Next if $z \in T^{-1}(B)$ that is $T(z) \in B$ then

$$T(z) = T(z_N + z_{N^\perp}) = T(z_{N^\perp}) = T_{|N^\perp}(z_{N^\perp}) \in B$$

and therefore:

$$T^{-1}(B) \subset T^{-1}_{|N^\perp}(B) + N$$

as desired.

It is at this point possible to define a function c_x on A_o by: $c_x = p_G \circ L^{-1}$ and a straightforward verification shows that c_x is a cylinder set measure. With the help of the Minlos-Sazanov theorem it is also possible to prove that it can be extended to a probability measure on A if and only if L is Hilbert-Schmidt, which is always the case in the finite dimensional case (that is if H is finite dimensional).

The next step, mooving parallel to the usual pattern, is to define the output transformation, expressing the circumstance that the state measurement is affected by additive white noise independent of that acting on the dynamics of the system.

Here comes a rather delicate point because the natural idea of defining products of cylinder set measures does not work. To appreciate the reason it is convenient to go back for a moment to the classical case, where everything runs because product topologies and product σ-algebras behave themselves in the sense that the σ-algebra generated by $T \times T$ is equal to the product of the σ-algebra generated by T, i.e. B_C, by itself. Thus the domain of the measure that makes the two noises independent, takes also care to make continuous maps measurable. In the white noise case, if on the product algebra $A_o \times A_o$ of subsets of $W \times W$ (that is the family of all finite disjoint unions of rectangles in $W \times W$ with cylinder sets as sides) the unique generalized probability measure $p_G \times p_G$ such that

$p_G \times p_G(C_1 \times C_2) = p_G(C_1) \cdot p_G(C_2)$, whenever C_1 and C_2 are cylinder subsets of W, is considered then "measurability" is missed because it is not necessarily true that the inverse image of a cylinder subset of W under a linear continuous transformation on $W \times W$ (with the product topology) into W is an element of $A_o \times A_o$.

The solution to the problem is to go the other way around and first endow $W \times W$ with the product topology. Such method works because this topology can be retrieved by means of a inner product (and is possible to operate similarly on topological quotients) and because of the crucial result that is stated and proved below. Thus endow $W \times W$ with the inner product $(\cdot,\cdot)_{W \times W}$ defined by $((v_1,v_2),(z_1,z_2))_{W \times W} = (v_1,z_1)_W + (v_2,z_2)_W$ for all (v_1,v_2) and (z_1,z_2) in $W \times W$ and observe that the topology of the corresponding pseudonorm is just the product topology. The topological product $W \times W$ is a complete, separable (reason: if D is a countable dense subset of W, then $D \times D$, which is countable, is dense in $W \times W$) non-Hausdorff linear topological space. The closure of the singleton of the origin in $W \times W$ is simply $\{0\}^-_W \times \{0\}^-_W$ where $\{0\}^-_W$ is the closure of $\{0\}$ in W. The usual loose identification of $W \times W$ with the topological quotient $W \times W/\{0\}^-_W \times \{0\}^-_W$, which is made onto a Hilbert space by the natural inner product, will be made. In this respect it is essential to make the simple remark that since the right starting space should have been the topological quotient $W/\{0\}^-_W$ the right ending space is the topological product $W/\{0\}^-_W \times W/\{0\}^-_W$ equipped with the inner product obtained as above from the inner product of $W/\{0\}^-_W$. However the ensuing Hilbert space is isomorphic to the Hilbert space $W \times W/\{0\}^-_W \times \{0\}^-_W$, as it is readily verified, so that the above procedure is correct.

Now consider the triplet $(W \times W, \hat{A}_o, \hat{p}_G)$, where $\hat{A}_o$ is the algebra of cylinder subsets of $W \times W$, and $\hat{p}_G$ is the cylinder set measure on $\hat{A}_o$ corresponding to the Gauss standard weak distribution. Then:

The algebra $\hat{A}_o$ properly contains the algebra $A_o \times A_o$ and the restriction of $\hat{p}_G$ to $A_o \times A_o$ is equal to $p_G \times p_G$.

To prove these assertions, begin with considering the product $C_1 \times C_2$ of any two cylinder subsets of W. Thus $C_1 = B_1 + F_1^\perp$ and $C_2 = B_2 + F_2^\perp$ where F_1 and F_2 are finite dimensional subspaces of W and B_1 and B_2 are Borel measurable subsets of F_1 and F_2 respectively. Next note that $C_1 \times C_2 = (B_1 + F_1^\perp) \times (B_2 + F_2^\perp) = (B_1 \times B_2) + (F_1^\perp \times F_2^\perp)$. On the other hand $B_1 \times B_2$ is a Borel subset of the finite dimensional subspace $F_1 \times F_2$ of $W \times W$ and furthermore $F_1^\perp \times F_2^\perp = (F_1 \times F_2)^\perp$ as it is easy to show, and this means that $C_1 \times C_2$ is actually a cylinder subset of $W \times W$. As to the "properly" part of the first statement it is readly seen that for example; if $\{e_i;\ i = 1,2,\ldots\}$ is any orthonormal basis for W the subset of $W \times W$

$$\{(v,z):\ (v,z) \in W \times W \text{ and } (v,e_1)^2_W + (z,e_1)^2_W = 1\}$$

is in $\hat{A}_o$ but not in $A_o \times A_o$.

It remains to be computed

$$\hat{p}_G((B_1 + F_1^{\perp}) \times (B_2 + F_2^{\perp})) = \hat{p}_{F_1 \times F_2}(B_1 \times B_2)$$

where $\hat{p}_{F_1 \times F_2}$ is the element of the Gauss standard weak distribution for $W \times W$ corresponding to the finite dimensional subspace $F_1 \times F_2$. To this purpose, consider any orthonormal bases $\{v_i : i=1,\dots,n\}$ for F_1 and $\{z_j : j=1,\dots,m\}$ for F_2, so that $\{(v_i,0)\} \cup \{(0,z_j)\}$ is an orthonormal basis for $F_1 \times F_2$, and let U be the isomorphism of R^{n+m} onto $F_1 \times F_2$ corresponding to this latter. Then:

$$\hat{p}_{F_1 \times F_2}(B_1 \times B_2) = (2\pi)^{\frac{n+m}{2}} \int_{B_1 \times B_2} e^{-\frac{1}{2}\|\cdot\|^2_{F_1 \times F_2}} d(\lambda^{n+m} \circ U^{-1}) = (2\pi)^{\frac{n+m}{2}} \int_{U^{-1}(B_1 \times B_2)} e^{-\frac{1}{2}\|\cdot\|^2_{R^{n+m}}} d\lambda^{n+m}$$

On the other hand by the way the basis for $F_1 \times F_2$ is choosen

$$U^{-1}(B_1 \times B_2) = U^{-1}(B_1 \times \{0\}) + U^{-1}(\{0\} \times B_2)$$

and this fact along with an application of the Fubini's theorem leads to the conclusion that

$$\hat{p}_{F_1 \times F_2}(B_1 \times B_2) = p_{F_1}(B_1) \cdot p_{F_2}(B_2) = p_G \times p_G((B_1 + F_1^{\perp}) \times (B_2 + F_2^{\perp}))$$

whereby the proof is completed.

At this point let Q_1 and Q_2 be the projections of $W \times W$ onto its factor spaces. Then keeping the assumption that $x_o = 0$, for each (ω_1, ω_2) in $W \times W$ the observation or output is the element of W given by (denoting explicitly the dipendence on (ω_1, ω_2))

$$y(\cdot)((\omega_1,\omega_2)) = L \circ Q_1((\omega_1,\omega_2)) + Q_2((\omega_1,\omega_2)) = (L \circ Q_1 + Q_2)((\omega_1,\omega_2))$$

Since the transformation $L \circ Q_1 + Q_2$ on $W \times W$ into W is linear and continuous, in view of a foregoing result, it is well defined the output-induced cylinder set measure, that is the cylinder set measure c_y on A_o given by;

$$c_y = \hat{p}_G \circ (L \circ Q_1 + Q_2)^{-1}$$

Turning to the specialization to the scalar valued case, H is henceforth assumed to be equal to R, so that $W = L_2$. Hence the foregoing stochastic differential equation reduces to

$$x(\cdot) = x_o \chi_{[0,1]}(\cdot) + \int_{[0,\cdot)} Ax(s)\,d\mu_L'(s) + \int_{[0,\cdot)} \omega(s)\,d\mu_L'(s)$$

where A is now any element of R (of course the constraint for the infinite dimensional case expressed by the celebrated Yosida-Hille-Feller-Miyadera-Phillips theorem disappears) and the integrals are simply Lebesgue integrals. The solution is now given by:

$$x(\cdot)(\omega) = T(\cdot)x_o + \int_{[0,\cdot)} T(\cdot - s)\,\omega(s)\,d\mu_L'(s)$$

where again the integral is a Lebesgue integral, and $\{T(t) : t \in [0,1]\}$ is the strongly continuous semigroup corresponding to A. Trivially now, in view of the integral equation, $x(\cdot)(\omega)$ is not only continuous, but also absolutely continuous for all ω in W as asserted previously.

The output equation is the same as before with the current specification of W. Therefore, because of the absolute continuity of $(L \circ Q_1)((\omega_1,\omega_2))$ for all ω_1 and ω_2 in W, y is a continuous function of bounded variation if and only if such is ω_2. It follows that:

$$\{(\omega_1,\omega_2): (\omega_1,\omega_2) \in W \times W \text{ and } y(\cdot)(\omega_1,\omega_2) \in B_V\} = W \times B_V$$

This latter set is a linear subspace of $W \times W$, and in fact a dense linear subspace because B_V is dense in W. On the other hand a cylinder subset of $W \times W$, say $B + F^{\perp}$, with F linear subspace of $W \times W$ and B Borel subset of F, is a linear subspace of $W \times W$ if and only if B is a linear subspace of F. Consequently whenever a cylinder subset of $W \times W$ is a linear subspace it is a closed linear subspace, which in turn implies that $W \times W$ is the only cylinder subset that contains $W \times B_V$.

As to the relationship between white noise and Brownian motion define the function $\hat{w}$ on $[0,1] \times W$ by: for all (t,ω) in $[0,1] \times W$

$$\hat{w}(t,\omega) = \int_{[0,t)} \omega d\mu_L'$$

To remove any doubt that the "stochastic process" $\{\hat{w}(t,\cdot): t \in [0,1]\}$ on the generalized probability space $\{W, A_o, p_G\}$ is the right model of the physical phenomenon Brownian motion, the section is concluded giving in detail all the relevant proofs.

First note that trivially for all ω in W, $\hat{w}(\cdot,\omega)$ is an absolutely continuous function and $w(0,\omega) = 0$.

Next for all t_1 and t_2 in $0,1$ with $t_2 > t_1$, clearly

$$\hat{w}(t_2,\omega) - \hat{w}(t_1,\omega) = \int \chi_{[t_1,t_2)} \omega d\mu_L' \text{ for all } \omega \in W$$

If F is the linear extension of $\{\chi_{[t_1,t_2)}\}$ in W, P_F the orthogonal projection of W onto F, and U the isomorphism of R onto F defined in the already specified manner, it is also obvious that:

$$\hat{w}(t_2,\cdot) - \hat{w}(t_1,\cdot) = \|\chi_{[t_1,t_2)}\|_W \, U^{-1} \circ P_F = (t_2 - t_1)^{\frac{1}{2}} U^{-1} \circ P_F$$

Thus, because $\hat{w}(t_2,\cdot) - \hat{w}(t_1,\cdot)/(t_2-t_1)^{\frac{1}{2}}$ is a cylinder function, its distribution, that is the probability measure $p_G \circ (\hat{w}(t_2,\cdot) - \hat{w}(t_1,\cdot)/(t_2-t_1)^{\frac{1}{2}})^{-1}$ is well defined and given by:

$$p_G \circ P_F^{-1} \circ U = p_F \circ U$$

(caution: with the same symbol U it is here and hereafter denoted the image map corresponding to U), so that for all B in $\mathcal{B}_R$ by definition of the Gauss standard weak distribution:

$$p_F \circ U(B) = (2\pi)^{-\frac{1}{2}} \int_{U(B)} e^{-\frac{1}{2}\|\cdot\|_F^2} d(\lambda \circ U^{-1}) = (2\pi)^{-\frac{1}{2}} \int_B e^{-\frac{1}{2}\|\cdot\|_R^2} d\lambda$$

which means that the distribution of $\hat{w}(t_2,\cdot) - \hat{w}(t_1,\cdot)/(t_2-t_1)^{\frac{1}{2}}$ is gaussian with zero mean and unit variance, and hence that $\hat{w}(t_2,\cdot) - \hat{w}(t_1,\cdot)$ has a gaussian distribution with mean zero and variance (t_2-t_1).

Finally suppose that $0 \le t_1 < t_2 < \ldots < t_n \le 1$ and consider the map $T_{t_1,\ldots,t_n}$ on W into R^{n-1} defined by

$T_{t_1,\ldots,t_n}(\omega)=(\hat{w}(t_2,\omega)-\hat{w}(t_1,\omega),\ldots,\hat{w}(t_n,\omega)-\hat{w}(t_{n-1},\omega))$ for all ω in W.

If F is now the linear extension in W of the set $\{\chi_{[t_1,t_2)},\ldots,\chi_{[t_{n-1},t_n)}\}$, P_F the orthogonal projection of W onto F and U the linear isomorphism of R^{n-1} onto F defined as usual in correspondence to the orthonormal base for F given by $\{\chi_{[t_1,t_2)}/(t_2-t_1)^{\frac{1}{2}},\ldots,\chi_{[t_{n-1},t_n)}/(t_n-t_{n-1})^{\frac{1}{2}}\}$, then it follows

$$T_{t_1,\ldots,t_n} = \mathrm{diag}\{(t_2-t_1)^{\frac{1}{2}},\ldots,(t_n-t_{n-1})^{\frac{1}{2}}\} \circ U^{-1} \circ P_F$$

and therefore it is actually true in the first place that the inverse image under $T_{t_1,\ldots,t_n}$ of any Borel subset of R^{n-1} is a cylinder subset of W with base in F and that the corresponding multidimensional distribution, that is the probability measure $p_G \circ T^{-1}_{t_1,\ldots,t_n}$ on $\mathcal{B}^{n-1}_R$, is well defined and given by

$$p_G \circ P_F^{-1} \circ U \circ (\mathrm{diag}\{(t_2-t_1)^{\frac{1}{2}},\ldots,(t_n-t_{n-1})^{\frac{1}{2}}\})^{-1} = p_F \circ U \circ \mathrm{diag}\{(t_2-t_1)^{-\frac{1}{2}},\ldots,(t_n-t_{n-1})^{-\frac{1}{2}}\}$$

Similarly to the preceding case, to make it easier, look first at the distribution of $U^{-1} \circ P_F$ which is given by

$$p_G \circ P_F^{-1} \circ U = p_F \circ U$$

so that for all B in $\mathcal{B}^{n-1}_R$

$$p_F \circ U(B) = (2\pi)^{-\frac{n-1}{2}} \int_{U(B)} e^{-\frac{1}{2}\|\cdot\|_F^2} d(\lambda^{n-1} \circ U^{-1}) = (2\pi)^{-\frac{n-1}{2}} \int_B e^{-\frac{1}{2}\|\cdot\|_{R^{n-1}}} d\lambda^{n-1}$$

hence the distribution of $(\mathrm{diag}\{(t_2-t_1)^{\frac{1}{2}},\ldots,(t_n-t_{n-1})^{\frac{1}{2}}\})^{-1} T_{t_1,\ldots t_n}$ is the product of zero mean unit variance one dimensional Gauss measures. It follows that the distribution of $T_{t_1,\ldots,t_n}$ is actually the product of the distributions of $\hat{w}(t_2,\cdot)-\hat{w}(t_1,\cdot),\ldots,\hat{w}(t_n,\cdot)-\hat{w}(t_{n-1},\cdot)$ that is that these "random variables" are actually independent. With this conclusion all physical properties defining Brownian motion have been retrieved.

4. FINAL REMARKS

The foregoing discussion shows that in the classical probabilistic model all observed results are a priori known to belong to a fixed zero measure event. This foundational difficulty gives in turn rise to technical problems in applications. For instance suppose to attempt the construction of a likelinood functional via Radon-Nikodym derivatives. Thus for each given values of the unknown parameters (which will be some of the coefficient of the stochastic differential equation in question, in the foregoing simple case the value of A), consider the Radon-Nikodym derivative of the probability measure p_y with respect to the probability measure p_w. As is very well known such derivative is not unique, but each function equal to any of the derivatives modulo p_w is also Radon-Nikodym derivative. It is convenient here to use the symbol dp_y/dp_w to denote the family of all Radon-Nikodym of p_y with respect to p_w. Since any observed result y will necessarily belong to B_v, all is needed to know is the family of restrictions to B_v of members of dp_y/dp_w. But since $p_w(B_v) = 0$, this family is given by the collection of all restriction to B_V of arbitrary measurable functions on the measurable space $(C, \mathcal{B}_C)$. The analysis carried out in this note should also convince that this pathology cannot be bypassed because of its intrinsic nature and hence that there is no way out for constructing a likelihood functional.

In the white noise setting, filtering theory, the generalization of Radon-Nikodym derivatives, and white noise version of the Girzanov formula have already been given by Balakrishnan who also applied them to identification with apparently excellent result [9]. However it is enough here having introduced the approach and motivated the need of adopting the new theory instead of Ito theory, especially for engineering purposes. On the other hand as to filtering and identification as well as for extension to nonlinear cases (see e.g. [2] and [19]) white noise theory is still, in this early stage, a source of a great amount of challenging research problems, which deserve a primary interest both on mathematical and on applicative grounds; so that it might be premature to try now a complete assessement of it.

REFERENCES

[1] A.V.BALAKRISHNAN: *Stochastic differential systems*. New York, Springer Verlag, 1973.

[2] A.V.BALAKRISHNAN: *Stochastic bilinear partial differential equations*. Proc. 2nd USA-Italy Seminar on Variable Structure Systems, A. Ruberti and P.R.Mohler eds., New York, Springer Verlag, 1975.

[3] A.V.BALAKRISHNAN: *Applied Functional analysis*. New York, Springer Verlag, 1976.

[4] A.V.BALAKRISHNAN: *A white noise version of the Girsanov formula*. Proceedings of the internazional Symposium on stochastic differential equations, (Kyoto 1976) K.Ito Editor, Wiley Interscience 1978.

[5] A.V.BALAKRISHNAN: *Likelihood ratios for time continuous data models*. IRIA int. Symposium in new trends in system analysis, Dec. 1876, Versailles, France.

[6] A.V.BALAKRISHNAN: *Radon-Nikodym derivatives of a class of weak distributions on Hilbert spaces*. Applied Mathematics and Optimization, vol. 3, n. 2/3, 1977, pp. 209-225.

[7] A.V.BALAKRISHNAN: *Likelihood ratios for signals in additive white noise*. Applied Mathematics & Optimization, vol.3, n. 4, 1977.

[8] A.V.BALAKRISHNAN: *Stochastic optimization: time continuous data models*. Proceedings of the 8th IFIP Conference on Optimization technique, New York, Springer-Verlag, 1978.

[9] A.V.BALAKRISHNAN: *Parameter estimation in stochastic differential systems: theory and applications*. Developments in Statistics, vol. I, New York, Academic Press 1978.

[10] N.BOURBAKI: *Elements d'histoire des mathématiques*. Paris Hermann, 1974 (in french).

[11] BREIMAN: *Probability*. Reading (mass., Addison-Welsley, 1968.

[12] A.BROWN: *Additional remarks on active molecules*. Philosophical Pagazine n.s. 6, 1829, pp. 161-166.

[13] P.d'ALESSANDRO: *Vector topologies and the representation of spaces of random variables*. Applied Mathematics and Optimization, vol. 4, n. 1, 1977.

[14] P.d'ALESSANDRO: *Some remarks of topology and of measure theory applied to the definition of the Ito integral*. Applied Mathematics and Optimization, vol. 4, n. 3, 1978.

[15] P.d'ALESSANDRO, G.RINALDI and A.SASSANO: *A point of view in system theory*. Report 78-14, July 1978, Istituto di Automatica, University of Rome.

[16] P.d'ALESSANDRO, G.RINALDI and A.SASSANO: *A note on weak distributions*. Report 79-13, April 1979, Istituto di Automatica, University of Rome.

[17] M.H.A. DAVIS: *Linear estimation and stochastic control*. London, Chapman and Hall, 1977.

[18] A.DVORETSKI, P.ERDÖS and S.KAKUTANI: *Nonincrease everywhere of the Brownian motion process*. Proc. 4th Berkeley Symp. on Math. Stat. and Prob., vol. II, 1961, pp. 103-116.

[19] A.GERMANI and P.SEN: *White noise solutions for a class of distributed feedback systems with multiplicative noise*. Report Istituto di Automatica, University of Rome.

[20] R.S.LIPTSER and A.N.SHIRYAYEV: *Statistics of random process*. New York, Springer-Verlag, 1977.

[21] E.NELSON: *Dynamical theories of Brownian motion*. Princeton, Princeton University Press, 1967.

[22] J.NEVEU: *Bases mathématiques du calcul des probabilités*. Paris, Masson & C.ie, 1964.

[23] K.R.PARTHSSARATHY: *Probability measures on metric spaces*. New York, Academic Press, 1967.

[24] A.V.SKOROHOD: *Integration in Hilbert Space*, New York, Springer-Verlag, 1974.

Numerical Techniques for Stochastic Systems
F. Archetti and M. Cugiani (eds.)
© North-Holland Publishing Company, 1980

A CONTINUOUS-TIME APPROXIMATION TO OPTIMAL NONLINEAR FILTERING

Giovanni B. Di Masi
CNR-LADSEB and
Istituto di Elettrotecnica
Università di Padova
Padova, Italy

Wolfgang J. Runggaldier
Seminario Matematico and
Istituto di Statistica
Università di Padova,
Padova, Italy

An approximation to the nonlinear filtering problem for diffusion models, given in terms of functionals of a continuous-time Markov chain, is provided. The approximating functionals satisfy a recursive relation.

INTRODUCTION

In the present work a recursive approximate solution for the nonlinear filtering problem is derived.

Consider the n-dimensional diffusion process given by the Itô equation

$$dx(t) = a(x(t),t)dt + b(x(t),t)dw(t), \quad 0 \le t \le T, \quad x(0) = x_o , \tag{1.1}$$

where a and b are R^n- and $R^{n\times m}$- valued functions respectively on $R^n\times[0,T]$ and w is an m-dimensional standard Wiener process.

The process x(t) is partially observed through the p-dimensional process y(t) given by

$$dy(t) = c(x(t),t)dt + dv(t) , \quad y(0) = y_o , \tag{1.2}$$

where c is an R^p- valued function on $R^n\times[0,T]$ and v is a p-dimensional standard Wiener process independent of x(t).

Let $\mathcal{Y}_t$ be the σ-algebra generated by $\{y(s), s\le t\}$ and let f be a real valued measurable function on R^n. The nonlinear filtering problem consists in the evaluation for each t of the conditional expectation

$$\hat{f}(x(t)) = E\{f(x(t)) \mid \mathcal{Y}_t\} \tag{1.3}$$

if it exists.

Our results will be based on the following assumptions:

A1. <u>The functions a, b, c and f in (1.1), (1.2) and (1.3) are continuous and bounded</u>.

A2. <u>For each initial condition x_o, equation (1.1) has on [0,T] a unique solution in the weak sense /1/</u>.

Work partially supported by the Consiglio Nazionale delle Ricerche under Contract 79.00700.01

Furthermore, for simplicity of exposition, we will consider deterministic initial conditions x_o and y_o and we will limit ourselves to the scalar case, i.e., n=m=p=1.
In the literature equations have been derived both for the conditional expectation $\hat{f}(x(t))$ and for the conditional density of x(t). A rather comprehensive exposition of these results can be found in /1/. However, the actual application of such results is still a problem and therefore attempts have been made to obtain approximate solutions that can be recursively computed.

The starting point of our approximation is given by the so-called Kallianpur-Striebel formula of which two versions will be given below. The first one is slightly less general but it allows a better understanding of its connection with Girsanov's theorem /2/ on measure-transformations in function spaces. The following result /3; Prop.VI.5.1/ will clarify such connection.

Let the scalar processes x(t) and y(t) given, for n=m=p=1, by (1.1) and (1.2) be defined on the probability space $\{\Omega, \mathcal{F}, P\}$. Define

$$\frac{d P_{0,t}}{dP} = \exp\left[- \int_0^t c(x(s),s)dv(s) - 1/2 \int_0^t c^2(x(s),s)ds \right], \ 0 \le t \le T, \quad (1.4)$$

then, under our assumptions, we have

i) $P_{0,t}$ is a probability measure on $\{\Omega, \mathcal{F}\}$. (1.5a)

ii) Under $P_{0,t}$, $\{y(s), s \le t\}$ is a standard Wiener process. (1.5b)

iii) Under $P_{0,t}$, $\{c(x(s),s), s \le t\}$ and $\{y(s), s \le t\}$ are independent. (1.5c)

iv) The restriction of $P_{0,t}$ on $\sigma\{c(x(s)), s \le t\}$ is the same as the corresponding restriction of P. (1.5d)

v) $P \ll P_{0,t}$ and

$$\frac{dP}{dP_{0,t}} = \exp\left[\int_0^t c(x(s),s)dy(s) - 1/2 \int_0^t c^2(x(s),s)ds \right]. \quad (1.5e)$$

We can now state a first version of the Kallianpur-Striebel formula /3;Prop.VI.5.2/.
Let f be a measurable function such that $E\{|f(c(x(t),t)|\} < \infty$ for all $t \le T$, then we have

$$E\{f(c(x(t),t)) | \mathcal{Y}_t\} = \frac{E_o\{f(c(x(t),t)) \, dP/dP_{0,t} | \mathcal{Y}_t\}}{E_o\{dP/dP_{0,t} | \mathcal{Y}_t\}}$$

$$= \frac{E_o\left\{ f(c(x(t),t))\exp\left[\int_0^t c(x(s),s)dy(s) - 1/2 \int_0^t c^2(x(s),s)ds\right] | \mathcal{Y}_t \right\}}{E_o\left\{ \exp\left[\int_0^t c(x(s),s)dy(s) - 1/2 \int_0^t c^2(x(s),s)ds\right] | \mathcal{Y}_t \right\}} \quad (1.6)$$

where E_o denotes the expectation with respect to the measure $P_{0,t}$.
Notice that, because of (1,5c), the conditioning in (1.6) fixes the process $\{y(s), s \le t\}$ while the expectation is actually only over the process $\{x(s), s \le t\}$.

A second versionof the Kallianpur-Striebel formula will be given in (1.7). There the random variable to be estimated will not necessarily be a function of the process $c(x(t),t)$; furthermore the independence of the processes $c(x(t),t)$ and $y(t)$ will be made more explicit by defining $x(t)$ and $y(t)$ on a product space. Let the processes $x(t)$ and $y(t)$ be defined on $\{\Omega', \mathscr{F}', P'\}$ and $\{\Omega, \mathscr{F}, P\}$ respectively, where $\{\Omega', \mathscr{F}', P'\}$ is a copy of $\{\Omega, \mathscr{F}, P\}$. Lift $x(t)$ and $y(t)$ to the product space $\{\Omega'\times\Omega, \mathscr{F}'\times\mathscr{F}, P'\times P\}$ by setting $x(t,\omega',\omega)=x(t,\omega')$ and $y(t,\omega',\omega) = y(t,\omega)$. Then for each integrable random variable f on $\{\Omega', \mathscr{F}', P'\}$ we have /4/, /5; Th.1/.

$$E\{f\,|\,\mathscr{Y}_t\} = \frac{\int f(\omega')\exp\left[\int_0^t c(x(s,\omega'),s)dy(s,\omega)-1/2\int_0^t c^2(x(s,\omega'),s)ds\right]P(d\omega')}{\int \exp\left[\int_0^t c(x(s,\omega'),s)dy(s,\omega)-1/2\int_0^t c^2(x(s,\omega'),s)ds\right]P(d\omega')} \quad \text{a.s.,} \tag{1.7}$$

This second version will be used in the sequel in the following form /6; Sec.7.5/. Let $\mathscr{C}[0,T]$ be the Borel σ-algebra induced on $C[0,T]$ by the sup norm topology. Let $\bar{x}(t)$ be a version of $x(t)$ independent of $y(t)$ in the sense that $\bar{x}(t)$ induces on $\mathscr{C}[0,T]$ the same probability measure as $x(t)$ and $\{\bar{x}(s),s\leq t\}$ and $\{y(s),s\leq t\}$ are independent. Let f be a measurable function such that $E\{|f(x(t))|\}<\infty$ for all $t\leq T$ and define

$$V_f(t) = E\left\{f(\bar{x}(t))\exp\left[\int_0^t c(\bar{x}(s),s)dy(s)-1/2\int_0^t c^2(\bar{x}(s),s)ds\right] \,\middle|\, \mathscr{Y}_t\right\}. \tag{1.8}$$

Then we have

$$E\{f(x(t))\,|\,\mathscr{Y}_t\} = V_f(t)/V_1(t) \quad \text{a.s.,} \tag{1.9}$$

where $V_1(t)$ is given by (1.8) with $f=1$.

In the following sections we will approximate (1.9) by approximating its numerator and denominator by analogous functionals of a continuous-time Markov chain which will be obtained by spatial discretization of the original diffusion process $x(t)$. As the discretization step goes to zero, the chain will converge weakly to the diffusion and, as a consequence, the corresponding functionals of the chains will converge to the functional of the diffusion. For the actual computation of the approximating functionals the problem will be considered in a bounded region by stopping the process $x(t)$ at its first exit from such a region; in this way an approximating chain with a finite number of states will be obtained. Finally a recursive procedure for the computation of the approximating functionals will be given.

In /6; Sec.7.5/ an analogous approximation is given, where discrete-time approximating chains are used. This allows a much easier derivation of the recursive relation for the approximating functionals but has the disadvantage that only sampled values of the observation are used. In our approach, due to the continuous-time nature of the approximation procedure, there is no loss of information as far as the observation process is concerned.

DISCRETIZATION AND APPROXIMATING FUNCTIONALS

Consider the scalar diffusion process $x(t)$ given for $n=m=1$ by (1.1), fix a positive and real discretization step h and define

$$\lambda^h_+(x,t)=a^+(x,t)/h + b^2(x,t)/2h^2, \tag{2.1a}$$

$$\lambda_-^h(x,t)=a^-(x,t)/h+ b^2(x,t)/2h^2 \quad , \tag{2.1b}$$

$$\lambda^h(x,t)=- |a(x,t)|/h-b^2(x,t)/ h^2 \quad , \tag{2.1c}$$

where $a^+(x,t)=\max\{a(x,t),0\}$ and $a^-(x,t)=[-a(x,t)]^+$.

Observe that $\lambda_+^h(x,t)\geq 0$, $\lambda_-^h(x,t)\geq 0$ and $\lambda_+^h(x,t)+ \lambda_-^h(x,t)+\lambda^h(x,t)=0$, so that we can define a continuous-time Markov chain $\{x^h(t),0\leq t\leq T\}$ with state space $R^h=\{z\in R: z=x_o+nh,\ n \text{ integer}\}$, transition rates given by (2.1) and such that $P\{x^h(0)=x_o\}=1$. As transition probabilities we can therefore choose

$$P\{x^h(t+\Delta)=x+h \mid x^h(t)=x\} = \int_t^{t+\Delta} \lambda_+^h(x,s)ds \tag{2.2a}$$

$$P\{x^h(t+\Delta)=x-h|x^h(t)=x\} = \int_t^{t+\Delta} \lambda_-^h(x,s)ds \tag{2.2b}$$

$$\sum_{\substack{n\in Z\\ n\neq -1,0,+1}} P\{x^h(t+\Delta)=x+nh|x^h(t)=x\}=o(\Delta) \tag{2.2c}$$

$$P\{x^h(t+\Delta)=x|x^h(t)=x\}=1+ \int_t^{t+\Delta} \lambda^h(t,s)ds-o(\Delta) \tag{2.2d}$$

As mentioned in the previous section, in order to get an actually computable approximation we have to consider the problem on a bounded region. For this purpose let $G\subset R$ be open and bounded, let $G^h=G\cap R^h$ and define the stopping times

$$\tau = \begin{cases} T & \text{if } x(t)\in G \text{ for all } t\in[0,T] \\ \inf\{t:0\leq t\leq T,\ x(t)\notin G\} & \text{otherwise,} \end{cases} \tag{2.3}$$

$$\tau^h = \begin{cases} T & \text{if } x^h(t)\in G^h \text{ for all } t\in[0,T] \\ \inf\{t:0\leq t\leq T;\ x^h(t)\notin G^h\} & \text{otherwise} \end{cases} \tag{2.4}$$

as well as the corresponding stopped processes

$$\tilde{x}(t) = x(t\wedge\tau), \quad \tilde{x}^h(t) = x^h(t\wedge\tau^h). \tag{2.5}$$

Let $\tilde{y}(t)$ be the observation corresponding to the stopped process $\tilde{x}(t)$, namely

$$\tilde{y}(t) = y_0 + \int_0^t c(\tilde{x}(s),s)ds+v(t) \quad , \tag{2.6}$$

and $\tilde{\mathcal{Y}}_t$ be the σ-algebra generated by $\{\tilde{y}(s),\ 0\leq s\leq t\}$. The problem then becomes that of approximating

$$E\{f(\tilde{x}(t)) \mid \tilde{\mathcal{Y}}_t\} = \tilde{V}_f(t)/\tilde{V}_1(t) \qquad \text{a.s.} , \tag{2.7}$$

where

$$\tilde{V}_f(t) = E\{f(\bar{\tilde{x}}(t))\exp\left[\int_0^t c(\bar{\tilde{x}}(s),s)d\tilde{y}(s) - 1/2 \int_0^t c^2(\bar{\tilde{x}}(s),s)ds\right] \mid \tilde{\mathcal{Y}}_t\} \tag{2.8}$$

with $\bar{\tilde{x}}(t)$ a version of $\tilde{x}(t)$ independent of $\tilde{y}(t)$. Notice that because of the boundedness of $\tilde{x}(t)$, the boundedness of f is not a particularly restrictive assumption.

Let $\bar{\tilde{x}}^h(t)$ be a version of $\tilde{x}^h(t)$ independent of $\tilde{y}(t)$ (as $\bar{\tilde{x}}^h(t)$ has discontinuous trajectories, instead of the σ-algebra $\mathcal{C}[0,T]$ we have to consider the σ-algebra $\mathcal{D}[0,T]$ which will be defined in the following section). In analogy to (2.8) we will then choose the approximating functional as given by

$$\tilde{V}_f^h(t) = E\{f(\bar{\tilde{x}}^h(t))\exp\left[\int_0^t c(\bar{\tilde{x}}^h(s),s)d\tilde{y}(s) - 1/2 \int_0^t c^2(\bar{\tilde{x}}^h(s),s)ds\right] \mid \tilde{\mathcal{Y}}_t\} \tag{2.9}$$

CONVERGENCE OF THE DISCRETIZED PROCESSES AND OF THE APPROXIMATING FUNCTIONALS

We will first show that $x^h(t)$ converges weakly to $x(t)$ as $h \downarrow 0$. Following a standard technique in weak convergence theory, this will be done in two steps.

Step 1: we will show that every sequence $\{x^{h_n}(t),\ h_n \downarrow 0\}$ is weakly sequentially compact, namely there exists a weakly converging subsequence. To this end, let $D^n[0,T]$ be the space of R^n- valued functions on $[0,T]$ that are right continuous, left continuous at T and have left hand limits, endowed with a metric /8;p.111/ (equivalent to the Skorokhod metric) with respect to which $D^n[0,T]$ is separable and complete. We will denote by $\mathcal{D}^n[0,T]$ the corresponding Borel σ-algebra and, for simplicity, we will often use the symbol $D^n[0,T]$ to denote the corresponding measurable space $\{D^n[0,T], \mathcal{D}^n[0,T]\}$. It is known /8; Th.6.1 and 6.2/ that on $D^n[0,T]$ weak sequential compactness is equivalent to tightness.
Define

$$A^h(t) = \int_0^t a(x^h(s),s)ds \quad , \tag{3.1a}$$

$$B^h(t) = x^h(t) - x_o - \int_0^t a(x^h(s),)ds \quad , \tag{3.1b}$$

so that $x^h(t) = x_o + A^h(t) + B^h(t)$. We then have the following result whose proof can be found in /7/.

Proposition 1: Every sequence $\{(x^{h_n}(t), A^{h_n}(t), B^{h_n}(t)),\ h_n \downarrow 0\}$ is tight in $D^3[0,T]$ and the limit of every converging subsequence has trajectories in $C^3[0,T]$ a.s.

Step 2: we will show that every weakly converging sequence $\{x^{h_n}(t),\ h_n \downarrow 0\}$ converges to the same limit, which is precisely the original diffusion x(t). Taking into account assumption A2., this is accomplished by the following

Theorem 1: for each weakly converging sequence $\{(x^{h_n}(t), A^{h_n}(t), B^{h_n}(t)),\ h_n \downarrow 0\}$ there exists a standard Wiener process $\tilde{w}(t)$ such that the limiting process is given by $(\xi(t), A(t), B(t))$, with

$$\xi(t) = x_o + A(t) + B(t) \,, \tag{3.2a}$$

$$A(t) = \int_0^t a(\xi(s),s)ds \,, \tag{3.2b}$$

$$B(t) = \int_0^t b(\xi(s),s)dw(s). \tag{3.2c}$$

For the proof see /7/.

In the following Theorems 2 and 3, we will show the convergence of the stopped processes and of the approximating functionals. The proofs can again be found in /7/.

Let $\bar{G}$ be the closure of G and define the stopping time

$$\tau' = \begin{cases} T \text{ if } x(t)\in \bar{G} \quad \text{for all } t\in[0,T] \\ \\ \inf\{t: 0\leq t\leq T,\ x(t)\notin \bar{G}\} \text{ otherwise.}\end{cases} \tag{3.3}$$

Then we have

Theorem 2: if $P\{\tau'=\tau\}=1$, then $\tilde{x}^h(t)$ converges weakly to $\tilde{x}(t)$ as $h\downarrow 0$.

Theorem 3: under the assumption of Theorem 2, $\tilde{V}_f^h(t)$ converges to $\tilde{V}_f(t)$ a.s. for each $t\in[0,T]$, as $h\downarrow 0$.

RECURSIVE COMPUTATION OF THE APPROXIMATING FUNCTIONALS

Rewrite the approximating functional $V_f^h(t)$ given by (2.9) as

$$\tilde{V}_f^h(t) = E\{f(\bar{\tilde{x}}^h(t))\, R^h(t))\,|\,\tilde{\mathcal{Y}}_t\} \,, \tag{4.1}$$

where

$$R^h(t) = \exp\left[\int_0^t c(\bar{\tilde{x}}^h(s),s)d\tilde{y}(s) - 1/2\int_0^t c^2(\bar{\tilde{x}}^h(s),s)ds\right] \tag{4.2}$$

and define

$$u^h(t,\bar{\tilde{x}}^h(t),\tilde{y}^t) = E\{R^h(t)\ |\ \sigma\{\bar{\tilde{x}}^h(t)\}\times\tilde{\mathcal{Y}}_t\} \,, \tag{4.3}$$

where $\tilde{y}^t$ denotes the set of random variables $\{y(s), s\leq t\}$. Using lower-case Greek letters to denote the states of the chain $\tilde{x}^h(t)$ and defining

$$p_\alpha^h(t) = P\{\tilde{x}^h(t) = \alpha\} \,, \tag{4.4}$$

we have

$$\tilde{V}_f^h(t) = E\{ f(\bar{\bar{x}}^h(t)R^h(t) \mid \tilde{\mathcal{Y}}_t\} = E\{ E\{ f(\bar{\bar{x}}^h(t))R^h(t) \mid \{\bar{\bar{x}}^h(t)\}\times\tilde{\mathcal{Y}}_t\} \mid \tilde{\mathcal{Y}}_t\}$$

$$= E\{ f(\bar{\bar{x}}^h(t))\, U^h(t,\bar{\bar{x}}^h(t),\, \tilde{y}^t) \mid \tilde{\mathcal{Y}}_t\} = E\{ f(\bar{\bar{x}}^h(t))\, U^h(t,\bar{\bar{x}}^h(t),\tilde{y}^t)\} \quad (4.5)$$

$$= \sum_\alpha f(\alpha)\, U^h(t,\alpha,\tilde{y}^t)\, p_\alpha^h(t).$$

Since the probabilities $p^h(t)$ can be calculated a priori, the problem of getting a recursive relation for $\tilde{V}_f^h(t)$ is equivalent to that of getting one for $U^h(t,\bar{\bar{x}}(t), \tilde{y}^t)$. From (4.2) we have by Itô's lemma

$$R^h(t) = 1 + \int_0^t R^h(s)c(\bar{\bar{x}}^h(s),s)d\tilde{y}(s) \quad (4.6)$$

and therefore

$$U^h(t,\bar{\bar{x}}^h(t),\tilde{y}^t) = 1+E\left\{\int_0^t R^h(s)c(\bar{\bar{x}}(s),s)d\tilde{y}(s) \mid \sigma\{\bar{\bar{x}}(t)\} \times \tilde{\mathcal{Y}}_t\right\} \quad (4.7)$$

Using a particular stochastic version of Fubini's theorem /7; Th.4/, we get

$$U^h(t,\bar{\bar{x}}^h(t),\tilde{y}^t) = 1 + \int_0^t E\{ R^h(s)c(\bar{\bar{x}}^h(s),s) \mid \sigma\{\bar{\bar{x}}^h(t)\} \times \tilde{\mathcal{Y}}_s\}\; d\tilde{y}(s), \quad (4.8)$$

where the conditioning σ-algebra is now $\sigma\{\bar{\bar{x}}^h(t)\} \times \tilde{\mathcal{Y}}_s$, and by the markovianity of the process $\bar{\bar{x}}^h(t)$ (past and future are independent given the present) we have

$$E\{ R^h(s)c(\bar{\bar{x}}^h(s),s) \mid \sigma\{\bar{\bar{x}}^h(t)\} \times \tilde{\mathcal{Y}}_s\} = E\{ E\{ R^h(s)c(\bar{\bar{x}}^h(s),s) \mid \sigma\{\bar{\bar{x}}^h(t),\, \bar{\bar{x}}^h(s)\}\times$$
$$\tilde{\mathcal{Y}}_s\} \mid \sigma\{\bar{\bar{x}}^h(t)\} \times \tilde{\mathcal{Y}}_s\} = E\{ c(\bar{\bar{x}}^h(s),s)E\{ R^h(s) \mid \sigma\{\bar{\bar{x}}^h(s)\} \times \tilde{\mathcal{Y}}_s\} \mid \sigma\{\bar{\bar{x}}^h(t)\}\times$$
$$\tilde{\mathcal{Y}}_s\} = E\{ c(\bar{\bar{x}}^h(s),s)\, U^h(s,\bar{\bar{x}}^h(s),\tilde{y}^s) \mid \sigma\{\bar{\bar{x}}^h(t)\}\times \tilde{\mathcal{Y}}_s\} \quad (4.9)$$
$$= \sum_\beta c(\beta,s)\, U^h(s,\beta,\tilde{y}^s)P\{\bar{\bar{x}}(s)=\beta \mid \bar{\bar{x}}(t)\}\;.$$

Defining

$$q_{\beta\alpha}^h(s,t) = P\{\bar{x}^h(s) = \beta \mid \bar{x}^h(t)=\alpha\}\;, \qquad s\leq t, \quad (4.10)$$

we get from (4.8) and (4.9) the recursive relation

$$U^h(t,\alpha,\tilde{y}^t) = 1+ \int_0^t \sum_\beta c(\beta,s)\, U^h(s,\beta,\tilde{y}^s)q_{\beta\alpha}^h(s,t)d\tilde{y}(s). \quad (4.11)$$

ACKNOWLEDGEMENT

The authors are grateful to Prof. S.K. Mitter of M.I.T. for his continuing encouragement and helpful comments.

REFERENCES

/1/ Liptser, R.S. and Shiryayev, A.N., Statistics of random processes I (Springer Verlag, New York, 1977).

/2/ Girsanov, I.V., On transforming a certain class of stochastic processes by absolutely continuous substitution of measures, Theory of Prob. and its Applic. 5 (1960) 285-301.

/3/ Wong, E., Stochastic processes in information and dynamical systems (Mc Graw-Hill, New York, 1971).

/4/ Kallianpur, G. and Striebel, C., Estimation of stochastic systems: arbitrary system process with additive white noise observation errors, Ann. Math. Statistics 39 (1968) 785-801.

/5/ Clark, J.M.C., The design of robust approximations to the stochastic differential equations of nonlinear filtering, in: Skwirzynski, J.K. (ed.), Communication systems and random process theory (Sijthoff and Noordhoff, Alphen aan den Rijn, 1978).

/6/ Kushner, H.J., Probability methods for approximations in stochastic control and for elliptic equations (Ac.Press, New York, 1977).

/7/ Di Masi, G.B. and Runggaldier, W.J., Continuous-time approximations for the nonlinear filtering problem, C.N.R.-LADSEB Report N.79/04 (1979). Submitted for pubblication.

/8/ Billingsley, P., Convergence of probability measures (Wiley, New York, 1968).

Numerical Techniques for Stochastic Systems
F. Archetti and M. Cugiani (eds.)
© North-Holland Publishing Company, 1980

THE USE OF STOCHASTIC MODELS IN AIR POLLUTION PREDICTION AND CONTROL: PROBLEMS AND NUMERICAL RESULTS

C.Bonivento
A.Tonielli

Istituto di Automatica
Università di Bologna
Italia

G.Fronza
A.Spirito

Centro Teoria dei Sistemi
Politecnico di Milano
Italia

This paper is a part of the work done during the past and the current year at I.I.A.S.A. (Laxemburg - Austria) by a group of researchers from the Centro Teoria dei Sistemi (Milano), the Centro Scientifico IBM (Venezia) and the Istituto di Automatica (Bologna), in fulfillment of the research project no. REN-3003 on "Modeling and Control of Industrial and Urban Air Pollution".

1. INTRODUCTION

From the viewpoint of air quality control, pollution episodes (= rapid enhancements and subsequent decreases of ground level pollutant concentrations) are the most relevant aspect. The dynamics of episodes must be usually represented at hourly (or less) time scale, so that Gaussian models (see for instance Sutton |1|, Pooler |2|, Calder |3|, Runca et al. |4|) usually offer an inadequate description.

In principle, a most suitable mathematical representation of episodes should be obtained by integrating the advection-diffusion equation, which gives a rather accurate description (both in time and space) of the dispersion of a pollutant (see for instance Randerson |5|, Shir and Shieh |6|). Unfortunately, the approach requires a high computational effort and furthermore the input of the advection-diffusion model consists of extremely detailed information about meteorology and emission. In practical situations such detail of information is not available, namely the model inputs are affected by strong uncertainties. This circumstance explains why advection-diffusion models seldom offer a satisfactory performance, in particular when used for real-time forecast, namely for predicting at each time step future concentration levels on the basis of current information about concentrations, meteorology and emission.

The present paper, which summarizes the work by Runca et al. |7| and Fronza et al. |8| and Bonivento et al. |9|, describes a different approach to the real-time forecast problem. It basically consists of embedding a numerical solution scheme of the advection-diffusion equation into a stochastic framework and subsequently applying to the well-known Kalman prediction techniques (see also Bankoff and Hanzevack |10|, Desalu et al. |11|, Sawaragi and Ikeda |12|).

Following this approach, the control scheme results as in Fig. 1. Specifically the paper is organized as follows. Section 2 describes the numerical scheme as well as its interpretation as a discrete dynamical system, which is subsequently turned into a stochastic system by introducing properly defined noise terms.

Then (Section 3), an adaptive Kalman predictor, derived from the stochastic system is described, together with its real-time forecast of sulphur dioxide concentrations in the Venetian lagoon region (Section 4). In particular, the relevant four-hour-ahead forecast improvement with respect to the "deterministic" predictor, illustrated in Section 2 is pointed out.

In Section 5 the performance of different stochastic models of hourly wind (at a given height) are compared. Precisely, the forecast quality of all the predictors derived from the models is tested on a real case (wind at a meteorological station near Sassuolo, in Northern Italy).

Three models are respectively developed:

a) An AutoRegressive (AR) model of wind speed;
b) A bivariate AR model of the wind components (western and southern) in the horizontal plane;
c) A bivariate AutoRegressive Moving Average (ARMA) model, identified as a "canonical" input-output relationship derived from the innovation filter state representation.

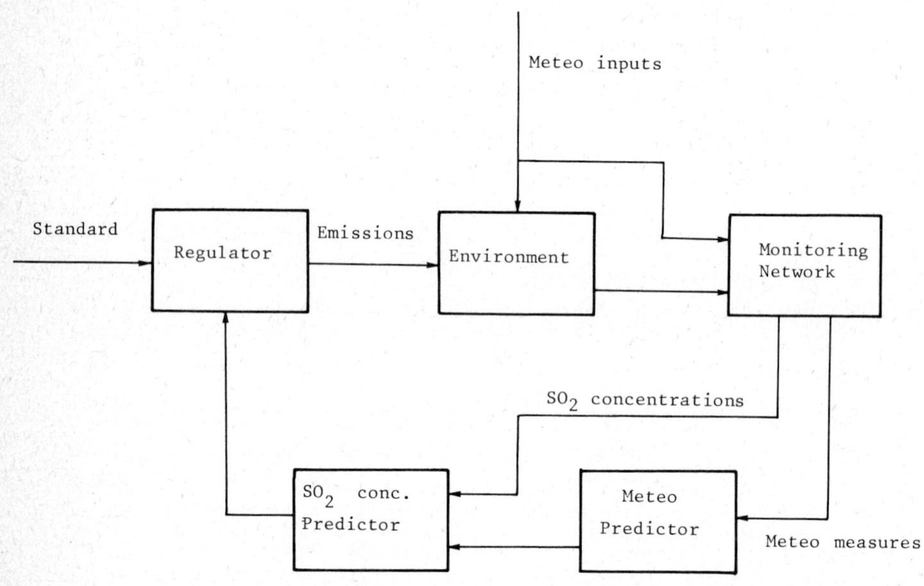

Fig. 1 - The control scheme

2. NUMERICAL SCHEME AND DISCRETE STOCHASTIC SYSTEM

For simplicity, the illustration of the scheme is carried out on a particular two dimensional-single source form of the advection-diffusion equation. However the extension to the three dimensional multisource case, which was actually considered in the study on pollution in the Venetian lagoon is straightforward (see Section 4).

The above mentioned two dimensional-single source form is the following (see |13|).

$$\frac{\partial c}{\partial t} + v(z)\frac{\partial c}{\partial x} = \frac{\partial}{\partial z}[K^z(z)\frac{\partial c}{\partial z}] + Q\,\delta(z-h)\,\delta x \qquad (1)$$

with initial and boundary conditions

$$c(x,z,0) = 0 \qquad (2)$$

$$c(x_W,z,t) = c(x_E,z,t) = 0, \qquad \forall z,\ \forall t, \qquad (3)$$

$$K^z\frac{\partial c}{\partial z} = 0, \quad z = 0, H, \qquad \forall x,\ \forall t, \qquad (4)$$

where

- h = source height
- H = height of the inversion layer base,
- $v(z)$ = wind speed (blowing in the x-direction at level z),
- $K^z(z)$ = vertical diffusion coefficient at level z,
- x_W = abscissa of western boundary of the integration region (just west of the source),
- x_E = abscissa of eastern boundary,
- Q = source emission rate,
- $\partial(\cdot)$ = Dirac's function.

In (1)-(4) input (source and wind speed) and parameter (diffusion coefficient) dependence upon time has been neglected in order to simplify the notation in the description of the solution algorithm.

According to the method of fractional steps (see for instance Yanenko |14|), eq. (1) is first splitted into three differential equations which separately take into account the contributions of source, advection and diffusion terms respectively:

$$\frac{\partial c}{\partial t} = Q\,\delta(z-h)\,\delta(x) \qquad (5)$$

$$\frac{\partial c}{\partial t} + v(z)\frac{\partial c}{\partial x} = 0 \qquad (6)$$

$$\frac{\partial c}{\partial t} = \frac{\partial}{\partial z}(K^z(z)\frac{\partial c}{\partial z}). \qquad (7)$$

Introduce the grid notation: Δx_i = grid spacing between points (i,m) and $(i+1,m)$, Δz_m = grid spacing between (i,m) and $(i,m+1)$. Index i takes values $0,1,\ldots,I+1$, in particular $(1,m h)$ coincides with the source. Time will be denoted by index k, specifically $c^k_{i,m}$ will represent concentration at: time $k\Delta t$ at point (i,m). Index m takes values $0,1,\ldots,M+1$; in particular ground level and mixing layer height correspond to $m=1$ and $m=M$ respectively. Moreover $m=0$ and $m=M+1$ represent fictious layers, which allow to satisfy boundary condition (5) through setting

$$\Delta z_0 = \Delta z_1,\ \Delta z_{M-1} = \Delta z_M,\ K^z_0 = K^z_1,\ K^z_{M-1} = K^z_M,\ c^k_{i,0} = c^k_{i,1},\ c^k_{i,M} = c^k_{i,M+1},\ \forall k,\ \forall i.$$

Then consider the following algorithm for integrating eqs. (5)-(7) in the k-th time interval.

a) Contribution of the source term (solution of eq.(5)).

Solving eq.(5) simply consists of adding the contribution of the source to the concentration field at time $k\Delta t$. Precisely the step is given by:

$$c^{*}_{1,mh} = c^{k}_{1,mh} + \frac{4Q\,\Delta t}{(\Delta x_1 + \Delta x_0)(\Delta z_{mh} + \Delta z_{mh-1})}$$

$$c^{*}_{i,m} = c^{k}_{i,m} \qquad \text{for } (i,m) \neq (1,mh)$$

where $c^{*}_{i,m}$ denotes the concentration field after the step.

b) <u>Contribution of the advection term</u> (solution of eq.(6)).

First consider the forward time, backward space finite difference approximation to eq.(6)

$$c^{**}_{i,m} = c^{*}_{i,m} - \frac{v_m \Delta t}{\Delta x_i} (c^{*}_{i,m} - c^{*}_{i-1,m}) \tag{8}$$

where $c^{**}_{i,m}$ is the concentration field after the step. Scheme (8) has first order accuracy and the Courant condition for numerical stability requires

$$v_m \Delta t < \Delta x_i \tag{9}$$

To first order, eq.(8) is an approximation of

$$\frac{\partial c}{\partial t} + v_m \frac{\partial c}{\partial x} = \frac{V_m}{2} (\Delta x_i - v_m \Delta t) \frac{\partial^2 c}{\partial x^2} .$$

After many time steps, the numerical errors associated to (8) can be quantified by the pseudo-diffusion coefficient

$$P = \frac{V_m}{2} (\Delta x_i - v_m \Delta t)$$

which reduces to zero for

$$v_m \Delta t = \Delta x_i . \tag{10}$$

Since the wind is a function of the vertical coordinate, namely v_m is not constant with m, in view of (9), condition (10) can be satisfied only at the height where wind speed is maximum.

On the other hand P should obviously be zero at least at source height ($m = m\,h$) where the highest concentration gradients occur.

A solution to this problem has been provided by Carlson (see for instance Richtmyer and Norton |15|). His scheme is illustrated in Fig. 2 in the (x,t) plane (m = constant), where the dashed trajectories I and II respectively correspond to situations $v_m \Delta t < \Delta x_i$ and $v_m \Delta t \geq \Delta x_i$. Specifically, according to Carlson's scheme if situation II of Fig.2 occurs, eq.(8) is replaced by

$$c^{**}_{i,m} = c^{*}_{i-1,m} - \frac{\Delta x_i}{v_m \Delta t} (c^{**}_{i-1,m} - c^{*}_{i-1,m}) . \tag{11}$$

The use of eq.(11) instead of eq.(8) for $v_m \Delta t > \Delta x_i$ makes Carlson's scheme unconditionally stable and allows to reduce the truncation error to zero at source height by choosing $\Delta x_1 = v_{mh} \Delta t$.

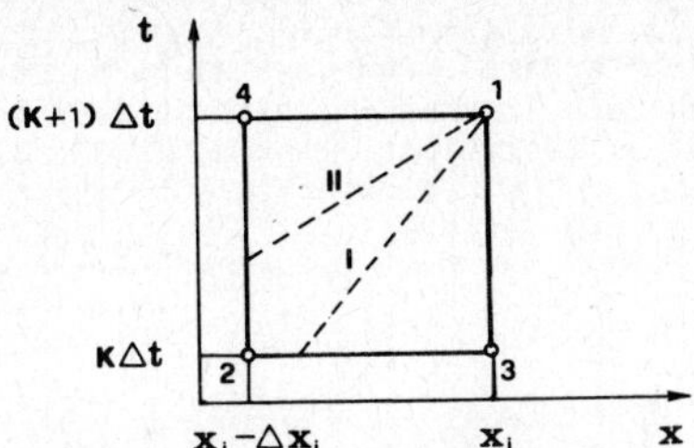

Fig. 2 - Carlson's scheme in the (x,t) plane:
I) Trajectory in the case $v_m\,\Delta t < \Delta x_i$,
II) Trajectory in the case $v_m\,\Delta t \geq \Delta x_i$

c) Contribution of the diffusion term (solution of eq.(7))

Numerical approximation to eq.(7) can be obtained by the method due to Crank and Nicolson |16| :

$$c^{k+1}_{i,m} = c^{**}_{i,m} + \frac{1}{2}\{ D_z\,[c^{k+1}_{i,m}] + D_z\,[c^{**}_{i,m}]\} \tag{12}$$

where

$$D_z\,[c^{k+1}_{i,m}] = \frac{\Delta t}{\Delta z^2_m}\,[K^z_{m+1}\,c^{k+1}_{i,m+1} - (K^z_{m+1} + K^z_{m-1})c^{k+1}_{i,m} + K^z_{m-1}c^{k+1}_{i,m-1}]$$

is the standard difference operator for centered diffusion and c^{k+1} denotes the concentration field at time $(k+1)\,\Delta t$.

The overall algorithm is here summed up for commodity

$$c^{*}_{1,mh} = c^{k}_{1,mh} + \frac{4Q\,\Delta t}{(\Delta x_1 + \Delta x_0)(\Delta z_{mh} + \Delta z_{mh-1})} \tag{13'}$$

$$c^{*}_{i,m} = c^{k}_{i,m}\,, \qquad (i,m) \neq (1,m\,h) \tag{13"}$$

$$c^{**}_{i,m} = \begin{cases} c^{*}_{i,m} - \dfrac{v_m \Delta t}{\Delta x_i}\,(c^{*}_{i,m} - c^{*}_{i-1,m}), & v_m\,\Delta t < \Delta x_i \quad (14') \\[2ex] c^{*}_{i-1,m} - \dfrac{\Delta x_i}{v_m\,\Delta t}\,(c^{**}_{i-1,m} - c^{*}_{i-1,m}), & V_m\,\Delta t \geq \Delta x_i \quad (14") \end{cases}$$

$$c^{k+1}_{i,m} = c^{**}_{i,m} + \frac{1}{2}\{ D_z\,[c^{k+1}_{i,m}] + D_z\,[c^{**}_{i,m}]\}. \tag{15}$$

Though algorithm (13)-(15) is first order, it is more accurate than other more complex first and second order schemes. Such better performance depends not only upon the possibility of having zero truncation error at source height, but also upon the method itself of fractional steps. This clearly turns out from the following considerations. For a while assume for simplicity a uniform grid ($\Delta x_i = \Delta x$, $\Delta z_m = \Delta z$). Then by eliminating the intermediate fields c^* and c^{**} , eqs.(13)-(15) yield for $v_m \Delta t = \Delta x$:

$$c^{k+1}_{i,m} = c^{k}_{i-1,m} + \frac{1}{2}\{ D_z [c^{k+1}_{i,m}] + D_z [c^{k}_{i,m} + \frac{Q\,\Delta t}{\Delta x\,\Delta z}] \} + \frac{Q\Delta t}{\Delta x\,\Delta z} . \quad (16)$$

The corresponding approximation to eq.(1) by the direct scheme (i.e. the scheme obtained by using the same finite difference operator but without splitting) is

$$c^{k+1}_{i,m} = c^{k}_{i-1,m} + \frac{1}{2}\{ D_z [c^{k+1}_{1,m}] + D_z [c^{k}_{i,m}]\} + \frac{Q\Delta t}{\Delta x \Delta z} . \quad (17)$$

By comparing eqs.(16),(17), the following conclusions can be drawn:

- After k time intervals, while the fractional step scheme moves the pollutant front up to $v_m\, k\,\Delta t$, the direct scheme has a time lag of Δt, namely moves the front up to $v_m(k-1)\,\Delta t$.
- In the region $x > o$, where no source exists, eq.(17) approximates the diffusion operator along $x = \text{constant}$, while eq.(16) approximates it along the trajectory segment $\Delta x = v_m \Delta t$. By choosing $\Delta x = v_{mh} \Delta t$, eq.(16) holds at source heights so that transport as well as diffusion are properly described in the region of maximum concentration gradients.
- The fractional step scheme computes the new concentration values by exploiting the three "most relevant" point of the mesh, namely (see Fig.2) points 1,2,3 or 1,2,4 (depending upon the trajectory through point 1 precisely upon whether it intersects Δx or Δt).

As a further comment, it must be pointed out that the suitability of Carlson's scheme to the advection step derives from the circumstance that in the region of the largest concentration gradients, in view of the choice $\Delta x_i = v_{mh} \Delta t$, the Courant number is closed to unity. Hence all Fourier modes are transported with velocity closed to wind speed and with negligible dissipation. Far from this region, as far as the concentration field gets smoother, only low frequency harmonics are required for its representation.

For such harmonics, even if the Courant number is small, dispersion and dissipation errors are limited.

Finally, it must be recalled that there are other fractional step procedures illustrated in air pollution literature (see for instance Shir and Shieh |6|, Bankoff and Hanzevack |10|). The basic difference between such schemes and algorithm (13)-(15) is due to the circumstance that they treat only the vertical terms with an implicit formulation, while the remaining steps are solved via explicit approach. Hence the horizontal grid spacing, limited by stability conditions, is uniform and subsequently the computational effort is higher. As pointed out in the Introduction, the choice is justified by the uniform distribution of the sources (domestic source instead of the industrial ones considered in the present study).

A part from conspicuous computational saving, a non-uniform grid in the horizontal plane allows to locate grid points exactly in monitoring stations, thus avoiding any interpolation when comparing forecast and observed data.

Through straightforward but cumbersome computations (see for instance Fronza et al.

8), it can be shown that algorithm (13)-(15) can be written in the compact form (where dependence upon time as well as dependence of the diffusion coefficients upon the stability class is made explicit):

$$X(k+1) = \phi(v(k),\ s(k))X(k) + \phi\,(v(k),\ s(k))E(k) \qquad (18)$$

where

$X(k)$	= vector of concentrations in all grid points at time k;
$v(k)$	= vector representing the average wind field in the k-th period;
$s(k)$	= stability class in the k-th period;
$\phi(v(k),s(k))$	= suitable matrix, whose entries only depend upon wind field and stability class.

Equation (18) can be considered as the state equation of a discrete dynamical system having state $X(k)$ and inputs $E(k)$, $v(k)$, $s(k)$.

Moreover if $y(k)$ denotes the vector of concentration measurements at time k (taken in monitoring stations all coincident with grid points), an output transformation of the discrete system can be defined as

$$y(k) = H\,X\,(k) \qquad (19)$$

where H is a matrix whose elements are either 0 or 1.

Embedding the discrete system (18)-(19) into a stochastic environment simply means adding random terms in its two relationships in order to represent the "inaccuracies" of the model itself. Sfecifically, the stochastic version of system (18),(19) is given by

$$X(k+1) = \emptyset\,(v(k),\ s(k))X(k) + \emptyset\,(v(k),s(k))E(k) + n(k) \qquad (18)$$

$$y(k) \quad = H\ X(k) + w(k) \qquad (19)$$

where

$n(k)$ = stochastic term ("process noise"), which accounts for all the sources of disagreement between the model and the actual dynamics of the pollution phenomenon (i.e. for neglected physical inputs in the advection-diffusion equation such as rain or chemical reactions, for errors introduced by the assignment of parameter values, for errors due to the model structure, to numerical inaccuracies and so on).

$w(k)$ = stochastic term ("measurement noise") which accounts for errors in measurements.

The random processes $\{n(k)\}_k$ and $\{w(k)\}_k$ are commonly assumed to be zero-mean white noises, namely to have a correlation structure of the following type (E(.)= = expectation operator).

$$E\ [n(k)n^T(k+\tau)] = \begin{cases} Q(k) & \tau = 0 \\ 0 & \tau \neq 0 \end{cases}$$

$$E\ [w(k)w^T(k+\tau)] = \begin{cases} R(k) & \tau = 0 \\ 0 & \tau \neq 0 \end{cases}$$

3. THE KALMAN PREDICTOR

The stocnastic model (18)-(19) can be used for real time pollution forecast, namely for predicting at the beginning of each time interval, future concentration levels on the basis of current information about concentrations, emission and meteorology. Specifically, the recursive one-step ahead forecast algorithm (Kalman predictor), derived from model (18),(19) is given by (see for instance Kalman |17|, Jazwinski |18|):

$$\hat{X}(k|k) = \hat{X}(k|k-1) + G(k)\ [Y(k) - H\hat{X}(k|k-1)] \qquad (20)$$

$$\hat{X}(k+1|K) = \emptyset\,(v(k),s(k))\ \hat{X}(k|k) + \emptyset\ (v(k),s(k))\ E\,(k) \qquad (21)$$

where

$\hat{X}(k+1|K)$ = prediction of $X(k+1)$ made at time k, namely at the beginning of the k-th interval of time;

$\hat{X}(k|k)$ = filtered state, namely a posteriori (i.e. at time k) estimation of $X(k)$ on the basis of the new available datum $y(k)$. Precisely such estimation is given by eq.(20) as a correction of the previous fore cast $\hat{X}(k|k-1)$ and is introduced in eq.(21) instead of $\hat{X}(k|k-1)$ in order to give a better prediction of $X(k+1)$;

$$G(k) = P(k|k-1)H^T[HP(k|k-1)H^T + R(k)]^{-1} \quad \text{(Kalman gain)} \qquad (22)$$

$$P(k|k-1) = E[(\hat{X}(k|k-1) - X(k))(\hat{X}(k|k-1) - X(k))^T] = \text{forecast error covariance matrix.}$$

In turn, such covariance matrix is recursively evaluated through the following e-quations.

$$P(k|k) = P(k|k-1)\,\{I - H^T[HP(k|k-1)\,H^T + R(k)]^{-1}HP(k|k-1)\} \quad (23)$$

$$P(k+1|k) = \emptyset\ (v(k),\ s(k))\ P(k|k)\emptyset^T(v(k),\ s(k)) + Q(k+1) \quad (24)$$

The r-step ahead prediction $(r = 2,3,\ldots)$ is obtained recursively by

$$\hat{X}(k+r|k) = \emptyset(v(k+r-1),\ s(k+r-1))\ \hat{X}(k+r-1|k) + \\ + \emptyset(v(k+r-1),\ s(k+r-1))\ E(k+r-1) \qquad (25)$$

The actual implementation pf the Kalman predictor (20)-(25) raises a number of conceptual and practical problems which are now discussed in detail.

a. Assignment of Q(k) and R(k)

The correction a posteriori of the previous forecast $\hat{X}(k|k-1)$ into $\hat{X}(k|k)$ (a better initial state for the new prediction step (21)) is made in eq.(20) by weighting the new datum $y(k)$ through the Kalman gain $G(k)$. In turn, such weight matrix depend, in view of eqs.(22)-(24), upon the noise intensities $Q(k)$ and $R(k)$. In conclusion, at every forecast step, the Kalman predictor corrects ("filters") the initial state of the step by taking into account the noise intensities.

Obviously, $Q(k)$ and $R(k)$ must be regarded as input data to the filter (22)-(25), namely must be evaluated before the filtering is made. In principle, $R(k)$ should turn out from an analysis of the accuracy of the measurement system, while $Q(k)$ should turn out from considerations of fitting between the numerical scheme and the real word.

This is not accomplished, in practice. Rather, Q(k) and R(k) are usually estimated by recursive algorithms, based on a posteriori analysis at each time step about the performance of the predictor at previous time steps.

Specifically, there are different types of adaptive Kalman predictors, namely Kalman predictors supplied with a recursive algorithm for a posteriori estimation of Q(k) and R(k). Unfortunately, the most rigorous ones (see for instance Mehra |19|) cannot be applied to the present case, because of the non-stationarity of system (18),(19), precisely because the matrix Ø(v(k), s(k)) is not the same for every k.

Thus, in the application described in Section 4, a heuristic adaptive approach has been followed, which represents a slight generalization of a procedure by Jazwinski |20|.

b. Treatment of emission uncertainties

The adaptive mechanisms mentioned above (noise intensities updated at each time step) are commonly too weak to allow a good forecast of pollution episodes, when these are due to conspicuous (but unknown to the predictor) emission enhancements. Usually, a more robust correction is obtained by the so-called state enlargement (extended Kalman predictor, see for instance Jazwinski |18|). Through often effective (Bankoff and Hanzevack |10|), extended (adaptive) Kalman predictors require the dimension of the state of the stochastic system to be doubled and correspondingly the computational burden heavily increases.

Hence, a simpler and less expensive, through rougher and heuristic, recursive adjustment of emission inputs has been considered in the present work. Precisely, eq.(18) has been modified as follows:

$$X(k+1) = \emptyset\,(v(k),\, s(k))\,X(k) + \emptyset\,(v(k), s(k))\,\theta\,(k)E(k) + n(k) \qquad (18)$$

The scalar $\theta(k)$ is defined as

$$\theta\,(k) = \sum_{\ell}\hat{x}_{\ell}(k|k) / \sum_{\ell}\hat{x}_{\ell}(k|k-1) \qquad (26)$$

where

$\hat{x}_{\ell}(k|k)$ and $\hat{x}_{\ell}(k|k-1)$ are the ℓ-components of $\hat{X}(k|k)$ and $\hat{X}(k\ k-1)$ respectively and the sums are made over all components.

From eq.(26), Θ (k) is a ratio between the total "mass" of pollutant estimated a posteriori (namely, after the arrival of the new measurement vector y(k)) and the "mass" previously forecast. If the two "masses" do not coincide, then an emission variation, occurred between time (k-1) and time k and do not revealed to the predictor, is assumed and the emission for the subsequent step (prediction of X(k+1) made at time k) is correspondingly modified.

c. Computational effort considerations

From eqs.(20)-(25), the implementation of the Kalman predictor implies at each iteration step the manipulation of square matrices having order equal to the number of grid points. Through the numerical scheme described in Sect. 2 admits non-uniform grid-spacing, the order of such matrices may easily be one thousand, corresponding to an untolerable computational burden. In order to cut such burden down, the procedure suggested by Bankoff and Hanzevack |10| has been used. Such procedure basically consists of filtering only part of the state, precisely the concentrations in the grid points of some subregions of intereset in the area under consideration.

There is clearly a danger inherent to the method. In fact, the initial state of the forecast step (21) may now turn out to be a very "irregular" concentration field, since some state components have been filtered and some haven't. In particular, there may result strong variations between the components corresponding to subregion boundaries and components corresponding to grid points immediately outside. Hence, "artificial" high gradients may be introduced, with negative effects on the forecast step. Whether the distortion caused by partial filtering is relevant or not, it can be ascertained only by simulation of the Kalman predictor on the real case (see next section).

d. Meteorological input to the predictor

From eqs.(21), (25), the forecast of r-step ahead concentration fields, made at the beginning of the k-th time interval requires the knowledge of emission, wind field and stability inputs for the time intervals $\{k\Delta t, (k+1)\Delta t\}$, $\{(k+1)\Delta t, (k+2)\Delta t\}$, $\{(k+r-1)\Delta t, (k+r)\Delta t\}$.

Such needed information about emission in these future periods, should in principle be available, since emission is a decision variable of polluters, who are assumed to collaborate or to be forced to collaborate to the prediction (anyway, if wrong information is supplied, the forecast quality is safeguarded by the correction mechanism described in B).

On the contrary, future meteorology is not obviously known.

There are basically two possible approaches for supplying with the wind and stability inputs required by the concentration predictor.

1. Set up a meteorological predictor, whose forecasts are introduced as inputs into eqs.(21),(25).
2. Simply postulate a persistent meteorology (future wind stability will be the same as the present ones).

Clearly, the forecast performance of the concentration predictor under approach 2) is a lower bound, since it corresponds to the roughest treatment of the meteorological inputs. On the contrary, the upper bound of such performance corresponds to a situation where wind and stability inputs are supplied to eqs.(21), (25) by a perfect meteorological predictor (forecast wind and stability always equal to their future time values).

Moreover, the performance under approach 1) is expected to lie within the two bounds, the distance to the upper bound obviously depending upon the quality of the metereological forecast.

The above analysis about the sensitivity of concentration prediction to the treatment of meteorological inputs has actually been done in the Venetian lagoon study, as illustrated in detail in the next section.

Furthermore, a deeper study on wind prediction has been carried out in the Sassuolo industrial area |9|, as illustrated in Section 5. The application, however, of the obtained wind predictors to the pollution concentration forecast is in progress at the present.

4. THE APPLICATION TO THE VENETIAN CASE

a. Description of the area and pollution problem

The area under consideration (Fig. 3) is located in the northeastern part of Italy, at the upper shore of the Adriatic Sea. It includes part of the extreme end of the Padana Plain and part of the Venetian lagoon, precisely it consists of the urban centers of Mestre, Marghera and Venice and the industrial area of Porto Marghera (one of the largest in Europe). The urban centers of Mestre and Marghera, situated on the mainland, have developed very rapidly in the last three decades and have now a surface area of about 10 km^2. The industrial area is about 15 km^2 and its main activities include oil-refining, petrochemical production, metallurgical processing of iron and other metals, and electric power production. Six km from the mainland, in the middle of the Lagoon, is the historical center of Venice, covering an area of 6 km^2 and standing on a cluster of small islands.

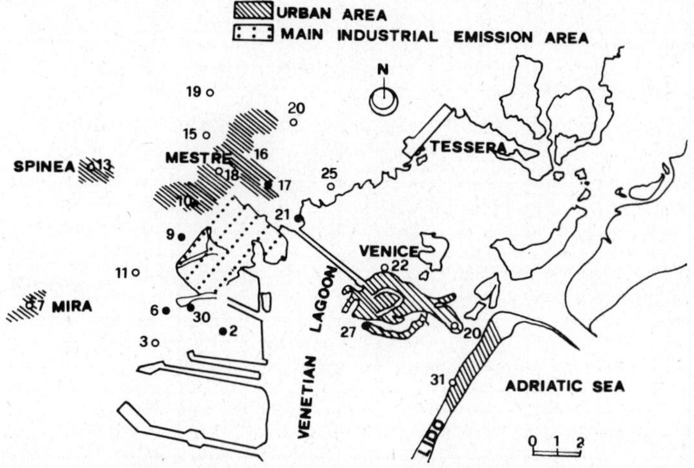

Fig. 3 - Venetian lagoon area and monitoring network (● = station working since Feb. 1973, o = station working since Feb. 1974)

The air pollution problem in the region has already been analyzed in a number of studies. For instance the influence of meteorological factors on pollution levels has been investigated by Zannetti et al. |21|. Moreover mathematical models, both deterministic (Runca et al.) and "black box" stochastic (Zannetti et al. |22|, Finzi et al. |23|) have been used for simulating or predicting long-term or short-term average SO_2-concentrations in the area.

In particular, Zannetti et al. |22| analyzed the occurrence of episodes both in the historical center of Venice and in the mainland. It turned out that most of the pollution episodes occur in the industrial area, while the few ones concerning Venice correspond to winter winds blowing from the industrial area and have smaller intensity. Hence, using concentration measurements in the historical center would not add much information to a control scheme of air quality. Thus the present model has been implemented only in correspondence with the region shown in Fig. 4. Of course, this choice has significantly reduced the computational effort.

b. Input data set

The available emission, meteorological and SO_2 concentration data were the following.

Data concerning each industrial source were directly obtained by 1971 National Census figures. To have an idea of the overall emission in the region, the estimated pollutant released from industries totalled 160,000 tons per year, in front of approximately 10,000 tons per year due to domestic heating.

Both meteorological and concentration data used in the present study have been provided by the monitoring network (see Fig. 4 again) installed by Tecneco on behalf of the Governmental Department of Health. This network consists of one meteorological station and 24 SO_2 - monitoring sensors.

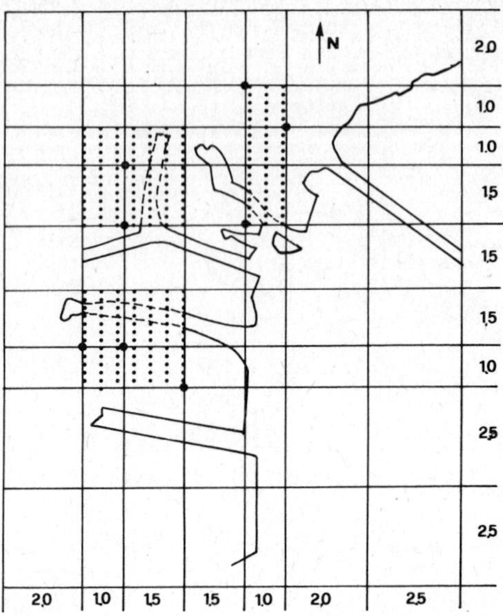

Fig. 4 - Geometry of the grid in the horizontal plane with distances in km (● = stations working since Feb. 1973, labelled in accordance with Tecneco classification)

The meteorological station, 15 m above the ground, records hourly wind speed and direction, temperature, pressure, humidity, rainfall, cloudiness and fog. Wind direction is recorded according to the eight sectors of the compass, thus introducing an indetermination of $\pm$ 22°30'.

Concentration data recorded by the 24 monitoring sensors are transmitted to a small computer which elaborates the data and records the hourly average values as well as daily statistics. In 1973, the year to which this study refers only ten stations were put into operation. Two out of them are located in Venice, the remaining ones are the stations considered in this application.

In general terms concentration data exhibit satisfactory reliability, while this is not true for the other types of data. In fact emission data are only average rates, a very rough input when modelling an episode, partly or mainly due to extra-release. As for meteorological measures, one station cannot obviously point out the spatial variation of wind and diffusion parameters. Just these input uncertainties explained the poor performance of the advection-diffusion model (see point d) below) and the subsequent reformulation in stochastic terms.

c. Model specifications

The region of interest (16.5 km x 18.0 km wide) has been discretized by means of 10 x 12 x 7 grid points. The horizontal grid spacing ranges from a minimum of 1 km to a maximum of 2.5 km (see Fig. 4, where the monitoring stations, all coincident with grid points are also reported). The vertical grid size were specified as follows:

$$\Delta z_m = \begin{cases} 50 \text{ meters}, & m = 1,2 \\ 75 \text{ meters}, & m = 3,4 \\ (H-250)/2 & m = 5,6 \end{cases}$$

Since no measurement of the mixing depth H was available, H was kept equal to 500 meters in all simulations.

On the above mesh, the following form of eq.(1) has been used (differently from eq.(1) here input and parameter dependence upon time is explicitly pointed out)

$$\frac{\partial c}{\partial t} + v_x(z,s(t)) \frac{\partial c}{\partial x} + v_y(z,s(t)) \frac{\partial c}{\partial y} = \tag{27}$$

$$= K^x(s(t)) \frac{\partial^2 c}{\partial x^2} + K^y(s(t)) \frac{\partial^2 c}{\partial y^2} + \frac{\partial}{\partial z} [K^z(z,s(t)) \frac{\partial c}{\partial z}] + S(x,y,z)$$

with initial and boundary conditions

$$c(x,y,z,t) = 0 , \qquad t = 0$$

$$K^z \frac{\partial c}{\partial z} = 0, \qquad z = 0, H$$

$$c(x,y,z,t) = 0 , \qquad \forall t \text{ at side boundaries.} \tag{28}$$

In eq.(27), $S(x,y,z,t)$ is the source term, $v_x(z,s(t))$ and $v_y(z,s(t))$ are the wind speed components in the horizontal plane, $s(t)$ denotes atmospheric stability at time t in accordance with the classification by Pasquill |24| . Specifically, Pasquill characterizes air turbolence bz six classes ranging from A (strong stability) to F (extreme stability).

Boundary condition (28) is acceptable since the integration region is extended enough in the horizontal plane so that at its boundaries pollution is actually negligible.

The wind field, as clear from eq.(27) has only horizontal components, a standard assumption in advection-diffusion modelling (though somehow questionable in an area characterized by sea breezes).

Moreover, as also specified by eq.(27) the wind field has been considered as uniform in each horizontal plane. Such uniform field at level z has been evaluated via the power law

$$\vec{v}_m(t) = \vec{v}_R(t)\left(\frac{z_m}{z_R}\right)^{\alpha(s(t))} \tag{29}$$

where

$\vec{v}_m(t)$ = wind vector at the m-th level ($z = z_m$) at time t;

$\vec{v}_R(t)$ = wind vector supplied at time t by the meteorological situation (see b.) in the present section), which is situated at quote z_R (15 meters);

$\alpha(s(t))$= given function of stability reported in Table 1 (first column).

As for $K^z(z,s(t))$, the classic formula by Shir and Shieh has been modified in the following way

$$K^z(z,s(t)) = K^D(s(t))\ z \exp(-\rho(s(t)).z/H)$$

The values of $\rho(s(t))$ are reported in Table 1 , second column. As for $K^D(s(t))$, it has been obtained by

$$K^D(s(t)) = z_R^{-1} K^z(z_R, s(t)) \exp(\rho(s(t)) z_R/H)$$

where $K^z(z_R, s(t))$ is the vertical diffusion coefficient at the level of the meteorological station and is reported in Table 1 (third column).

Finally, as for horizontal diffusion coefficients, it has been assumed $K^x(s(t)) = K^y(s(t))$ = values reported in Table 1 (fourth column).

TABLE 1 - Wind and diffusion parameters versus Pasquill stability class

$s(t)$	$\alpha(s(t))$	$\rho(s(t))$	$K^z(z_R,s(t))$ (m^2/sec)	$K^x(s(t)) = K^y(s(t))$ (m^2/sec)
A	0.05	6	45.	250.
B	0.1	6	15.	100.
C	0.2	4	6.	30.
D	0.3	4	2.	10.
E	0.4	2	0.4	3.
F	0.5	2	0.2	1.

d. Results of numerical scheme simulation

The fractional step procedure (13)-(15) described in Section 2 for the two dimensional eq.(1), has been applied to the three dimensional advection-diffusion model (27) (considered period March-October 1973). The extension of algorithm (13)-(15) to the solution of eq.(27) is straightforward and consists of a source step of form (13), two advection steps of form (14) in x and y-direction respectively and three diffusion steps of form (15) in the x,y and z-direction respectively.

The comparison between simulation results and real data has been favourable in correspondence with low or medium pollution periods but disappointing in presence of episodes. For example Fig. 4 shows the dynamics of a relevant episode (peak more than twice admissible standard) together with model simulation figures. It is worthwhile adding that the episode practically regarded only the station (n. 9), whose record is reported in Fig. 5. Other stations did not measure "extraordinary" pollution and there model fitting has been satisfactory.

A similar result has been obtained in correspondence with another relevant 1973 episode (Fig. 6).

In both cases the discrepancies between model and reality can be reasonably ascribed to input uncertainty, mainly to lack of information about extra-release by some polluter. Namely, Fig. 5 and Fig. 6 prove the statement made in the Introduction: advection-diffusion models are accurate representations of the phenomenon but usually their input uncertainties do not allow an acceptable description of pollution episodes. Hence it is no use looking for sophisticated numerical accurate solution algorithms, they can give only negligible improvement in front of a possibly relevant enhancement of computation times.

e. Kalman prediction

Of course, a "stochastic embedding" procedure, quite similar to the one illustrated in Section 2, holds for the numerical solution scheme of the three dimensional advection-diffusion equation. The resulting stochastic model is here rewritten by putting into evidence that wind speed has two (horizontal) components in this case:

$$X(k+1) = \emptyset\ (v(k),\ d(k),\ s(k))\ X(k) +$$
$$+ \emptyset\ (v(k),\ d(k),\ s(k))\ E(k) + n(k) \qquad (30)$$

$$y(k) \quad = HX(k) + w(k) \qquad (31)$$

where v(k) and d(k) respectively denote the vector of wind intensities at different levels and wind direction. As for the Kalman predictor derived from system (30), (31), specifications concerning points C) and D) of the previous section must be supplied.

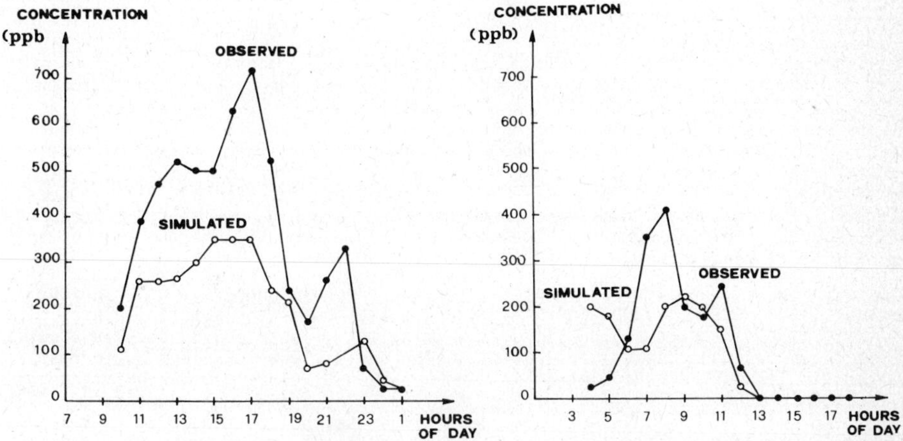

Fig. 5 - Simulation and measures for episode in station no 9 on April 7, 1973

Fig. 6 - Simulation and measures for episode in station no 30 on August 2, 1973

The locations of the grid subregions mentioned in C) (= subregions where the concentration field is filtered) are illustrated in Fig. 4. Precisely, the dashed areas of Fig. 4 represent the bases of the three subregions at ground level (m = 1) and correspond to the most polluted zones. In the vertical axis, each subregion reaches the level m = 3, namely has an extension of two layers above the ground. As for point D) of Section 3, the Kalman predictor of future concentrations has been run in correspondence with the three different types of meteorological inputs.

1) Inputs given by a mathematical meteorological predictor. Precisely, for f = 1,2,...,r-1 a forecast $\hat{s}(k+f|k)$ (made at time k for the stability class at time k + f) has been obtained in accordance with the following simple probabilistic criterion:

$$\hat{s}(k+f|k) \rightarrow \max_{s(k+f)} \mathrm{Prob}(s(k+f)|s(k-1),q)$$

where Prob(s(k+f))|s(k-1),q) is the probability of having class s(k+f) at time k+f, given the information that the class has been s(k-1) in the interval $\{(k-1)\,\Delta t,\ k\,\Delta t\}$ and that time $k\,\Delta t$ is the q-th hour of the day

(q = 1,2,...,24).
Similarly, the wind direction sector has been forecast in accordance with the criterion.

$$\hat{d}(k+f|k) \to \max_{d(k+f)} \text{Prob}(d(k+f)|d(k-1),q)$$

Finally wind intensity in the meteorological station has been forecast by means of an ARMA predictor (see Box and Jenkins |25|). From such forecast, predictions $\hat{v}(k+f|k)$ of the whole future profiles of wind intensities have been obtained through the power law already illustrated in c.).

2) Inputs given by assuming persistent meteorology ($v(k+f) = v(k-1)$, $d(k+f) = d(k-1)$, $s(k+f) = s(k-1)$).

3) Inputs given by assuming a perfect meteorological predictor, namely true inputs.

The four-hour ahead forecast performance under the three conditions of input treatment is shown in Fig. 7 in correspondence with the episode of April 7, 1973, and in Fig. 8 in correspondence with the episode of August 2, 1973. As expected, approach I) gives an intermediate performance between II) and III), but very near to the ideal situation III). In fact the correlation between forecast and true concentration data has been 0.90, 0.32, 0.92 (for the cases of Fig. 7I, 7II, 7III respectively, and 0.76, 0.50 and 0.77 (for the cases of Fig. 8I, 8II, 8III respectively). As clear from a comparison between Fig. 7 and Fig. 8 with the corresponding performance of the numerical scheme alone (Fig. 5 and Fig. 6) the improvement of quality is conspicuous.

The remainder of the forecast concentration field results non-distorted in both episode situations, through generally an overstimation of the future fields. This is due to the correction mechanism of the emissions illustrated in Section 3,point B) which is based on the multiplicative scalar coefficient $\Theta(k)$ and hence results in the simultaneous enhancement of all the emissions. Precisely, the existence of an episode around some station (n. 9 on April 7, 1973 and n. 30 on August 2, 1973) leads the correction mechanism to increase all emissions and subsequently to raise all the forecast field. Of course this effect could be avoided by setting up a suitable selective mechanism of correction of emission information.

In non-episode situations, the predictor has also given good performance, but this is not a particularly significant result. Finally, for computational times, each four-hour-ahead forecast required approximately three minutes on an IBM 370 computer.

5. THE APPLICATION TO THE SASSUOLO CASE

a. An air model of hourly wind speed

The area under consideration is shown in Fig. 9, where the wind monitoring station (located 50m above the ground) is also pointed out. Hourly wind measurements during the period March 1 - May 31, 1974 have taken into account.

First, hourly wind speed has been considered. Its record exhibited a pronounced daily periodicity, as shown in Fig. 10, where means μ_r and standard deviations σ_r corresponding to the different hours of the day are reported ($r = 1,2,\ldots,24$). Hence, the data have been manipulated by the following cyclic standardization (Δt = 1 hour; $k = 1,2,\ldots,92$).

$$v(24*k+r) = (V(24*k+r) - \mu_r)/\sigma_r \tag{32}$$

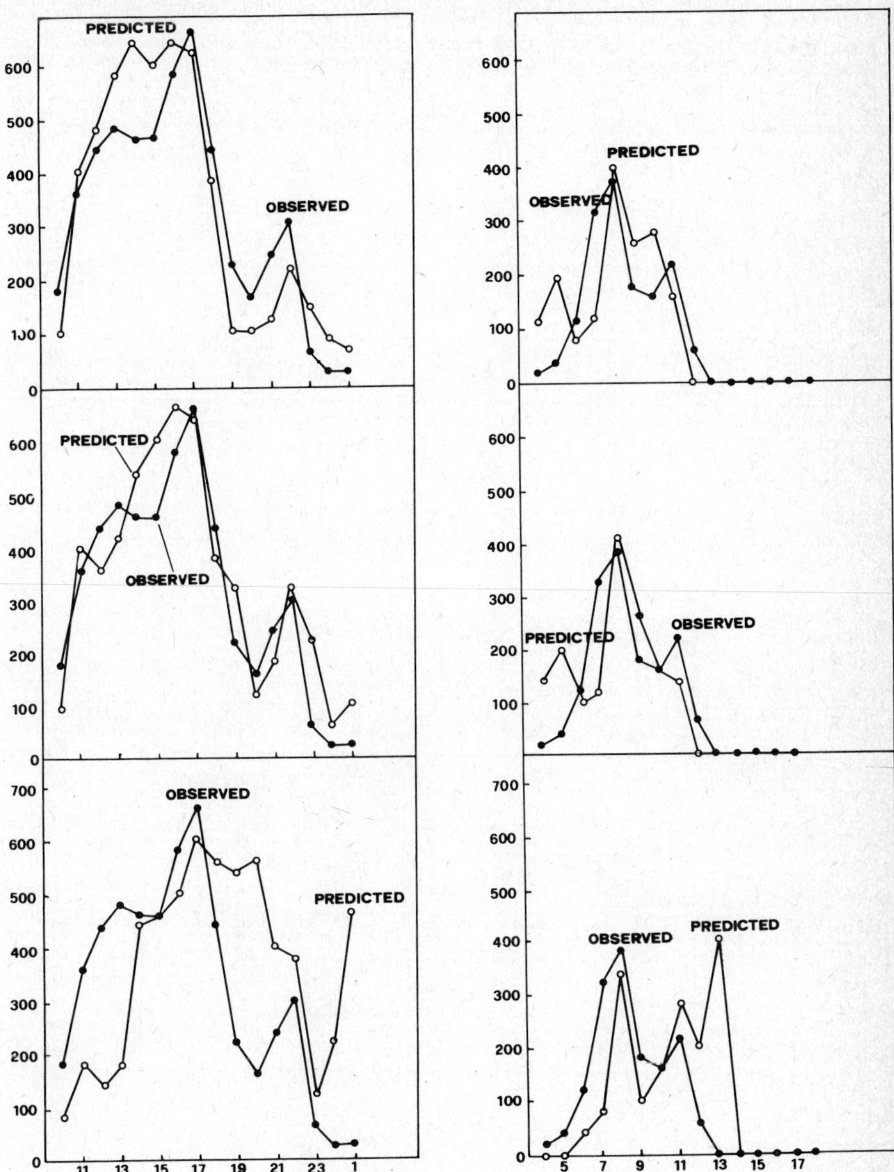

Fig. 7 - Four hour-ahead episode Kalman prediction with I) Forecast, II) Persistent, III) True meteo rological inputs (April 7, 1973)

Fig. 8 - Four hour-ahead episode Kalman prediction with I) Forecast, II) Persistent III) True meteorological in puts (August 2, 1973)

where

$V(24*k+r)$ = k-th day hourly wind speed, i.e. instantaneous wind speed, averaged over the time interval $(r-1)\Delta t$, $r\Delta t$, which is the r-th hour of a day.

Then, the time series of v(k), derived from the record of wind measurements through eq.(32) has been assumed to be the (partial) realization of a stationary stocastic process $\{v(k)\}$.

Such process has been described by means of an autoregressive model:

$$v(k+1) = \sum_{j=1}^{p} \Phi_j \, v(k-j+1) + \varepsilon(k) \tag{33}$$

where $\{\varepsilon(k)\}$ is zero mean white noise.

From the previous defined record of data, during the period March 1 - April 30, 1974, the parameters of model (33) have been estimated through standard least-square algorithm, in correspondence with different values of the model order p. Using the sum of the squares of modeling errors, as the order test index, since it practically flattens for $p = 4$, such value has been retained as p . The corresponding parameter estimates are shown in Table 2.

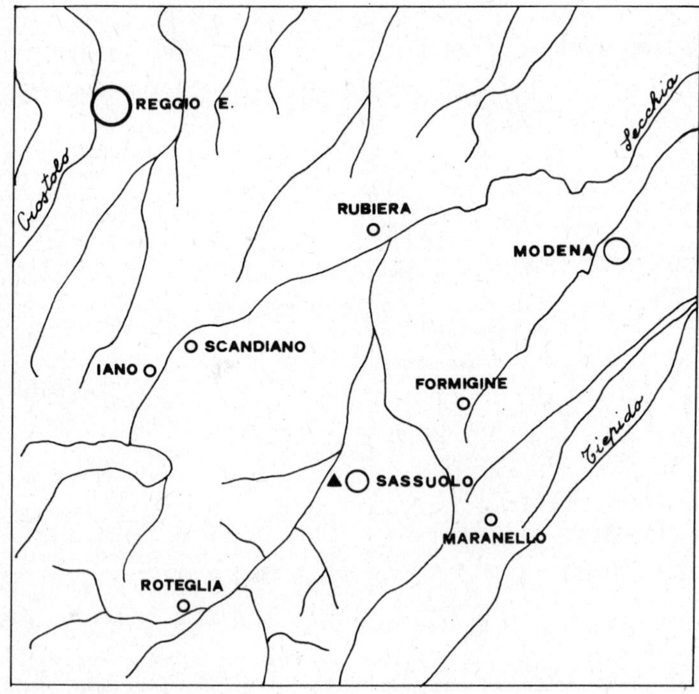

Fig. 9 - Area under consideration

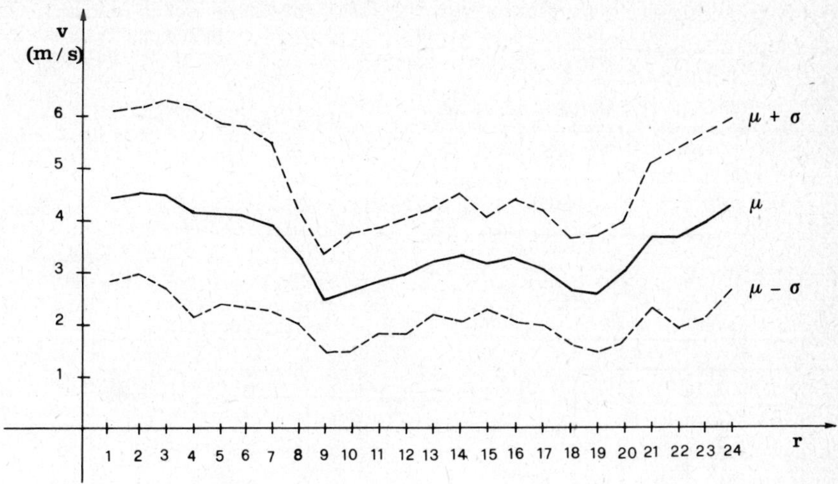

Fig. 10 - Mean μ_r and standard deviation σ_r of hourly wind speed versus hour r of the day

The τ-step ahead real-time predictor derived from eq.(33) is simply given by (trivial case of Kalman predictor):

$$\hat{v}(k+\tau \mid k) = \sum_{j=1}^{p} \Phi_j \tilde{v}(k+\tau-j) \tag{34}$$

where, for $\tau \leqslant p$

$$\tilde{v}(k+\tau-j) = \begin{cases} v(k+\tau-j) & \tau \leqslant j \leqslant p \\ \hat{v}(k+\tau-j \mid k) & 1 \leqslant j < \tau \end{cases}$$

while for $\tau > p$.

$$\tilde{v}(k+\tau-j) = \hat{v}(k+\tau-j \mid k) \qquad j = 1,2,\dots,p.$$

Of course, in view of eq.(32) the forecast of the variable of interest can be obtained from eq.(34) as

$$\hat{V}(k+\tau \mid k) = \mu_r + \sigma_r \hat{v}(k+\tau \mid k) \tag{35}$$

Predictor (34-35) has been tested on 1 month of wind data from May 1 to May 30 1974, using means μ_r and standard deviations σ_r computed on March 1 - April 30 1974, and the overall result is summarized in Table 3. The indexes of the Table are, respectively: ρ = correlation between observed and predicted data, σ_e =standard deviation of the prediction error. The result of Table 3 must be compared with the forecast performance of the trivial persistent predictor ("future wind speed equal to present one") reported in Table 4.

TABLE 2 - Parameter estimates of the AR wind speed model

Φ_1	Φ_2	Φ_3	Φ_4
+ 0.872	- 0.094	- 0.011	+ 0.035

TABLE 3 - Forecast performance of the univariate AR predictor of wind speed

	ϱ	σ_e
1	0.8663	0.775
2	0.7357	1.053
3	0.6345	1.206
4	0.5601	1.299

TABLE 4 - Forecast performance of the persistent predictor of wind speed

	ϱ	σ_e
1	0.8378	0.880
2	0.6618	1.274
3	0.5186	1.522
4	0.4039	1.698

b. A bivariate AR model of hourly wind components

A bivariate model of hourly wind components in the horizontal plane is described in this section. Precisely, let (T means transposition symbol)

$v_1(k)$ = k-th (cyclically standardized) hourly western component of wind;

$v_2(k)$ = k-th (cyclically standardized) hourly southern component of wind;

$\underline{v}(k) = |v_1(k) \; v_2(k)|^T$.

Then consider the following autoregressive representation of the standardized wind vector $\underline{v}(k)$:

$$\underline{v}(k+1) = \sum_{j=1}^{p} \psi_j \underline{v}(k-j+1) + \underline{\varepsilon}(k) \tag{36}$$

Model (36) has also been identified (Table 5) by following the procedure mentioned in Sec. 5.A and using the same data record. In this case, the most suitable value of model order has turned out to be $p^* = 3$.

Apart from the vector form, the predictor derived from (36) is quite similar to (34)-(35). Its component-by-component performance on the same data record as in 5.A is summarized in Table 6 and must be compared with the forecast quality of the component-by-component trivial persistent predictor (Table 7).

Moreover, from component forecasts $\hat{V}_1(k+\tau \mid k)$ and $\hat{V}_2(k+\tau \mid k)$, it is straightforward to derive a prediction of wind speed:

$$\hat{V}(k+\tau \mid k) = (\hat{V}_1^2(k+\tau \mid k) + \hat{V}_2^2(k+\tau \mid k))^{\frac{1}{2}} \tag{37}$$

The quality of forecast (37) (Table 8) must be compared with the performance of the previous wind speed predictor (34)-(35) (Table 3). The comparison points out a significant improvement when turning from eqs.(34)-(35) to eq.(37). This means that cross correlations between the two wind components are a valuable information for the forecast.

Finally, from each component prediction it is possible to derive a forecast of wind direction

$$\hat{d}(k+\tau \mid k) = \text{arctg}\ \hat{V}_2(k+\tau \mid k)/\hat{V}_1(k+\tau \mid k) \tag{38}$$

whose quality is summarized in Table 8b.

TABLE 5 - Parameter estimates of the Bivariate AR model

Ψ_1	Ψ_2	Ψ_3
ψ_{11} = + 0.956	ψ_{11} = - 0.256	ψ_{11} = + 0.096
ψ_{12} = + 0.069	ψ_{12} = - 0.044	ψ_{12} = + 0.041
ψ_{21} = + 0.123	ψ_{21} = + 0.016	ψ_{21} = - 0.060
ψ_{22} = + 0.841	ψ_{22} = - 0.019	ψ_{22} = + 0.028

TABLE 6 - Forecast performance of the bivariate AR predictor: a) western, b) southern component

	ϱ	σ_e
1	0.9517	0.974
2	0.8958	1.409
3	0.8644	1.591
4	0.8424	1.703

- a -

	ϱ	σ_e
1	0.9033	0.709
2	0.8152	0.957
3	0.7368	1.117
4	0.6700	1.226

- b -

TABLE 7 - Forecast performance of the component by component persistent predictor: a) western, b) southern component

	ϱ	σ_e
1	0.9178	1.284
2	0.7782	2.109
3	0.6371	2.696
4	0.4917	3.189

- a -

	ϱ	σ_e
1	0.8831	0.798
2	0.7638	1.135
3	0.6505	1.380
4	0.5451	1.575

- b -

TABLE 8 - Forecast performance of the bivariate AR predictor: a) wind speed, b) wind direction

	ϱ	σ_e
1	0.8688	0.859
2	0.7463	1.196
3	0.6641	1.365
4	0.6015	1.482

- a -

	ϱ	σ_e
1	0.8740	1.089
2	0.7954	1.373
3	0.7258	1.634
4	0.6663	1.832

- b -

c. A bivariate ARMA model

Let z denote the forward shift operator in the time domain. Then the following ARMA model of the wind vector has been considered.

$$\begin{bmatrix} f(z) - g_{11}(z) & - g_{12}(z) \\ - g_{21}(z) & f(z) - g_{22}(z) \end{bmatrix} \underline{v}(k) = f(z)\underline{\nu}(k) \qquad (39)$$

where

$$f(z) = z^n + a_1 z^{n-1} + \dots + a_n$$

$$g_{st}(z) = g_{st,1} z^{n-1} + g_{st,2} z^{n-2} + \dots + g_{st,n} \qquad s = 1,2; \quad t = 1,2$$

and $\{\underline{\nu}(k)\}$ is zero mean white noise.

Model (39) can be considered as an input-output relationship corresponding to the steady-state innovation filter representation:

$$\hat{\underline{x}}(k+1|k) = F\,\hat{\underline{x}}(k|k-1) + K\,\underline{\nu}(k) \qquad (40a)$$

$$\underline{v}(k) = H\,\hat{\underline{x}}(k|k-1) + \underline{\nu}(k) \qquad (40b)$$

where K is the steady-state Kalman gain and $\{\underline{\nu}(k)\}$ is the innovation sequence. Precisely, by letting $(m = 1,2,\dots,n)$

$$G_m = \begin{bmatrix} g_{11,m} & g_{12,m} \\ g_{21,m} & g_{22,m} \end{bmatrix}$$

The relationships between the parameters of the two representations (39) and (40) are the following (see for instance |26|):

f(z) is the minimal polynomial of F; (41)

$$G_m = \sum_{K=0}^{m-1} a_{m-k-1}\, H(F - KH)^k K \qquad (a_o = 1) \qquad (42)$$

From estimates of $\left\{ a_m, G_m \right\}_{m=1}^{n}$ and through (41),(42) it would be possible to determine an innovation filter state representation, namely a triplet F,K,H (see for instance |27|). However, if one is only interested into the forecast $\hat{v}(k+1|k) = H\hat{x}(k+1|k)$ of the variable of interest (see also |28|), it is straightforward to derive a prediction formula which makes use only of the estimates of $\left\{ a_m, G_m \right\}_{m=1}^{n}$. In fact, from (40b), it turns out to be

$$\underline{\nu}(k) = v(k) - \hat{\underline{v}}(k|k-1)$$

and hence by substitution into (39) the "canonical" predictor follows:

$$f(z)\hat{\underline{v}}(k|k-1) = G(z)\underline{v}(k) \qquad (43)$$

where

$$G(z) = \begin{bmatrix} g_{11}(z) & g_{12}(z) \\ g_{21}(z) & g_{22}(z) \end{bmatrix}$$

The explicit form of the one step ahead predictor (43) is

$$\hat{v}(k+n|k+n-1) = -\sum_{m=1}^{n} a_m \hat{\underline{v}}(k+n-m|k+n-m-1) + \sum_{m=1}^{n} G_m \underline{v}(k+n-m)$$

The parameters $\left\{ a_m, G_n \right\}_{m=1}^{n}$ of model (39)(and predictor (43) have been obtained via maximum - likelihood estimation applied to the same data of 5.A-B in correspondence with different values of the model order p. Specifically, the maximum of the likelihood function has been searched via Rosenbrock's algorithm |29|. Moreover the well-known AIC-test |30| has been performed in order to select the "true" order p. The value $p = 3$ has been obtained as p^*. The resulting parameter estimates are reported in Table 9.

The application of the predictor (43) to the same 30 days data record from May 1 to May 30 1974 as in 5.A-B has given the result summarized in Table 10 (component forecasts) and in Table 11 (forecast of wind speed and direction from component forecasts through eqs. (37) and (38). The derivation of the τ-th step ahead predictor ($\tau > 1$) from eq. (43) is straighforward and is omitted.

From a comparison between Table 11 and Table 8, the performance improvement of the bivariate ARMA with respect to the bivariate AR is very slight. In particular, the improvement seems unable to balance the extra-computational effort implied by the use of the more complex formula (43) in an air pollution prediction and control scheme.

TABLE 9 - Parameter estimates of the Bivariate ARMA model

a_1	a_2	a_3
+ 0.075	+ 0.093	+ 0.002

G_1	G_2	G_3
$g_{11} = +0.969$	$g_{11} = -0.244$	$g_{11} = +0.222$
$g_{12} = +0.174$	$g_{12} = -0.240$	$g_{12} = +0.128$
$g_{21} = +0.208$	$g_{21} = -0.102$	$g_{21} = -0.004$
$g_{22} = +0.994$	$g_{22} = +0.021$	$g_{22} = +0.022$

TABLE 10 - Forecast performance of the ARMA predictor: a) western, b) southern component

	ϱ	σ_e
1	0.9517	0.974
2	0.8958	1.408
3	0.8643	1.592
4	0.8424	1.703

- a -

	ϱ	σ_e
1	0.9838	0.707
2	0.8158	0.956
3	0.7373	1.116
4	0.6711	1.225

- b -

TABLE 11 - Forecast performance of the ARMA predictor: a) wind speed, b) wind direction

	ϱ	σ_e
1	0.8689	0.859
2	0.7477	1.194
3	0.6648	1.366
4	0.6020	1.483

- a -

	ϱ	σ_e
1	0.8669	1.113
2	0.7921	1.383
3	0.7254	1.635
4	0.6658	1.832

- b -

REFERENCES

|1| Sutton, O.G., Micrometeorology (Mc Graw Hill, New York, 1953)

|2| Pooler, F., A prediction model of mean urban pollution for use with standard wind roses, Intern. Jrnl. Air and Water Pollution, 4 (1961) 199-211

|3| Calder, K.L., A climatological model for multiple source urban air pollution, NATO/CCMS, Air Pollution, 5 (1971)

|4| Runca, E., Melli, P. and Zannetti, A., Computation of long-term average SO_2 concentration in the Venetian Area, Appl. Math. Mod., 1 (1976) 9-15

|5| Randerson, D., A numerical experiment in simulating the transport of sulfur dioxide through the atmosphere", Atmospheric Environment, 4 (1970) 615-632

|6| Shir, V.V., Shieh, A generalized urban air pollution model and its application to the study of SO_2 distribution in the St. Louis Metropolitan Area,Jrnl. of Applied Meteorology, 13 (1974) 185-204

|7| Runca, E., Melli, P. and Spirito, A., Real-time forecast of air pollution episodes in the Venetian region: Part 1: the advection-diffusion model, Appl. Math. Model (to appear)

|8| Fronza, G., Spirito, A. and Tonielli, A., Real-time forecast of air pollution episodes in the Venetian region: Part 2: the Kalman predictor, Appl. Math. Model (to appear)

|9| Bonivento, C., Fronza, G. and Tonielli, A., Stochastic Predictors of site wind, Tech. Rep., Istituto di Automatica, Bologna, Italy (July 1979)

|10| Bankoff, S.G. and Hanzevack, E.L., The adaptive filtering transport model for prediction and control of pollutant concentration in an urban airshed, Atmos. Envir., 9 (1975) 793-808

|11| Desalu, A.A., Gould, L.A. and Schweppe, F.C., Dynamic estimation of air pollution, IEEE Trans. on Autom. Contr., AC-19 (1974) 904-910

|12| Sawaragi, Y., Ikeda (Eds), Proceedings of the Symposium on Modelling of the control and prediction of air pollution, Kyoto, Japan (1974)

|13| Sardei, F., Runca, E., An efficient numerical scheme for solving time dependent problems of air pollution advection and diffusion, Proc. Seminar on Air Pollution Modelling, IBM Italy, Venice (1976)

|14| Yanenko, N.N., The method of fractional steps,(Springer-Verlag, Berlin,1971)

|15| Richtmyer, R.D., Norton, K.W., Difference methods for initial-value problems, (Interscience, New York, 1967)

|16| Crank, J. and Nicolson, P., A practical method for numerical solution of partial differential equations of heat conduction type, Pric. Cambridge Philos. Soc., 43 (1974)

|17| Kalman, R.E., A new approach to linear filtering and prediction theory, Trans. of the ASME, Sez. D, Jrnl. of Basic Engineering, 82, (1960) 17-25

|18| Jazwinski, A.H., Stochastic processes and filtering theory, (Academic Press, New York, 1970)

|19| Mehra, R.K., On the identification of variances and adaptive Kalman filtering, IEEE Trans. on Autom. Contr., AC-15, (1970) 175-184

|20| Jazwinski, A.H., Adaptive filtering, Automatica (1969) 475-495

|21| Zannetti, P., Melli, P. and Runca, E., Meteorological factors affecting SO_2 pollution levels in Venice, Atmospheric Env.,11 (1977) 605-616

|22| Zannetti, P., Finzi, G., Fronza, G. and Rinaldi, S., Time series analysis of Venice air quality data, Proc. of IFAC Symposium on Environmental Systems, Planning, Design and Control - Edited by Akashi and Sawaragi - (Pergamon Press,

1977)

|23| Finzi, G., Fronza, G., Rinaldi, S., Zannetti, P., Modelling and forecast of the dosage population product in Venice, Proc. of IFAC Symposium on Environ mental Systems, Planning, Design and Control, Edited by Akashi and Sawaragi, (Pergamon Press, 1978)

|24| Pasquill, F., Atmospheric Diffusion, (Van Nostrand, 1971)

|25| Box, G.E.P., Jenkins, G.M., Time series analysis, forecasting and control, Holden Day, San Francisco (1970)

|26| Spain, D.S., Identification and Modelling of discrete, stochastic, linear systems, Techn. Report n. 6302-10, Inf. Sys. Lab., Stanford Univ., August (1971)

|27| Tajima, K., Estimation of Steady-state Kalman Filtering Gain, IEEE Trans. on Autom. Contr., AC-23, n. 5, (1978) 944-945

|28| Aasnaes, H.B., Kailath, T., An Innovations Approach to Least Squares Estimation. Part VII: Some Applications of Vector Autoregressive Moving Average Models, IEEE Trans. on Autom. Contr., AC-18, n. 6, (1973) 601-607

|29| Rosenbrock, H.H., An automatic method for finding the greatest or teast value of a function, Comp. Jrnl., n. 3, (1960) 175-184

|30| Akaike, H., A new look at the statistical model identification, IEEE Trans. on Autom. Control, AC-19, (1974) 716-722

Numerical Techniques for Stochastic Systems
F. Archetti and M. Cugiani (eds.)
© North-Holland Publishing Company, 1980

LOWER ORDER MODEL REPRESENTATION OF MULTIVARIABLE SYSTEMS BY FACTORIZATION TECHNIQUES

E. Canuto[(o)], G. Menga[(+)], A. Villa[(o)]

[(o)] I.E.N. Galileo Ferraris - Corso Massimo d'Azeglio, 42- TORINO
CENS - Istituto Elettrotecnica Generale - Politecnico TORINO

[(+)] CENS - Istituto Elettrotecnica Generale - Politecnico TORINO

1. INTRODUCTION

It is a common engineering need to dispose of simple and yet adequate models both in simulating test and in designing low cost and effective control systems. A confirm ot this fact is the considerable attention which has been paid by systems and control theorists during the last decades to a well delimited but fundamental problem of model reduction: the approximation of alinear dynamic system by a reduced order model (for a comprehensive bibliografy see Genesio & Milanese [1976]).

A goal of this wide research effort is certainly to adapt the new system theory of the last twenties to the tight constrains posed by the identification and control practice.

The classical methods of approximation roughly divide in three streams:

a) aggregation methods (Aoki [1968]);
b) Padè approximation methods (Paynter [1957]) and continued fraction expansion (Chen & Shien [1968], Hutton & Friedland [1975]);
c) norm approximation methods (Heffes & Sarachik [1969]), Donati & Milanese [1970]).

The common characteristics of the above methods is that the approximation problem is approached by observing that the dynamical behaviour of a physical system may on occasion be dominated by the dynamics of a particular subsystem. Then model reduction may be based upon separating such dominant subsystem, neglecting other dynamics. The critical point in that a precise mathematical concept of dominance must be defined to decide a rational model reduction.

The goal of this work is a deeper analysis of the approximation criterion which is natural to be expressed in terms of the geometric properties of the system.

Let us recall the classical operation of discriminating within the system state space, a subspace of completely controllable and observable states (Wonham [1974]), i.e. the subspace which is responsible for the whole of the system input-output response. Then the approximation can be approached by relaxing observability and controllability concepts and defining a dominant subsystem in terms of strongly observable and contollable subspaces. This approach does not necessarily need a state variable description as a starting point, but input-output data as well, like impulse response or transfer function expansions.

In the light of the classical work of Kalman & Ho [1965] who related the geometric structure of the controllable/observable subspace to the algebraic structure of the impulse response Hankel matrix, several methods for representing linear systems through the analysis of the rank or the dependence relation of the system Hankel matrix were proposed. However due to lacking of efficient and robust numerical techniques for rank analysis they revealed unsuccesful for identification and non promising for solving the approximation problem.

Recently is has been shown (Menga & Perdon [1979]) that appropriate techniques of the numerical linear algebra (orthogonal factorization and singular value decomposition) play a significant role in the analysis of the structure of dynamical systems, and consequently in the determination of lower order models for approximating complex systems. Indeed through singular value decomposition a selection of a strongly observable/controllable subspace is easily obtained by discriminating between high and low singular values.

In this work both the classical and the recent results are revised, and a unifying theory is sketched. In Section 2 the relation between the impulse response Hankel matrix and the state space minimal realization is recalled. Section 3 proposes a unifyng view of the classical methods in the light of the state space decomposition concept. Section 4 is devoted to the recent result to obtain this state space decomposition by singular value analysis.

2. INPUT-OUTPUT EQUIVALENCE OF LINEAR DYNAMICAL SYSTEMS

2.1. Class of input-output equivalent dynamical systems

The input-output (external) properties of a causal, time-invariant, linear dynamical system are completely characterized by the impulse response matrix or the transfer function matrix, indicated with $G(t)$, $t \in \mathbb{R}^+$ (o) and $G(s)$, $s \in \mathbb{C}$ (o), respectively. Such representations are obtained through the relations

$$y(t) = \int_0^\infty G(r)\, u(t-r)\, dr \qquad (2.1)$$

$$Y(s) = G(s)\, U(s) \qquad (2.2)$$

where $y(t) \in \mathbb{R}^q$ and $u(t) \in \mathbb{R}^p$.
A wide and useful class of linear systems is characterized by an impulse response like

$$G(t) = G_o \delta(t) + \bar{G}(t) \qquad (2.3)$$

where $\delta(t)$ is the Dirac delta function and $G(.)$ is analytic over $\mathbb{R}^+$. In this case $G(t)$ admits the power expansion

$$G(t) = G_o \delta(t) + \sum_{i=1}^{\infty} G_i \frac{t^{i-1}}{(i-1)!} \qquad (2.4)$$

valid over $\mathbb{R}^+$, which under Laplace transform gives the expansion about $s=\infty$

$$G(s) = \sum_{i=o}^{\infty} G_i\, s^{-i} \qquad (2.5)$$

where G_i are called Markov parameters.

(o) $\mathbb{R}^+$ means the positive real line, or $(0,\infty)$ and $\mathbb{C}$ the complex plane.

A major problem in control and simulation practice is to transform an input-output description, like the sequence of Markov parameters $\{G_i\}_{i=o}^{\infty}$, in a more concise description, like rational transfer function or state equations, which may only approximate the input-output properties of the actual system. This procedure, denoted as <u>finite approximation</u>, is clearly an extension of the classical <u>realization problem</u>, which implies the exact recovering of the input -output properties.

As a starting point let us recall the classical realization problem from the Markov parameters (Ho & Kalman [1965]).

Let

$$\begin{aligned} \dot{x}(t) &= A\,x(t) + B\,u(t) \\ y(t) &= C\,x(t) + D\,u(t) \end{aligned} \tag{2.6}$$

be an n-dimensional state equations, with observability index r and controllability ubdex s. Denote also with T_i and U_j its truncated observability and controllability matrices.

Arrange the Markov parameters in a sequence of block-Hankel matrices

$$H_{ij} = \begin{bmatrix} G_1 & G_2 & \cdots\cdots & G_j \\ G_2 & G_3 & & \\ \vdots & & & \\ G_i & G_{i+1} & \cdots\cdots & G_{i+j} \end{bmatrix} \tag{2.7}$$

Let us recall some basic properties of Hankel matrices.

<u>Definition 1</u>

The infinite sequence $\{H_{ij}\}$ of Hankel matrices (2.7) is (n,r,s)-realizable if the integers r,s and n exist such that

$$\text{rank } H_{ij} = n<\infty \qquad i\leq r,\ j\geq s \tag{2.8}$$

Proposition 1

Any element H_{ij}, with $i \geq r$, $j \geq s$ of a (n,r,s,)-realizable Hankel sequence $\{H_{ij}\}$,can be factored as

$$H_{ij} = T_i U_j' \tag{2.9}$$

where T_i and U_j are n-columned matrices of full rank. The factorization is up to an n x n nonsingular matrix S, which defines an equivalence class of factors.

Proof

A factorization (2.9) of H_{ij}, under above assumptions, is easily obtained by selecting in H_{ij}, i.e. a row basis $h_1', \ldots, h_n'$ arranged in $\bar{H}_{ij}'$; then

$$\begin{aligned} U_j' &= \bar{H}_{ij}' \\ T_i &= H_{ij} \bar{H}_{ij} (\bar{H}_{ij}' \bar{H}_{ij})^{-1} \end{aligned} \tag{2.10}$$

The elements of the equivalence class correspond to difference choices of a basis for H_{ij}. ∎

Denote with $\mathcal{C}(N)$ and $\mathcal{R}(N)$ the colum and the row spaces of a matrix H. Let $\{N_i\}$ be an infinite sequence of n-columed block matrices where $N_i' = [M_1' \; M_2' \; \ldots \; M_i']$ and M_i is p-rowed.

Indicate with r the block index such that

$$\text{rank } N_j = n \qquad i \geq r \tag{2.11}$$

Definition 2

The matrices N_i, $i \geq r$, of the sequence $\{N_i\}_{i=}^{\infty}$ are (truncated) observability matrices with observability index r, if

$$\mathcal{C}(N_{i+1}) \subset \mathcal{C}\left(\begin{bmatrix} I_p & 0 \\ 0 & N_i \end{bmatrix}\right), \; i \geq r \tag{2.12}$$

∎

Proposition 2

The factors T_i and U'_j, $i \geq r$, $j \geq s$, in (2.9) are observability matrices.

Proof

The proof follows from the relations

$$\mathcal{C}(H_{ij}) = \mathcal{C}(T_i)$$
$$\mathcal{R}(H_{ij}) = \mathcal{R}(U'_j) \qquad (2.13)$$

and the structure of Hankel matrices. ∎

The fundamental result relating the Hankel sequence $\{H_{ij}\}$ to the minimal state space realization of G(s) is the following

Therem 1 (Silverman [1971])

A sequence of (n,r,s,)-realizable Hankel matrices $\{H_{ij}\}$ obtained by the Markov parameters of a transfer function G(s) univocally determines a class of n-dimensional realizations (2.6), characterized by

- r and s are the observability and controllability indices;
- the factors T_i and U'_j are the observability and controllability matrices;
- the class of realizations is one-to-one with the class of factors (T_i, U'_j). ∎

The procedure for obtaining the matrices (A,B,C,D) of the realization is summarized by the relations

$$U'_{j+1} = \left[\begin{array}{c|c} B & A \end{array}\right] \left[\begin{array}{c|c} I & 0 \\ \hline 0 & U_j \end{array}\right]$$
$$T'_{i+1} = \left[\begin{array}{c|c} C' & A' \end{array}\right] \left[\begin{array}{c|c} I & 0 \\ \hline 0 & T'_i \end{array}\right] \qquad (2.14)$$
$$D = G_o$$

for $i \geq r$, $j \geq s$

2.2. Structure of a dynamic system

Previous results relate the algebraic properties of the Hankel matrices of the Markov parameters to the geometry of the minimal realization. The factorization (V.2), selecting a basis for the controllability and observability subspaces, discriminates within the system state space a subspace of completely controllable and observable states, responsible for the system input-output properties. As a consequence this selection determines a unique minimal realization.

Definition 3

A structure for an input-output equivalent class of minimal state equations is a rule for selecting an element of the class; operatively it is a rule for performing the factorization (2.9).

Two types of rule has been proposed:

a) select a row or a colum basis of H_{ij} (algebraic rule);

b) factorize H_{ij} through numerical techniques (numerical rule).

Algebraic rule

The selection of a basis defines a set $\mathcal{L} = \{1_1, \ldots\ldots, 1\}$ or $\mathcal{R} = \{k_1, \ldots\ldots, k_n\}$ of position indices of the row or colum vectors in H_{ij}. If the set is nice (Denham [1974]) the minimal realization results in an algebraic canonical form. In the last case the rule is completely specified by the set of Kronecker indices (Popov [1972]), denoted as $\mu_1, \ldots\ldots, \mu_q$ for row selection and $\nu_1, \ldots\ldots, \nu_p$ for colum selection.

A row selection corresponds to the factorization

$$U_j' = \pi H_{ij} \tag{2.15}$$

where π is a selector matrix whose α-th colum is $[0 \ldots\ldots 010 \ldots 0]$ with the unity in the 1_α-th position. Moreover T_i maintains the same structure of π, apart the identically null rows (dependent rows), which now are representations with respect to the basis U_j' of the corresponding Hankel rows. A nice selection implies that the dependent rows are only combination of the preceeding ones.

Numerical procedures

In spite of the enormous bulk of literature on nice minimal realization of dynamic systems, this procedure, being only algebraic, leaves out of consideration the problem of really computing the factorization of the Hankel matrix. Yet the extension of the realization theory to the current practice, which is mainly concerned with realization problems from noisecorrupted data,can be made only on the basis of numerically efficient factorization techniques. To this end two different approaches, derived from linear algebra literature (Golub [1976], [1969]), has been proposed (Canuto & Menga [1977], Menga & Perdon [1979]), namely

1) orthogonal triangular factorization
2) singular value decomposition.

Orthogonal triangular factorization (with permutation)

If H_{ij} is (n,r,s)-realizable, it can be factorized as

$$H_{ij} = \Gamma \left[\begin{array}{c|c} R & 0 \\ \hline S & 0 \end{array}\right] \left[\begin{array}{c} Q_1' \\ \hline Q_2' \end{array}\right] \tag{2.16}$$

where $0_1'$ and $0_2'$ have orthogonal rows, $0_1'$ has n rows, and R is n x n lower triangular matrix; Γ is a reordering matrix which takes into account the selection of "robust" rows operated by the algorithm (Golub [1976], Canuto & Menga [1977]).

From (2.16) it results

$$U_j' = 0_1' \quad , \quad T_i = \Gamma \begin{bmatrix} R \\ S \end{bmatrix} \tag{2.17}$$

Introduce the position set $\ell = \{ 1_1, \ldots, 1_n$ of the rows of R in H_{ij}, derived from the matrix Γ , and the set $\bar{\ell} = \{ 1,2, \ldots, q, 1_{1+q}, \ldots, 1_{n+q} \}$, denote with $\bar{T}$ the matrix obtained by selecting in T_i the rows of position $\bar{\ell}$; then

$$\bar{T} = \left[\begin{array}{c|c} I_q & 0 \\ \hline 0 & R \end{array}\right] \left[\begin{array}{c} C \\ \hline A \end{array}\right] \tag{2.18}$$

The singular value decomposition

If H_{ij} is (n,r,s)- realizable, it can be factorized as

$$H_{ij} = \left[P_i \mid R_i \right] \left[\begin{array}{c|c} \Sigma & 0 \\ \hline 0 & 0 \end{array} \right] \left[\begin{array}{c} \Omega_j' \\ \hline S_j' \end{array} \right] \quad (2.19)$$

where Σ is the diagonal matrix of the nonzero singular values $\sigma_1 \geq \sigma_2 \geq \ldots \sigma_n > 0$ of H_{ij}; i.e. the non-negative square roots of the eigenvalues of $H_{ij}'\ H_{ij}$.

The matrices $[P_i\ \ R_i]$ and $[\Omega_j\ \ S_j]$ are both orthogonal; the first contains the eigenvector of $H_{ij}\ H_{ij}'$ and the second the eigenvectors of $H_{ij}'\ H_{ij}$.

If $i \geq s$ then from (2.19) the following realization results

$$U_j' = \Sigma\, \Omega_j \quad , \quad T_i = P_i \quad (2.20)$$

and

$$P_{r+1} = \left[\begin{array}{c|c} I & 0 \\ \hline 0 & P_r \end{array} \right] \left[\begin{array}{c} C \\ \hline A \end{array} \right] \quad (2.21)$$

$$\Sigma\ \Omega_1' = B$$

3. THE CLASSICAL APPROXIMATION METHODS

The classical approaches to the appoximation problem may be reviewed in the light of the results of Section 2 linking the Hankel matrix and the minimal realizations of systems.

Underlying the approximation of a linear dynamical system by a lower order model is a decomposition of the state space $\mathcal{X}$ into a direct sum of two subspace

$$\mathcal{X} = \hat{\mathcal{X}} + \tilde{\mathcal{X}} \quad (3.1)$$

where $\hat{\mathcal{X}}$ is the space of the lower order approximated model and $\hat{\mathcal{X}}$ represents the neglected dynamics.

Following (3.1), the state equations of the actual system (2.6) can be partitioned as

$$\dot{x} = \begin{bmatrix} A_{11} & A_{12} \\ A_{21} & A_{22} \end{bmatrix} x + \begin{bmatrix} B_1 \\ B_2 \end{bmatrix} u \tag{3.2}$$

$$y = \begin{bmatrix} C_1 & C_2 \end{bmatrix} x$$

The structure of the state matrix A discriminates between two types of state decomposition and approximation

a) modal type: $A_{12} = 0$ and or $A_{21} = 0$;

b) non modal type: $A_{21} \neq 0$ and $A_{12} \neq 0$.

In the modal case eigenvalues ofthe original system are partitioned between the dominant subsystem $(A_{11},B_1,C_1)$and the neglected subsystem (A_{22},B_2,C_2). In the modal case this partition is no more possible, and the approximated model (A_{11},B_1,C_1) does not recover in general the modes of the original system.

A deeper insight of the geometry of approximation is gained through the analysis of the observability and controllability matrices T_i and U_j'. Following the state space decomposition (3.1) and (3.2), a partition of T_i and U_j' results

$$T_i = \left[\hat{T}_i \,\middle|\, \tilde{T}_i\right], \quad U_j = \begin{bmatrix} \hat{U}_j & \tilde{U}_j \end{bmatrix} \tag{3.3}$$

By using the shifting property, the following relations are obtained

$$\left[\hat{T}_{i+1} \,\middle|\, \tilde{T}_{i+1}\right] = \left[\begin{array}{c|c} C_1 & C_2 \\ \hline \hat{T}_i A_{11} + \tilde{T}_i A_{21} & \hat{T}_i A_{12} + \tilde{T}_i A_{22} \end{array}\right] \quad i \geq r \tag{3.4}$$

$$\begin{bmatrix} \hat{U}'_{j+1} \\ \tilde{U}'_{j+1} \end{bmatrix} = \left[\begin{array}{c|c} B_1 & A_{11}\hat{U}'_j + A_{12}\tilde{U}'_j \\ \hline B_2 & A_{21}\hat{U}'_j + A_{22}\tilde{U}'_j \end{array}\right] \quad j \geq s$$

By inspection of (3.4) the following remarkable facts result

1) the matrix $\hat{T}_i$, related to the approximated model, is "observability" only if $A_{21}=0$;
2) the matrix $\hat{U}_j$, related to the approximated model, is "controllability" only if $A_{12}=0$.

Hence to obtain a model approximation, it is sufficient to bring the matrix A in an adequate block triangular form; however if the submatrix A_{11} has to be derived from the relation (3.4), care should be taken in selecting the right matrix ($\hat{T}_i$ or $\hat{U}'_j$).

In the case of non modal approximation, both $\hat{T}_i$ and $\hat{U}'_j$ are non "observability"; a way to obtain the hidden dynamics is to project, e.g.

$$\hat{T}_{i+1} \quad \text{on} \quad \left[\begin{array}{c|c} I_p & 0 \\ \hline 0 & \hat{T}_i \end{array}\right].$$

The norm of the residual gives a measure of the proximity of $\hat{T}_i$ to be dynamic, and the submatrix A_{11} which results corresponds to an orthogonal decomposition of $\mathcal{X}$.

Aggregation scheme of Aoki

It is easily shown that the modal approximation corresponds to the aggregation scheme of Aoki [1968],[1971],[1978], Chidambara & Skainker [1971], where the reduced order model

$$\dot{z} = A_{11} z + B_1 u \tag{3.5}$$

is obtained by the state aggregation

$$z = L x \tag{3.6}$$

satisfying

$$A_{11}L = L A \quad , \quad B_1 = L B \tag{3.7}$$

Thus the modal approximation is a perfect aggregation; while the non modal is a non perfect aggregation.

Classical modal approximations

The aggregation scheme of Aoki is subsequent to the classical modal approximation developed by several authors (Davison [1966],[1968], Marshall [1966], Mahaparta [1978], Chidambara [1967]). These schemes, which are perfect aggregation cases, decompose the state space by retaining the dominant modes of the system.

Non modal approximations

Non modal approximation methods are some geometric techniques developed by Wilson [1974], Mitra [1969], Anderson [1967], and Galiana [1973]. The reduced order model is obtained by projecting on a subspace of the state space the system response; the criterion of approximation is a projection error or an output approximation error. These methods generally do not maintain tha system eigenvalues, since the reduced model matrix A_m is related to the system matrix A by (Wilson [1974])

$$A_m = \theta_1 A \theta_2^{-1} \tag{3.8}$$

Only in the case of $\theta_1 = \theta_2$, the methods are perfect aggregation schemes: in this sense Mitra [1969] has shown that the modal approximation is a particular case of his own method.

Non perfect aggregation may be viewed also the Padè (or moment) approximation of Shamash [1975], and Zakian [1973], which operates on the system transfer function, and the continued fraction expansion of Chen & Shieh [1968], Chen [1974].

4. APPROXIMATION BY SINGULAR VALUE DECOMPOSITION

In this Section the approximation problem is restated in quite different terms with respect to the classical approaches. The results here reported are quite entirely contained in (Menga & Perdon [1979]). in this work techniques previously used in an identification context were firstly applied of the approximation problem. The method can be viewed as a relaxation of the classical realization theories and it has a strongly geometrical interpretation. Moreover the approximation is a non perfect aggregation scheme, which allows to measure precisely the proximity to a model scheme.

The criterion introduced carries an orthogonal decomposition of the state space which discriminates the strongly observable and controllable subspace. This decomposition is obtained with the aid of the singular value decomposition of the Hankel matrix of the system H_{ij}.

To formulate the criterion, suppose that in (2.19) a distinct gap exists between the m-th and the m+1-th singular value and that σ_{m+1} is small, i.e. $\sigma_m \geq \delta > \varepsilon \geq \sigma_{m+1}$.
According to Golub [1969], H has numerical rank m (o). This means that

> H_{ij} can be approximated by an m-ranked matrix K_{ij} with an error $\| H_{ij} - K_{ij} \|_2 \leq \varepsilon$

Let us rewrite the singular value decomposition of H_{ij} as

$$H_{ij} = \begin{bmatrix} \hat{P}_i & \tilde{P}_i \end{bmatrix} \left[\begin{array}{ccc|ccc} \sigma_1 & & & & & \\ & \ddots & & & 0 & \\ & & \sigma_m & & & \\ \hline & & & \sigma_{m+1} & & \\ & 0 & & & \ddots & \\ & & & & & \sigma_n \end{array}\right] \begin{bmatrix} \hat{Q}_j \\ \tilde{Q}_j \end{bmatrix} \tag{4.1}$$

where $\hat{P}_i$ and $\hat{Q}_j$ are the first m columns of P_i and Q_j; following Lemma Al an approximant matrix K_{ij} belonging to the neighbour of H_{ij} is obtained as

$$H_{ij} = P_i \begin{bmatrix} \sigma_1 & & & & \\ & \ddots & & & \\ & & \sigma_m & & \\ & & & 0 & \\ & & & & \ddots \end{bmatrix} Q'_j \tag{4.2}$$

with $\| H_{ij} - K_{ij} \|_2 = \sigma_{m+1} \leq \varepsilon$

Hence a distint gap in the sequence of the singular values of the

(o) See Appendix A.

system Hankel matrix is a <u>criterion</u> for selecting the order of the approximation.

The criterion has a geometric interpretation which enlights the procedure for obtaining the state realization of lower order. Denote with $\mathcal{C}(H_{ij})$ the column space of H_{ij}, and with $\mathcal{R}(H_{ij})$ its row space, wich coincide with the observability and controllability spaces of the underlying dynamical system. From (2.19) the columns of P_i are an orthonormal basis for $\mathcal{C}(H_{ij})$ and similarly the rows of Q_j' for $\mathcal{R}(H_{ij})$; following (4.1) $\mathcal{C}(H_{ij})$ and $\mathcal{R}(H_{ij})$ can be expressed as the direct sums

$$\begin{aligned} \mathcal{C}(H_{ij}) &= \mathcal{C}(\hat{P}_i) + \mathcal{C}(\tilde{P}_i) \\ \mathcal{R}(H_{ij}) &= \mathcal{R}(\hat{Q}_j') + \mathcal{R}(\tilde{Q}_j') \end{aligned} \tag{4.3}$$

Hence the approximation of the observability and controllability spaces of the system is possible by their $\mathcal{E}$ section (Golub [1969]) $\mathcal{C}(\hat{P}_i)$ and $\mathcal{R}(\hat{Q}_j')$. A measure of the approximation is given by the spectral norm of the projections residuals of the columns of H_{ij} on $\mathcal{C}(\hat{P}_i)$ which holds (Golub [1969])

$$\| \tilde{P}_i H \|_2 = \sigma_{m+1} \leq \mathcal{E} \tag{4.4}$$

The decompositions (4.1) and (4.3) induce a decomposition of the state space

$$\mathcal{X} = \hat{\mathcal{X}} + \tilde{\mathcal{X}} \tag{4.5}$$

where $\hat{\mathcal{X}}$, m-dimensioned, and $\tilde{\mathcal{X}}$, n-dimensioned, are respectively the strongly and weakly observable/controllable subspaces.

The above decomposition introduced in the realization equations (2.21) yields

$$\left[\hat{P}_{i+1} \;\middle|\; \tilde{P}_{i+1} \right] = \left[\begin{array}{c|cc} I_p & 0 & 0 \\ \hline 0 & \hat{P}_i & \tilde{P}_i \end{array} \right] \left[\begin{array}{c|c} C_1 & C_2 \\ \hline A_{11} & A_{12} \\ \hline A_{21} & A_{22} \end{array} \right] \tag{4.6}$$

and

$$\left[\begin{array}{ccc:ccc} \sigma_1 & & & & & \\ & \ddots & & & & \\ & & \sigma_m & & & \\ \hdashline & & & \sigma_{m+1} & & \\ & & & & \ddots & \\ & & & & & \sigma_n \end{array}\right] \begin{bmatrix} \hat{Q}_1' \\ Q_1' \end{bmatrix} = \begin{bmatrix} B_1 \\ B_2 \end{bmatrix} \tag{4.7}$$

<u>Definition 4</u>

The triple (A_{11}, B_1, C_1) corresponding to the state decomposition (4.5) is the ε - strongly observable-controllable realization of the actual system.

The selection of (A_{11}, B_1, C_1) amounts in general to a non perfect aggregation of the state variables, since the singular value decomposition (2.21) does not necessarily involves $A_{21} = 0$ or $A_{12} = 0$ (see Sect. 3).

To obtain a measure of the distance from a case of perfect aggregation let us investigate characteristics of the approximant observability (controllability) subspaces, $\mathcal{L}(\hat{P}_i)$ $(\mathcal{R}(Q_j'))$.

If $A_{21} \neq 0$ the matrix $\hat{P}_i$ does not enjoy the shifting property of the observability matrices, i.e.

$$\mathcal{L}(\hat{P}_{i+1}) \not\subset \mathcal{L}\left(\left|\begin{array}{cc} I & 0 \\ \hline 0 & P_i \end{array}\right|\right) \quad i \geq r \tag{4.8}$$

which is clear from

$$\hat{P}_{i+1} = \begin{bmatrix} C_1 \\ \hline \hat{P}_i A_{11} \end{bmatrix} + \begin{bmatrix} 0 \\ \hline \tilde{P}_i A_{21} \end{bmatrix} \tag{4.9}$$

The measure required can be expressed as the spectral norm of the projection residuals of $\hat{P}_{i+1}$ on $\begin{bmatrix} I & \\ 0 & \hat{P}_i \end{bmatrix}$, i.e.

$$\left[0 \;\vdots\; \tilde{P}_i'\right] \hat{P}_{i+1} \|_2 = \| A_{21} \|_2 \tag{4.10}$$

5. CONCLUSION

The need for a relaxation of the system geometric theory in order to meet the engineering aspects of the system pratice has been recently pointed out in different occasions (Moore & Laub [1978]). A very fundamental problem in every discipline is the modeling of phenomena which always is carried over with approximation criterions. In this paper recent results in the field of linear system approximation and a brief revision of the classical methods are reported.

APPENDIX A Numerical rank of a matrix

Let H be an s x r matrix of rank n, along with its singular value decomposition

$$H = P \Sigma Q' \tag{A.1}$$

where Σ is the diagonal matrix of its singular values $\sigma_1 \geq \sigma_2 \ldots \geq \sigma_r$ with $\sigma_{n+1} = \ldots = \sigma_r = 0$. Recall that

$$\sigma_1 = \|H\|_2 = \sup_{|x|_2 = 1} |Hx|_2 \tag{A.2}$$

where $|x|_2$ is the Euclidean vector norm and $\|H\|_2$ is the spectral matrix norm.

Definition A.1 (Golub [1969])

The s x r matrix H of rank n has numerical rank $(\delta, \varepsilon, m)_2$ with respect to the spectral norm if

$$\begin{aligned} &1)\quad r = \inf \{\text{rank } K \ : \|H - K\|_2 \leq \varepsilon\} \\ &2)\quad \delta = \sup \{\eta \ \ \|H - K\|_2 \leq \eta \rightarrow \text{rank } K \geq r\} \end{aligned} \tag{A.3}$$

This means that H can be approximated within an error ε by matrices of rank $\geq$r. Moreover this approximation cannot be improved (lowering the rank) within an error δ. A practical implementation of the approximat matrix is sugg sted by the following result

Lemma A.1 (Golub [1969])

Let $\sigma_1 \geq \sigma_2 \ldots\ldots \geq \sigma_n$ be the nonzero singular values of H. Then H has numerical rank $(\delta,\varepsilon,m)_2$ if and only if

$$\sigma_m \geq \delta > \varepsilon \geq \sigma_{m+1}$$

REFERENCES

1. ANDERSON J.H., 1967, Proc. iee. 1014.
2. AOKI M., 1968, Trans.on Automat.Contr., AC-13,246.
3. AOKI M., 1971, in "Optimization Methods for Large Scale Systems", Wismer Ed., McGraw-Hill.
4. AOKI M., 1978, Trans on Automat. Contr., AC-23,173.
5. CANUTO E. & MENGA G., 1977, IFAC Symp. on Discontinuous Computer Control Systems, Prague.
6. CHEN C. & SHIEH L., 1968, Trans. on Circuit Th., CT-16,197.
7. CHEN C., 1974, Int. J.Contr., 20,225.
8. CHIDAMBARA M.R., 1967, Trans; on Autom. Contr., AC-12,119,213,799.
9. CHIDAMBARA M.R. & SKAINKER R.B., 1971, Trans. on Automat. Contr., AC-16,175.
10. DAVISON E.J., 1966, Trans. on Automat. Contr., AC-11,93.
11. DAVISON E.J., 1968, Trans. on Automat. Contr., AC-13,214.
12. DENHAM M.J., 1974, Trans. on Automat. Cont., AC-19,646.
13. DONATI F. & MILANESE M., 1970, 2nd IFAC Symp., Prague.
14. GALIANA F.D., 1973, Int. J. Contr.,17,1313.
15. GENESIO R. & MILANESE M., 1976, Trans on Automat. Contr., AC 21,118.
16. GOLUB G.M., 1969, in "Statistical Computation", Springer, N.Y..
17. GOLUB G.H. & KLEMA V. & STEWARD G.W., 1976, Rep. Stan-CS-76-559, Stanford Univ.
18. GRUCA A. & BERTRAND P., 1978, Int. j .Contr., 28,953.
19. HEFFES H. & SARACHIK P.E., 1969, Bell Syst. Tech. J., 48,209.
20. HO B. L. & KALMAN R. E., 1965, Proc. 3rd Allerton Conf. on Circuit & System.
21. HUTTON M. & FRIEDLAND B., Trans. on Automat. Contr., AC-20,1975, 329.
22. MAHAPARTA, 1978, Trans. on Autom. Contr., AC-23,1106.
23. MARSHALL, 1966, Control, 642.
24. MENGA G.& PERDON A., 1979, 5th IFAC Symp. Identificat. Darmstadt.

25. MITRA R., 1969, Proc. IEE,1101.

26. MOORE B. C. & LAUB A. J., 1978, Trans. on Automat. Contr., AC-23,783.

27. PAYNTER H.M., Regelunstechnik, Moderne Theorien und ihre Verwendbarkeit, G. Muller Ed.R.O. Verlag, 1957.

28. POPOV V.M., 1972, SIAM J. Contr.10, no. 2.

29. SHAMASH Y., 1975, Int. J. Contr. 20,267.

30. SILVERMAN L. M. 1971, Trans. on Automat. Contr. AC-16,554.

31. WILSON B., 1974, Int. J. Contr. 20,57.

32. WONHAM W. M., Linear Multivariable Control: a geometric approach, 1974.

33. ZAKIAN V., 1973, Int. J. Contr., 18/455.

Numerical Techniques for Stochastic Systems
F. Archetti and M. Cugiani (eds.)
© North-Holland Publishing Company, 1980

STOCHASTIC PROCESSES IN THE MODELLING OF EARTHQUAKE OCCURRENCE

Elisa Grandori Guagenti

Istituto di Matematica della Facoltà di Architettura di Milano

1. Introduction

An appropriate analysis of the Seismic Risk requires stochastic models of

1 - magnitude of seismic events
2 - occurrence in time
3 - geometrical distribution of epicenters
4 - local seismicity at a site included in a seismic zone
5 - evaluation of damage.

This paper gives a brief panorama of the most common classic models and those recently proposed, expecially in regard to occurrence in time. Some numerical results will be discussed.

The fifth topic is typically a part of earthquake engineering: it requires an analysis of structural damage associated with alternative designs and of indirect damage. These subjects will not be treated in this paper, except for the probabilistic evaluation of a given level of damage.

Classic models will be considered in the first part; some recently proposed models in the second one.

The results of risk analysis are only an input for a seismic design decisions and, more generally, for public policy in risk prevention. A very short outline of design decisions will be given at the end.

The definitions of recurrent quantities in seismic risk analysis are:

energy, W, released in a single earthquake;

magnitude, M, $M = w_1 \ln W + w_2$;

local intensity, I_0, empiric scale;

waiting time, τ, from one event to the next one;

number of events, N_t, in time interval t;

epicentral distance, R, between epicenter and a given site whose seismicity has to be studied;

local intensity, I_{loc}, $I_{loc} = c_1 + c_2 I_o - c_3 \ln R$;

peak ground acceleration, Y, at the considered site, due to an earthquake occurring in the zone.

2. Classic Models.

The classic assumption for magnitude M is that its d.f. (distribution function) is exponential

$$(1) \qquad 1 - F_M(m) = e^{-\beta m}$$

with the limitation $m \geq 0$; this constraint is unimportant for earthquake engineering: indeed usually a lower bound m_o is imposed below which the seismic events are negligible for a seismic design. In this case the d.f. becomes:

$$(2) \qquad 1 - F_M(m) = e^{-\beta(m-m_o)} \qquad \text{with } m \geq m_o .$$

The classical model of earthquake arrivals is given by the Poisson process; then the distribution of the waiting time τ between one earthquake and the next one is exponential too:

$$(3) \qquad 1 - F_\tau(t) = e^{-\lambda t}$$

with mean μ and standard deviation s

$$\mu = s = \frac{1}{\lambda} .$$

The numbers of arrivals N_t in a time interval t has a Poisson distribution

$$(4) \qquad P\{N_t = n\} = \frac{e^{-\lambda t}(\lambda t)^n}{n!}$$

with mean

$$E(N_t) = \lambda t.$$

The r.v. (random variable) R depends on the geometrical distribution of earthquake sources. The most common models are based on linear fault,circular zone, rectangular zone.

In all three cases a homogeneous distribution of epicenters, with constant depth of focuses and symmetric location of focuses, is usually assumed , so that p.d.f. (probability density function) of r.v. R are respectively:

$$f_R(r) = \frac{r}{l\sqrt{r^2-d^2}} \; ; \qquad f_R(r) = \frac{2r}{r_1^2} \; ; \qquad f_R(r,x) = \frac{r}{ab\sqrt{r^2-x^2}}$$

where $r<r_1$ is the distance from the site to one focus; l is the lenght of linear fault and d its distance from the site; a,b are the dimensions of rectangular zone, and x is the distance from the site to one linear fault generating rectangular zone.

The extension to other cases with more than one fault or zone, differently located or shaped, is obvious.

An improvement of these hypotheses depends on an actual knowledge of fault geometry.

For engineering design it is important to predict the motion of the ground at the site where the buildings are, when one earthquake of a given size is produced at a given epiceter. A significant parameter is the peak ground acceleration Y. The attenuation law, based in part on theory but primarly on empirical correlation, is available to relate site parameter Y to epicentral distance R and magnitude M:

$$y = b_1 e^{b_2 m} \; r_o^{-b_3} \qquad\qquad r \leq r_o$$

$$y = b_1 e^{b_2 m} \; r^{-b_3} \qquad\qquad r_o \leq r \leq r_1$$

The r.v. Y is then function of two r.v. M and R. It is thus possible to evaluate the pdf of Y. Let us call p_y the probability that, at a given site, the peak ground acceleration Y is greater than y, when an earthquake occurs in the area.

Taking into account that Y is monotonic function of M we can write:

$$(5) \qquad 1 - F_Y(y) = p_y = \int_D \{1 - F_M[m(yr)]\} \; f_R(r) \; dr,$$

the integral being extended over seismogenetic area D.

The probability p_y defines the local effect due to an earthquake occurring at a random point in the area.

Local seismicity is actually defined taking into account the time occurrence too. Let us call N_{Yt} the number of earthquakes that, at a given site, during a time interval t, produce peak ground acceleration greater than y. If N_t is the total number of earthquakes, and assuming the independence among N_t and size and location of the events we have

$$(6) \qquad Pr\{N_{Yt} = n\} = \sum_{m=n}^{\infty} \binom{m}{n} p_Y^n (1-p_Y)^{m-n} Pr\{N_t = m\}.$$

It can be proved that, if N_t is Poisson distributed

$$(7) \qquad Pr\ \{N_{Yt} = n\} = \frac{e^{-p_Y \lambda t}(p_Y \lambda t)^n}{n!}.$$

This means that, in the above hypotheses, the local arrival process is Poisson process too.

3. Recently proposed models.

3.1 Magnitude M.

It has been observed by many Authors that the use of exponential distribution for magnitude M overestimates the frequency of large earthquakes. Moreover, for any reasonable energy-magnitude relation, the expected value of the energy released in a single earthquake becomes infinite.

An improvement of the exponential law can be obtained assuming a "quadratic" distribution

$$(8) \qquad 1-F_M(m) = e^{\beta_1 m+\beta_2 m^2} \qquad \text{with } \beta_2 < 0;\ m \geq 0.$$

This law generally fits the statistical data well even for large magnitudes, when it is used for zones with a rich set of data. Moreover the energy does not diverge. However it can be observed that we really need a reliable law when we are facing the problem of forecasting future earthquakes on the basis of a small set of statistical data. In this case the presence of two free coefficients β_1, β_2 makes the "quadratic" law very sensitive to statistical fluctuations and can lead to unacceptable errors in extrapolating to large magnitudes.

In order to overcome such difficulties it would be suitable to introduce a new d.f. with two main characteristics: a good fit for rich sets of data, even for large magnitudes; a rigid behaviour similar to the exponential distribution (one parameter distribution). A d.f. which seems to have these characteristics is the "double exponential"

$$(9) \qquad 1-F_M(m) = e^{-e^{\beta m}} \qquad -\infty < m < \infty$$

Besides it can be seen that, on the basis of this law, the expected value E(w) of energy is always limited.

The "goodness of fit" of the quadratic and double exponential laws has been compared over 60 years of seismic data: the double exponential distribution looks more stable than the quadratic law.

Another d.f. useful for fitting the data of large earthquakes is the extreme value distribution of the first type

$$(10) \qquad F_M(m) = e^{-e^{-m}}$$

It is interesting to note that all three proposed distributions have the same limiting distribution of their maximum values, the limiting distribution (10) of the first type. In fact all three d.f. satisfy the condition

$$\lim_{m\to\infty} \frac{d}{dm}\left[\frac{1-F_M(m)}{f_M(m)}\right] = 0.$$

In order to avoid the two mentioned unrealistic results, another criterion has been suggested: to impose an upper bound m_1 on magnitude values assuming that the truncated p.d.f. is proportional to relative non truncated:

$$f_M^{tr}(m) = k\ f_M(m)$$

with normalizing coefficient k, so as to satisfy

$$\int_{-\infty}^{m_1} f_M^{tr}(m) = 1.$$

This criterion is obviously available for every d.f.. But the deterministic choice of an upper bound m_1 is both debatable from the physical point of view and from a correct probabilistic approach. However in truncated cases the earthquakes beyond a distance r_{max} defined by solving

$$y = b_1 e^{b_2 m_1}\ r_{max}^{-b_3}$$

cannot cause a peak ground acceleration at the site greater then y since their magnitude cannot exceed m_1. Then the site influential field D(y) of sources becomes smaller and depending on y

$$D(y) = D_{(r\le r_{max})}$$

It can be proved that:

$$1 - F_y^{tr}(y) = p_y^{tr} = k[1-F_y(y)]_{r\leq r_{max}} + (1-k)\,\frac{\text{meas}D(y)}{\text{meas } D}$$

3.2 Waiting time τ.

In Poisson processes the waiting time τ is exponentially distributed. This implies two unacceptable consequencies: the immediat risk is constant in time and the origin time t_o is ininfluential. In fact, the probability of at most one event in the next time interval dt is

$$\Pr\{t<\tau\leq t+dt/_{\tau>t}\} = \frac{f_\tau(t)\,dt}{1-F_\tau(t)} = \phi(t)\,dt$$

We shall call "immediate risk" the function ϕ defined above.
In the Poisson hypothesis the risk function is constant irrispective of the time elapsed since occurrence of the last earthquake.
It is easily seen that, in the Poisson process, this is $\phi=\lambda$. This is not realistic: the previous history of the process is influential and the immediate risk ϕ must probably increase with the time elapsed from the last event.
The assumption that, for a given sismogenetic zone, the earthquake process takes into account the occurrence of past events seems more realistic than the assumption (3). With reference to large earthquakes (excluding after-shocks), it seems reasonable to propose one $f_\tau(t)$ that is not J-shaped as exponential one, i.e. with a maximum in the starting point, but a p.d.f. starting from zero and increasing until a single maximum.
Some tentative approaches leading to this kind of $f_\tau(t)$ have been actually proposed, for example:

(11) $$f_\tau(t) = \frac{\rho(\rho t)^{a-1} e^{-\rho t}}{\Gamma(a)}$$

or

(12) $$f_\tau(t) = a\rho(\rho t)^{a-1} e^{-(\rho t)^a}.$$

They are respectively a Γ or Weibull distribution with mean value

$$\mu = \frac{a}{\rho} \qquad \text{or} \qquad \mu = \frac{\Gamma(1+1/a)}{\rho}.$$

If $a > 1$ the risk function ϕ is increasing for both p.d.f. (11) and (12) as t increases: in Γ distribution from zero to ρ, in Weibull distribution from zero to infinity.

For example, with $\mu = 40$ years the assumption (3) leads to a constant value $\phi = 0.025$, while the eq. (11), with a = 2, leads to following values of

ϕ(10 years)=0,0167 ϕ(20 years)=0,025 ϕ(40 years)=0,035 and $\phi \to 0,05$ as $t \to \infty$. If in eq. (11) we suppose an integer value, we deal with Erlang distribution, and we have a very substantial advantage.

In order to obtain numerical answers for non integer a it may sometimes be preferable to interpolate numerically between solutions for integer a.

Let us now discuss the second question: the origin time. The Erlang (or Γ) distribution allows us to distinguish the origin time t_o. This is important in engineering applications: a typical case arises when t_o is the elapsed time from the last earthquake to the completion of a new construction at a given site.

What is the p.d.f. of the next earthquake waiting time, starting from t_o? Let us call f_1 this p.d.f. It is

$$f_1(t)dt = Pr\{t_o+t<\tau\leq t_o+t+dt/_{\tau>t_o}\}$$

If we assume the Poisson process $f_1 = f_\tau(t)$, i.e. the waiting time is independent of past history.

On the contrary if we assume the Erlang process we obtain

$$f_1(t) = \frac{\rho[\rho(t+t_o)]^{a-1}\ e^{-\rho t}}{(a-1)!\sum_0^{a-1} r\ \frac{(\rho t_o)^r}{r!}}$$

which depends on t_o.

3.2 Time up of r-th event.

We consider now not only the first event but the process of all future events. The problem arises of evaluating the waiting time

$$S_r = \tau_1+\tau_2+\tau_3+...+\tau_r$$

of r-th event; let k_r be its pdf. The random variables $\tau_2...\tau_r$ have all the same pdf, equal to $f_\tau(t)$; only the r.v. τ_1 has a different pdf equal to f_1, if $t_o \neq 0$. Such a process is called a "modified process". If $t_o = 0$ all the r.v. $\tau_1,\tau_2,...,\tau_r$ have the same pdf $f_\tau(t)$

and the process is called an "ordinary process".

The r.v. S_r is a sum of independent r.v.

The theory fournishes the Laplace transform k_r^* of pdf k_r of such r.v. S_r. It is

$$k_r^*(s) = f_1^*(s)\{f^*(s)\}^{r-1} \qquad \text{in "modified process"}$$

$$k_r^*(s) = [f^*(s)]^r \qquad \text{in "ordinary process"}$$

The knowledge of Laplace transform of $k_r(t)$ is useful also for damage evaluating.

3.4 Damage evaluation.

We deal only with material damage, not with loss of human lives.

If damage whose monetary cost is C will happen at a random future instant t with pdf $f_T(t)$ the present expected value $E(C_o)$ of C is

$$E(C_o) = \int_0^\infty Ce^{-\gamma t} f_T(t)\,dt = C\, f^*(\gamma)$$

where γ is discount rate for unit time: the Lapalce transform of a given waiting time T pdf is, multiplied by C, the expected present value of monetary damage C which will occur in random time T.

Assuming that the structure is repaired every time damage occurs, the expected present value of the r-th damage (whose value is C) is:

$$E(C_{o_r}) = C\, f_1^*(\gamma)[f^*(\gamma)]^{r-1}$$

In Erlang modified process

$$E(C_{o_r}) = C\,\frac{\sum_1^a i\,\dfrac{t_o^{a-1}}{(\rho+\gamma)^i\,(a-i)!}}{\sum_1^a i\,\dfrac{t_o^{a-i}}{\rho^i\,(a-i)!}}\left[\left(\frac{\rho}{\rho+\gamma}\right)^a\right]^r$$

In Erlang ordinary process, that is when the evaluation is done immediately after an earthquake (t_o=0),

$$E(C_{o_r}) = C\left[\left(\frac{\rho}{\rho+\gamma}\right)^a\right]^r$$

The ratio between the expected value of r-th damage and the first one is a useful parameter. Even the second damage may be much smaller than the first one, if local seismicity at a single site is considered, for example if a=3 μ=50 (years) γ=0,07 (per year)

$$\frac{E(C_{o_2})}{E(C_{o_1})} = 0{,}098 \qquad \frac{E(C_{o_3})}{E(C_{o_1})} = 0{,}0096$$

3.5 Different levels of damage.

Until now one particular damage of monetary value C has been considered. But we have seen that local effects are defined, to a first approximation, by peak ground acceleration Y whose probability law is p_y. Therefore we need to distinguish the different levels of damage, which we call y-damages; the waiting time of y-damage is τ_y.

As before we can evaluate, trough the Laplace transforms, the three waiting time pdf:

$h_1^*(s)$ from origin time to the first y-damage;

$h^*(s)$ from one y-damage to the next one;

$h_r^*(s)$ from origin time to the r-th y-damage.

They are:

$$h_1(t) = f_1(t)p_y + p_y(1-p_y)^2 k_2(t) + \ldots$$

$$h_1^*(s) = p_y f_1^*(s) \frac{1}{1-(1-p_y)f^*(s)}$$

$$h^*(s) = p_y \frac{f^*(s)}{1-(1-p_y)f^*(s)}$$

$$h_r^*(s) = h_1^*(s)[h^*(s)]^{r-1}$$

The present expected value of all the future y-damages is:

$$E(C_y) = C_y \sum_1^\infty r \int_0^\infty e^{-\gamma t} h_r(t)\,dt = C_y p_y \frac{f_1^*(\gamma)}{1-f^*(\gamma)}$$

Thus the stochastic sequence of seismic y-damages can be treated, for monetary damage evaluation, as the damage occurring at the completion of the structure, provided we multiply the probability p_y by a "time discount" factor δ

$$\delta = \frac{f_1^*(\gamma)}{1-f^*(\gamma)}$$

If we deal with Poisson process this is:

$$\delta = \frac{\lambda}{\gamma}$$

If we deal with ordinary Erlang process

$$\delta = \frac{\rho^a}{(\rho+\gamma)^a - \rho^a}.$$

The knowledege of waiting time allows us to evaluate the distribution of numbers N_t of events in time interval t. In fact it is:

$$Pr\{N_t < r\} = Pr\{S_r > t\} = 1-K_r(t)$$

where $K_r(t)$ is d.f. of S_r; and

$$Pr\{N_t = r\} = K_r(t) - K_{r+1}(t)$$

In ordinary Erlang process it is simply:

$$Pr\{N_t = r\} = \sum_{m=ra}^{ra+a-1} \frac{(\rho t)^m e^{-\rho t}}{m!}$$

The number N_t^y of events corresponding to y-damage is evaluated in an analogous way.

4. Some words about decision problems.

When applying these mathematical models to engineering decisions many problems arise to which the previous analysis furnishes the input data.

The main question is: what is the criterion of choice among alternative designs? Is it an optimization problem?

In order to give some answers to this question we recall that in the seismic codes, the formula giving the intensity of the lateral forces contains in general many coefficients to which the same intensity is proportional. Let c_i be the coefficient, at the site i, depending on local seismicity. As a first approximation we consider an area where only local seismicity differs from point to point, while all the remaining conditions (local soil conditions, shape of response spectrum, social and economic conditions, characteristic of builddings,...) are assumed as constant. The coefficients c_i define the severity of building codes, i.e. the aseismic design.

As we have seen above it is possible to evaluate the economic value of a given material damage; but how to evaluate the value of expected losses of human lives? The human life has not a monetary cost. Although some Authors propose to define the economical consequences of human losses, such approach seems not correct. It is morally, socially unacceptable and it leads in some cases to absurd consequences.

Then it is impossible to give the seismic building codes as results of optimization problem.

But different criteria of choice are acceptable. For example: let D be the total monetary cost in a time interval t of seismic risk prevention over a given area (considering both the cost of seismic design and the cost of future damage, direct and indirect, due to earthquakes). The following principle seems acceptable: the distribution of the coefficients c_i must minimize the expected number of victims E(V) over the area for a given value of D; i.e. starting from optimum distribution of the coefficients c_i it must be impossible to reduce the expected number of victims maintaining constant the total cost D.

If the area is divided into n small portions which can be considered as points, we face the problem

$$\text{(13)}\qquad \begin{aligned} &\min E[V(c_i)] \qquad \text{with} \\ &E[D(c_i)] = \text{const} \end{aligned}$$

where $V(c_i) = \Sigma V_i(c_i)$ and $D(c_i) = \Sigma D_i(c_i)$ are functions of code coefficients c_i through the above laws of earthquake occurrence and the following correlations, proposed by earthquake engineering,

$$\text{(14)}\qquad \begin{aligned} y_c &= y_c(c_i) \\ V_i &= V_i(N_t^{y_c}) \\ D_i &= D_i(y, c_i) \end{aligned}$$

where y_c is the peak ground acceleration corresponding to the collapse of the building.

The study of (14) goes beyond our purposes, more about can be found in the references; here we simply point out that (13) furnishes the coefficients c_i, i.e. they solve the problem of seismic zoning as an optimization problem.

Naturally how large the funds D must be is not a mathematical problem. The choice of the appropriate level of protection against the seismic risk depends on political choice.

Nevertheless in order to aid this choice it is possible to perform a correct cost-benefit analysis if we adopte the concept of marginal cost of a saved life. This concept is useful when dealing with comparison between the prevention costs related to different risks and it is

$$\frac{\Delta D}{\Delta L} = - \frac{dD(c)}{dV(c)},$$

obviously its evaluation depends on the occurrence models together with (14).

REFERENCES

1. Cox, D.R.: Renewal Theory, Science Paperbacks, 1967.
2. Esteva, L.: Consideraciones praticas en la estimativa Bayesiana de riesgo sismico, UNAM, Instituto de Ingenieria, n. 248, 1970.
3. Shlien, S. and Toksöz, M.N.: Frquency-magnitude statistics of earthquake occurrence, Earthquake Notes (Eastern Section of the Seismological Society of America), 1970, 41, 5-18.
4. Cornell, C.A.: Probabilistic analysis of damage to structures under seismic loads, in Dynamic Waves in Civil Engineering, D.A. Howells, I.P. Haigh and C. Taylor, eds., John Wiley, London 1971.
5. Merz, H.A. and Cornell, C.A.: Seismic risk analysis based on a quadratic magnitude-frequency law, B.S.S.A., 1973, vol. 63, n.6.
6. Hasofer, A.M.: Design for infrequent overloads, Earthquake Engineering and Structural Dynamics, 1974.
7. Oliveira, C.S.: Seismic risk analysis for a site and a metropolitan area, Report EERC, 75-3, 1975.
8. Lommitz-Adler, J. and Lommitz, C.: A new magnitude-frequency relation, Tectonophysics, 49, 1978, 237-245.
9. Grandori, E. and Grandori, G.: An application of decision theory to seismic zoning, 6 WCEE, New Delhi, 1977.
10. Grandori G.: Seismic zoning as a problem of optimization, 2nd Int. Conf. on Structural Safety and Reliability, Münich, 1977.
11. Grandori, E., Grandori, G. and Petrini, V.: Engineering decisions and seismic risk prevention, IABSE Seminar, Bergamo 1978.
12. Grandori, E., Grandori, G. and Petrini, V.: A discussion of "non-linear" magnitude-frequency laws, 2nd U.S.National Conference on Earthquake engineering, Stanford, 1979.
13. Grandori Guagenti,E., Gianni Molina,C.: Valutazione probabilistica di danni con rischio variabile, Acc. Ist. Lomb. di Scienze e Lettere, 1979.
14. Newmark, N.M., Rosenblueth,E.: Fundamental of Earthquake engineering, Prentice Hall, 1971.

Numerical Techniques for Stochastic Systems
F. Archetti and M. Cugiani (eds.)
© North-Holland Publishing Company, 1980

GENERAL PURPOSE VERSUS SPECIALIZED ALGORITHMS FOR AN ILL-CONDITIONED ESTIMATION PROBLEM

Barbara Bacchelli and Marina Di Natale
Istituto di Matematica - Università di Milano

In this paper the performarces of a variable metric algorithm and a specialized algorithm for minimizing functions that are sums of squared terms are compared in solving an unconstrained nonlinear least squares problem arising in the earthquake hypocentre location. For this practical problem, which is in general ill-conditioned, the mathematical features are analyzed to point out the circumstances under which the superiority of the specialized methods might be not maintained. Several numerical experiments, performed to solve the question of the comparative efficiency, are reported.

1. Introduction

In this paper the question of the comparative efficiency between a Gauss Newton Based method and a Variable Metric method is considered in solving an ill-conditioned parameter estimation problem of the form

$$\min_{\underline{\theta} \in \mathbb{R}^n} \sum_{i=1}^{m} R_i(\underline{\theta})^2$$

where the residuals R_i are nonlinear functions. This least squares problem arises from the practical one of earthquake hypocentre location under suitable statistical assumptions on the distribution of the random errors in the data.

It has been suggested that problems of the considered form exist for which Variable Metric methods may be more efficient than Gauss Newton Based ones. For the hypocentre location problem the mathematical features are analyzed in order to point out those circumstances under which the performances of algorithms belonging to the above families are unpredictable. Two particular algorithms, OPLS and OPVM(°), were employed for numerical experiments based on Monte Carlo simulation. The numerical results show that there exist values of the standard deviation of the errors in the data for which the Variable Metric algorithm OPVM is more efficient than the Gauss Newton Based algorithm OPLS. A particular attention is paid to the behaviour of the algorithms for increasing ill-conditioning.

(°) Both belonging to the OPTIMA Package of the Numerical Optimisation Centre of Hatfield.

2. The earthquake hypocentre location problem as a parameter estimation problem.

The earthquake hypocentre location problem can be considered as a parameter estimation problem under the traditional assumption of a point-source to represent the limited portion of earth where seismic energy is released.
The seismic event is to be located in time and space using the data provided by the recording stations and assuming a physical model to characterize the path of the seismic waves.
If the model is correct, the following nonlinear regression scheme can be assumed for the observations:

$$\tilde{t}_i = f_i(\underline{\theta}) + \varepsilon_i \qquad (i = 1,\dots,m)$$

where
m = number of recording stations
$\tilde{t}_i$ = observed first arrival time at the i-th station
$\underline{\theta} = (\theta_1, \dots, \theta_n)$ vector of focal parameters
f_i = function which assignes the first arrival time for the i-th station according to the assumed physical model
ε_i = random error in the observation for the i-th station.

The errors ε_i are usually considered as independent normally distributed random variables with mean zero and known standard deviations σ_i. Then the vector of the focal parameters can be estimated via the Maximum Likelihood Principle (Mood et al., 1972)

$$\max_{\underline{\theta} \in \mathbb{R}^n} \prod_{i=1}^{m} \exp\left\{ -\frac{1}{2} \frac{(\tilde{t}_i - f_i(\underline{\theta}))^2}{\sigma_i^2} \right\}$$

which is equivalent to the classical Least Squares Criterion with weighted residuals

$$\min_{\underline{\theta} \in \mathbb{R}^n} \sum_{i=1}^{m} \frac{(\tilde{t}_i - f_i(\underline{\theta}))^2}{\sigma_i^2} .$$

In the following the standard deviations σ_i are assumed to be all equal to σ.

Four orthogonal axes are considered as coordinate axes: two axes lying on the earth surface, one directed downward and the time axis oriented in the direction of increasing time. The spatial coordinates are measured in kilometers and the time coordinate in

seconds.
Four local parameters are considered:

θ_1, θ_2 = coordinates of the epicentre

θ_3 = depth of the hypocentre

θ_4 = origin time of the seismic event.

As physical model, the homogeneous half-space model with constant velocity v is assumed in which the stations lie all at the same level, that is

$$f_i(\underline{\theta}) = \theta_4 + \frac{\sqrt{(x_i-\theta_1)^2+(y_i-\theta_2)^2+\theta_3^2}}{v} \tag{2.1}$$

where x_i, y_i are the coordinates of the i-th station. This model represents only a first approximation to real structures but has the advantage that the functions f_i can be explicitely formulated and analitic derivatives can be easily computed.
Hence, under the above assumptions, an unconstrained nonlinear least squares problem of the form

$$\min_{\underline{\theta} \in \mathbb{R}^4} \sum_{i=1}^{m} \left(\tilde{t}_i - \theta_4 - \frac{\sqrt{(x_i-\theta_1)^2+(y_i-\theta_2)^2+\theta_3^2}}{v} \right)^2$$

will be considered.
We intend to deal with a local problem which means that m is not greater than 10 and hypocentre and stations are in a region with a diameter of few kilometers.

3. The mathematical features.

In this Section we consider the question of the comparative efficiency of Gauss Newton Based methods, specialized for sum of squares problems, versus Variable Metric methods and we analyze the mathematical aspects in relation to the hypocentre location problem. Indeed Mc Keown (1975) showed that in general the performance is strictly related to the characteristics of the objective function and that there exist problems of sum of squares form for which the Variable Metric algorithms may be more efficient than the Gauss Newton Based algorithms.
According to a common notation, the function to be minimized is of

the form

$$F(\underline{\theta}) = \sum_{i=1}^{m} R_i^2(\underline{\theta})$$

The following quantities are to be recalled:

$$H = \nabla^2 F = 2\{J^T J + B\}$$

where $H_{j,k} \equiv \dfrac{\partial^2 F}{\partial\theta_j \, \partial\theta_k}$ $\qquad j, k = 1,\dots,n$

$$J_{j,k} \equiv \frac{\partial R_j}{\partial \theta_k} \qquad j = 1,\dots,m \qquad k = 1,\dots,n$$

$$B \equiv \sum_{i=1}^{m} H^i R_i \tag{3.1}$$

$$H^i_{j,k} \equiv \frac{\partial^2 R_i}{\partial\theta_j \partial\theta_k} \qquad j, k = 1,\dots,n \qquad i = 1,\dots,m$$

As well known, the Gauss Newton Based methods approximate the Hessian matrix by $2J^TJ$ neglecting B, while the Variable Metric methods use suitable updating formulas.

We observe that the rate of convergence of an algorithm from the Gauss Newton Based family may be poor for large values of the quantity

$$\rho = \frac{|\lambda_{max}|\ (B)}{\lambda_{min}(J^TJ)} , \tag{3.2}$$

where λ_{min} and $|\lambda_{max}|$ are the minimum and the maximum absolute eigenvalue respectively, a quantity ρ which can be considered as a measure of the greatest error in H induced by ignoring B (Mc Keown, 1975) .

On the other hand the efficiency of a Variable Metric algorithm in solving a least squares problem seems to be more sensitive to general nonquadraticity in the function rather than to a particular term such as B (Biggs, 1971). For problems characterized by a large value of ρ , it is reasonable to think that Variable Metric methods may be more efficient than specialized Gauss Newton Based methods; anyway a final answer can only be obtained by numerical experiments.

In the following the interest is focused on those situations in which the performance of the algorithms may not be predicted in advance: for the hypocentre location problem the quadraticity of the objective function and the circumstances under which the rate ρ is likely to be large are studied. A particular attention is paid to the singularity of J^TJ which causes the well known difficul-

ties of ill-conditioning for the Gauss Newton Based algorithms in addition to the effect it has on the rate of convergence.
Let us study the linearity of the residuals. The Hessian matrices of the residuals are of the form:

$$H^i = \frac{1}{v}\begin{bmatrix} \frac{(x_i-\theta_1)^2-d_i^2}{d_i^3} & \frac{(x_i-\theta_1)(y_i-\theta_2)}{d_i^3} & \frac{(x_i-\theta_1)(-\theta_3)}{d_i^3} & 0 \\ & \frac{(y_i-\theta_2)^2-d_i^2}{d_i^3} & \frac{(y_i-\theta_2)(-\theta_3)}{d_i^3} & 0 \\ & & \frac{\theta_3^2-d_i^2}{d_i^3} & 0 \\ & & & 0 \end{bmatrix}$$

where $d_i = \sqrt{(x_i-\theta_1)^2+(y_i-\theta_2)^2+\theta_3^2}$.
The euclidean norm of H^i can be easily obtained:

$$(3.3) \qquad \|H^i\|_E = \frac{\sqrt{2}}{v\, d_i} \qquad i = 1,\ldots,m.$$

Since for the local problem the distances are bounded (Sect. 2), for any value of the parameters in the real range the matrices H^i are far to be null, that is the objective function cannot be considered nearly quadratic. As already noticed, this characteristic is relevant for the performance of the Variable Metric algorithms. Furthermore, the matrix B cannot be considered negligible for the near-linearity of the residuals.
From definition (3.2), the quantity ρ is large if the maximum absolute eigenvalue of B is large compared with the minimum eigenvalue of J^TJ. Then we study the numerator $|\lambda_{max}|$ (B) of the ratio ρ and the singularity of J^TJ.
We recall (Ralston, 1965) that for an nxn simmetric matrix A

$$(3.4) \qquad \begin{aligned} \|A\|_S &= |\lambda_{max}| \ (A) \\ \|A\|_S &\le \|A\|_E \end{aligned}$$

where $\|\cdot\|_S$ and $\|\cdot\|_E$ are respectively the spectral and the euclidean norm. Applying (3.4) to the matrix B, from (3.1), (3.3) and from the properties of the norm, it is easy to obtain the bound

$$(3.5) \qquad |\lambda_{max}| \ (B) \le \frac{\sqrt{2}}{v} \sum_{i=1}^{m} \frac{|R_i|}{d_i} .$$

In the local problem the distances d_i are in general away from zero, hence $|\lambda_{max}|(B)$ might be large if some residual is large. Since the residuals are of the form $R_i(\underline{\theta})=f_i(\underline{\theta}^*)+\varepsilon_i-f_i(\theta)$, where θ^* is the true focal vector, recalling the probabilistic relation.

$$Pr\{-2\sigma \leq \varepsilon_i \leq 2\sigma\} \cong 0.95 ,$$

where σ is the standard deviation of the reading errors, we have that their magnitude at the minimum is bounded by 2σ in a large number of cases. Hence if the standard deviation produces large residuals the ratio ρ is likely to be large. Furthermore, the denominator $\lambda_{min}(J^TJ)$ has to be considered small as the hypocentre location problem is in general ill-conditioned independently on σ. Indeed, according to the model (2.1), the matrix J^TJ is of the form

$$J=\begin{bmatrix} \sum_{i=1}^{m} \frac{(x_i-\theta_1)^2}{v^2 d_i^2} & \sum_{i=1}^{m} \frac{(x_i-\theta_1)(y_i-\theta_2)}{v^2 d_i^2} & -\sum_{i=1}^{m} \frac{(x_i-\theta_1)\theta_3}{v^2 d_i^2} & -\sum_{i=1}^{m} \frac{(x_i-\theta_1)}{v d_i} \\ & \sum_{i=1}^{m} \frac{(y_i-\theta_2)^2}{v^2 d_i^2} & -\sum_{i=1}^{m} \frac{(y_i-\theta_2)\theta_3}{v^2 d_i^2} & -\sum_{i=1}^{m} \frac{(y_i-\theta_2)}{v d_i} \\ & & \sum_{i=1}^{m} \frac{\theta_3^2}{v^2 d_i^2} & \sum_{i=1}^{m} \frac{\theta_3}{v d_i} \\ & & & m \end{bmatrix}$$

It is clear that the matrix is nearly-singular if the ratios of the hypocentral depth to the hypocentre-station distance are close to zero. In fact:

$$|J^TJ_{K,3}| \leq \frac{1}{v} \sum_{i=1}^{m} \frac{|\theta_3|}{d_i} \qquad K = 1,\dots,4. \quad (°)$$

For a local problem the distances d_i are in a bounded range, hence the ratios are close to zero only when the hypocentral depth is close to zero. This fact commonly happens in real situations, causing the problem to be ill-conditioned.

Recalling that the residuals are affected by random errors, we finally want to consider the limit case of no perturbations in the

(°) In general $v \geq 1$. If $v<1$, $\frac{1}{v}$ is replaced by $\frac{1}{v^2}$.

data, which of course never occurs in the practice. In this case, at the minimum $\underline{\theta}^*$

$$R_i(\underline{\theta}^*) = 0 \qquad i = 1,\ldots,m$$

$$J^T(\underline{\theta}^*)J(\underline{\theta}^*) = H(\underline{\theta}^*) .$$

Hence the G.N.B. methods employ approximations of H which are very good, but nevertheless their performance is not easily predictable when J^TJ is singular at the minimum. In presence of ill-conditioning even the performance of a V.M. algorithm cannot be expected because the singularity regards H too.

4. Numerical experiments.

The efficiency of two algorithms, belonging to the considered families, OPLS, a safeguarded Gauss Newton Based algorithm and OPVM, a Variable Metric algorithm, both from the OPTIMA Package, was tested in solving the earthquake hypocentre location problem.
The following procedure was employed for the trials.
Assuming a set of focal parameters $\underline{\theta}^*$, the quantities

$$t_i = f_i(\underline{\theta}^*) \qquad i = 1,\ldots,m$$

were generated according to the model (2.1) with fixed constant velocity of 5 km/sec and stations placed as in the six stations network of Ancona (Fig. 1). The origin time θ^*_4 was always assumed to be 10 sec. without loss of generality (*); four simulation epicentres A,B,C,D were placed in different sites of the region in order to represent some really observed situations (Fig. 1).
Following a practice commonly used in the hypocentre location, the starting point of the algorithms was chosen as the smallest reading time for the origin time, the coordinates of the corresponding station for the epicentre, an arbitrarily fixed nonzero value for the depth.
The numerical experiments were performed according to the mathematical analysis of Section 3 and the results are displayed in two subsections regarding the limit case of no perturbations in the data and the case of data affected by random errors. The comparison of

(*) The value 10 allows to deal with quantities of the same order of magnitude.

the efficiency of the two algorithms is made in terms of number of iterations and function calls.

4.1 No perturbations in the data.

Assuming no perturbations in the data, the residuals are of the form

$$R_i(\underline{\theta}) = t_i - f_i(\underline{\theta}) \qquad i = 1,\dots,m$$

and the simulation parameter $\underline{\theta}^*$ is the minimum of the objective function. Hence it is easy to compare the accuracy of the final estimates and to produce situations of ill-conditioning.
A set of decreasing values for the hypocentral simulation depth was considered. The numerical results are shown in Table 1. We notice that both algorithms display a satisfactory performance in spite of ill-conditioning: the accuracy of the solution in terms of distance between the actual minimum and the final estimate is of order 10^{-2} for the null depth and 10^{-6} or 10^{-5} for the other depths (see column "error") and the final value of the objective function is in the range 10^{-15}- 10^{-12} (see column "function value").
As expected, in general increasing the ill-conditioning the number of iterations and function calls increases too for both algorithms; anyway the number required by OPVM is nearly three times the number required by OPLS. Hence the superior efficiency of the specialized algorithm OPLS versus the general one OPVM is clearly stressed by the numerical evidence for any value of the depth.
The case of null simulation depth is considered in more detail because the matrices J^TJ and H are singular at the minimum. In Table 2 the sixth column lists the number of iterations needed by OPVM to reach the best function value obtained by OPLS; the numerical superiority of the least squares algorithm is confirmed.

4.2 Data affected by random errors.

According to the assumptions of Sect.2., the residuals are of the form

$$R_i(\underline{\theta}) = t_i + \varepsilon_i - f_i(\underline{\theta}) \qquad i = 1,\dots,m.$$

For the trials, the random errors ε_i were generated according to

the normal distribution $N(0,\sigma)$.
Recalling the observations related to the inequality(3.5) we remark that the quantity ρ may be large in connection with large values of the standard deviation σ. Several values of σ were considered; for any value of θ_1^*, θ_2^* two values of simulation depth were fixed, a positive one which well pertains the physical situation and the null one which causes the singularity of J^TJ. Having fixed the simulation parameter $\underline{\theta}^*$, 1000 sets of perturbations $(\varepsilon_1, \ldots, \varepsilon_m)$ were generated for any value of σ.
The numerical results are shown in Tables 3 and 4. From Table 3 we notice that both algorithms tend to converge at points unacceptably far from the solution for the largest values of σ (see column "DIV"). For these cases of "divergence" the final estimate $\hat{\underline{\theta}}$ given by OPLS is nearer to the simulation point $\underline{\theta}^*$ than the one given by OPVM (see column "$d(\hat{\underline{\theta}}, \underline{\theta}^*)$ of Table 3 reporting the euclidean distance averaged on 1000 trials). For the smallest values of σ the two algorithms reach the same points. As expected, as σ decreases the distance $d(\hat{\underline{\theta}}, \underline{\theta}^*)$ decreases too.
Table 4 displays the performance of the algorithms in terms of average number of iterations and function calls on 1000 trials. In particular the ratio RI of the average number of iterations of the algorithms

$$RI = \frac{\overline{n^\circ} \text{ Iterations (OPLS)}}{\overline{n^\circ} \text{ Iterations (OPVM)}}$$

is plotted versus σ in Fig. 2.
Taking the results as a whole, it turns out that there exist critical values of σ above which the general Variable Metric algorithm OPVM is more efficient than the specialized one OPLS. According to the mathematical analysis of Sect. 3 the superiority of OPLS is lost when the standard deviation produces residuals which are too large to consider B negligible with respect to J^TJ.
Obviously ill-conditioning allows the critical values to be lower.
It has to be pointed out that in the hypocentre location problem the reading errors which are observed in the practice are usually not greater than about 0.4 sec., that is the values of σ above 0.2 less pertain the real problem. Hence, in real situations, the specialized algorithm is more efficient than the general one.
A particular attention has to be paid to the performance of the algorithms when the simulation point has null depth. As σ tends to

zero, according to the fact that the matrix B becomes negligible and that the final estimate $\tilde{\underline{\theta}}$ approaches the simulation point $\underline{\theta}^*$ (see columns $d(\hat{\underline{\theta}},\underline{\theta}^*)$ of Table 3), the ill-conditioning of the problem increases for both algorithms. From Table 4 it turns out that the number of iterations and the number of function calls increase for OPVM while they unexpectedly decrease for OPLS. Hence the increasing ill-conditioning of the problem seems to influence more the performance of the Variable Metric algorithm than the performance of the specialized one. This probably happens because of the contemporaneous improvement of the approximations employed by the Gauss Newton Based method OPLS.

Acknowledgments:
We thank J.J. Mc Keown for fruitful suggestions and encouragement.

References

[1] Bard, J., Nonlinear parameter estimation (Academic Press, 1977).

[2] Biggs, M.C., JIMA, Vol. 8, N. 3, (1971) 315-328.

[3] Dennis, I.E. and Morè, J.J., Quasi-Newton methods, motivation and theory, SIAM Review, Vol. 19, n. 1 (1977) 46-89.

[4] Mc Keown, J.J., On algorithms for sums of squares problems, in: Dixon, L.C.W. and Szegö, G.P. (eds.), Towards global optimisation (North Holland, Amsterdam, 1975).

[5] Manual "OPTIMA" Package (1976) of the Numerical Optimisation Centre of the Hatfield Polytechnic, Hatfield, Hertfordshire.

[6] Ralston, A., A first course in numerical analysis (Mc Graw - Hill, 1975).

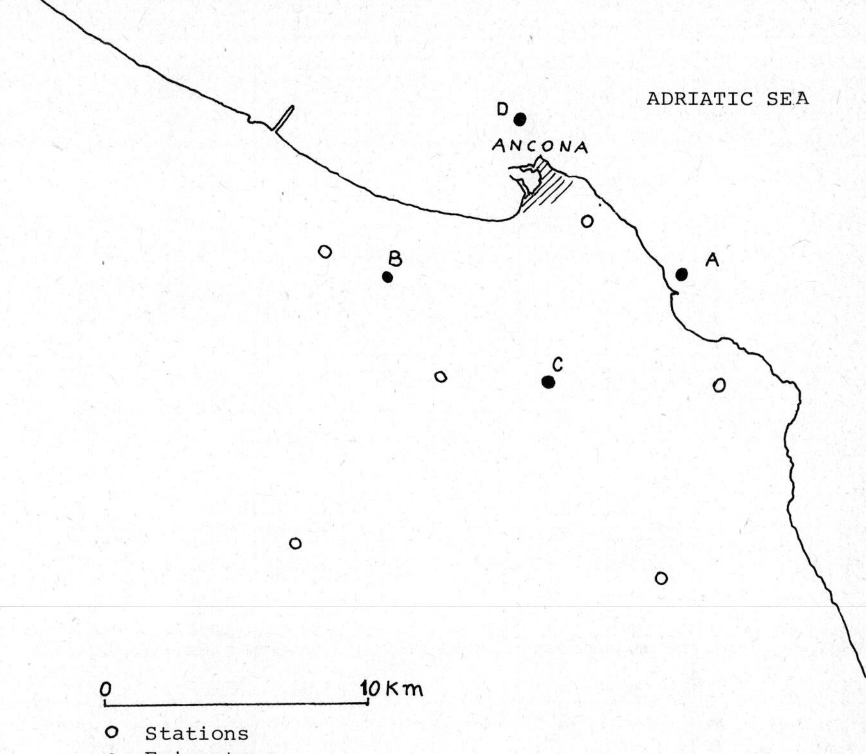

Figure 1. - Six stations network of Ancona

Table 1.

Epicentre	Depth θ_3^*	Number of iterations		Number of function calls		Error		Function value	
		OPLS	OPVM	OPLS	OPVM	OPLS	OPVM	OPLS	OPVM
A	0.	14	43	20	65	.2-02	.9-02	.6-14	.4-12
	.25	10	36	19	61	.1-05	.2-05	.5-13	.4-14
	.50	9	26	13	40	.3-05	.7-06	.3-14	.2-14
	.75	7	26	12	38	.1-05	.7-06	.9-14	.6-14
	1.	10	25	10	44	.2-05	.9-06	.9-15	.4-14
B	0.	13	44	21	61	.1-02	.2-02	.4-13	.1-13
	.25	9	26	15	36	.2-05	.5-05	.2-13	.5-14
	.50	7	24	13	42	.2-06	.1-05	.2-13	.9-14
	.75	7	22	13	36	.7-06	.1-05	.3-13	.7-14
	1.	8	20	12	29	.1-05	.4-06	.5-14	.8-14
C	0.	13	33	21	47	.2-02	.8-02	.7-13	.8-12
	.25	8	27	14	49	.4-05	.5-05	.3-13	.3-13
	.50	9	24	16	37	.7-06	.3-05	.3-13	.4-14
	.75	8	22	14	42	.1-05	.2-05	.2-13	.2-13
	1.	7	22	15	30	.1-05	.2-05	.6-13	.1-14
D	0.	13	36	21	65	.3-02	.4-02	.6-13	.6-13
	.25	10	25	13	43	.2-05	.1-05	.5-14	.8-14
	.50	10	23	10	35	.3-05	.3-05	.1-14	.1-14
	.75	8	21	11	39	.2-05	.2-05	.1-14	.7-14
	1.	8	22	13	45	.1-05	.2-05	.5-14	.1-13

Stopping criterion: $|\nabla F| < 0.000\,0001$, or predicted function reduction $< 0.000\,0001$

Table 2.

Epicentre	OPLS		OPVM		Iterations to attain OPLS function value
	function value	n° of iterations	function value	n° of iterations	
A	.6 -14	14	.4 -12	43	43
B	.4 -13	13	.1 -13	44	42
C	.7 -13	13	.8 -12	33	33
D	.67-13	13	.64-13	36	34

Table 3.

Epicentre	σ	$\theta_3^* = 0$				$\theta_3^* > 0$			
		DIV		$d(\underline{\hat{\theta}},\underline{\theta}^*)$		DIV		$d(\underline{\hat{\theta}},\underline{\theta}^*)$	
		OPLS	OPVM	OPLS	OPVM	OPLS	OPVM	OPLS	OPVM
A	1.	294	302	639.04	**	310	315	779.95	**
	.50	180	181	332.25	3761.25	221	221	517.27	5342.69
	.40	123	112	181.68	2358.42	164	163	348.41	3565.82
	.30	50	50	80.92	816.20	90	90	139.10	1622.29
	.20	11	12	13.33	318.23	23	23	26.17	170.72
	.10	0	0	2.36	2.36	0	0	3.10	3.07
	.05	0	0	1.39	1.39	0	0	1.46	1.46
	.01	0	0	0.53	0.53	0	0	0.29	0.29
B	1.	242	240	617.79	9388.76	308	310	904.96	**
	.50	96	95	165.80	1802.67	188	187	446.13	4434.40
	.40	35	36	66.61	866.09	125	125	272.22	2471.13
	.30	10	10	14.60	122.99	66	66	77.27	681.12
	.20	1	1	3.67	3.67	5	5	8.91	43.07
	.10	0	0	1.33	1.33	0	0	1.95	1.91
	.05	0	0	0.87	0.87	0	0	0.93	0.93
	.01	0	0	0.38	0.38	0	0	0.19	0.19
C	1.	206	206	613.65	6623.49	244	244	768.86	8507.62
	.50	53	53	104.44	699.75	86	86	186.44	1795.74
	.40	19	19	30.25	246.36	47	47	75.89	513.88
	.30	1	1	3.81	3.81	12	12	9.91	46.28
	.20	0	0	2.46	2.46	0	0	3.27	3.24
	.10	0	0	1.41	1.41	0	0	1.62	1.62
	.05	0	0	0.97	0.97	0	0	0.76	0.76
	.01	0	0	0.39	0.39	0	0	0.15	0.15
D	1.	321	322	664.70	**	365	365	931.07	**
	.50	189	191	323.69	4293.46	278	278	635.12	8016.36
	.40	136	138	207.85	2642.07	227	230	478.43	6447.60
	.30	66	69	87.07	1348.76	154	159	287.34	3618.85
	.20	13	16	13.43	271.59	64	65	95.17	1305.51
	.10	0	0	2.51	2.51	1	1	4.42	4.36
	.05	0	0	1.38	1.38	0	0	1.81	1.81
	.01	0	0	0.45	0.45	0	0	0.37	0.37

$\theta_3^* > 0$ denotes the following depths for the epicentres: $\theta_3^*(A)=5$ Km. $\theta_3^*(B)= 7$ Km., $\theta_3^*(C)=4$ Km., $\theta_3^*(D)=9$ Km.

DIV denotes the number of cases for which the final estimate $\underline{\hat{\theta}}$ satisfies $\sqrt{(\hat{\theta}_1 - \theta_1^*)^2 + (\hat{\theta}_2 - \theta_2^*)^2 + (\hat{\theta}_3 - \theta_3^*)^2} > 50$ Km.

$d(\underline{\hat{\theta}},\underline{\theta}^*)$ denotes the average on 1000 trials of the euclidean distance between the final estimates $\underline{\hat{\theta}}$ and the simulation point $\underline{\theta}^*$

** denotes the cases in which $d(\underline{\hat{\theta}},\underline{\theta}^*) > 10^5$

Table 4.

Epicentre	σ	$\theta_3^* = 0$				$\theta_3^* > 0$			
		$\bar{n}^\circ$ Iterations		$\bar{n}^\circ$ Function calls		$\bar{n}^\circ$ Iterations		$\bar{n}^\circ$ Function calls	
		OPLS	OPVM	OPLS	OPVM	OPLS	OPVM	OPLS	OPVM
A	1.	43.46	22.08	80.77	41.23	41.28	22.52	75.35	41.78
	.5	27.20	22.38	49.69	40.17	24.02	23.48	42.45	42.33
	.4	22.13	22.34	40.13	39.90	19.10	23.15	33.38	41.35
	.3	17.95	22.04	32.53	39.13	15.19	22.90	26.26	40.56
	.2	15.12	21.51	27.64	38.01	11.78	21.80	20.35	38.32
	.1	12.27	21.02	22.13	38.01	8.32	20.06	14.34	35.10
	.05	10.76	21.87	18.92	38.54	6.99	18.70	12.18	32.88
	.01	10.19	26.45	17.59	45.92	6.37	18.68	11.29	33.15
B	1.	37.63	21.32	69.85	39.90	35.36	22.34	63.47	41.93
	.5	20.37	19.39	37.31	34.81	15.81	20.96	27.15	38.11
	.4	16.65	18.64	30.29	33.26	12.10	20.16	20.36	36.56
	.3	13.84	17.92	25.24	31.70	9.10	19.25	15.00	34.52
	.2	12.23	17.95	22.42	31.70	7.19	18.50	12.10	33.15
	.1	10.39	18.98	18.94	33.06	6.44	17.38	11.16	31.74
	.05	9.43	20.66	16.83	35.53	6.24	17.57	11.08	32.28
	.01	9.09	24.43	15.56	40.94	6.24	17.36	10.75	32.46
C	1.	29.17	20.41	52.76	36.85	30.18	21.30	54.25	38.82
	.5	15.34	18.50	27.47	32.37	14.32	19.70	25.01	34.76
	.4	13.65	17.98	24.75	31.34	12.06	19.03	21.20	33.46
	.3	12.19	17.58	22.26	30.71	10.26	18.15	17.99	32.00
	.2	11.03	17.52	20.14	30.76	8.41	18.33	14.51	32.21
	.1	9.98	18.70	17.94	32.84	6.88	17.68	11.76	31.40
	.05	9.44	19.97	16.22	34.77	6.43	17.11	11.10	30.84
	.01	9.25	23.42	15.96	39.91	6.41	17.59	11.44	31.85
D	1.	44.50	22.92	81.60	43.00	40.43	23.22	72.19	43.26
	.5	27.20	22.27	49.50	40.44	22.01	23.66	37.74	43.23
	.4	23.00	22.08	41.82	39.78	17.53	24.23	29.59	44.16
	.3	18.08	21.59	32.92	38.74	12.64	23.52	20.91	42.62
	.2	13.90	21.24	25.30	37.66	8.75	21.67	14.36	38.44
	.1	11.12	21.06	19.63	37.04	6.88	19.30	11.98	34.36
	.05	10.92	20.98	19.04	37.23	6.52	19.11	11.52	33.90
	.01	10.78	22.89	18.33	41.04	6.23	19.33	10.83	35.24

$\theta_3^* > 0$ denotes the following depths for the epicentres: $\theta_3^*(A)=5$ Km. $\theta_3^*(B)=7$ Km., $\theta_3^*(C)=4$ Km., $\theta_3^*(D)=9$ Km.

$\bar{n}^\circ$ denotes the average number on 1000 trials.

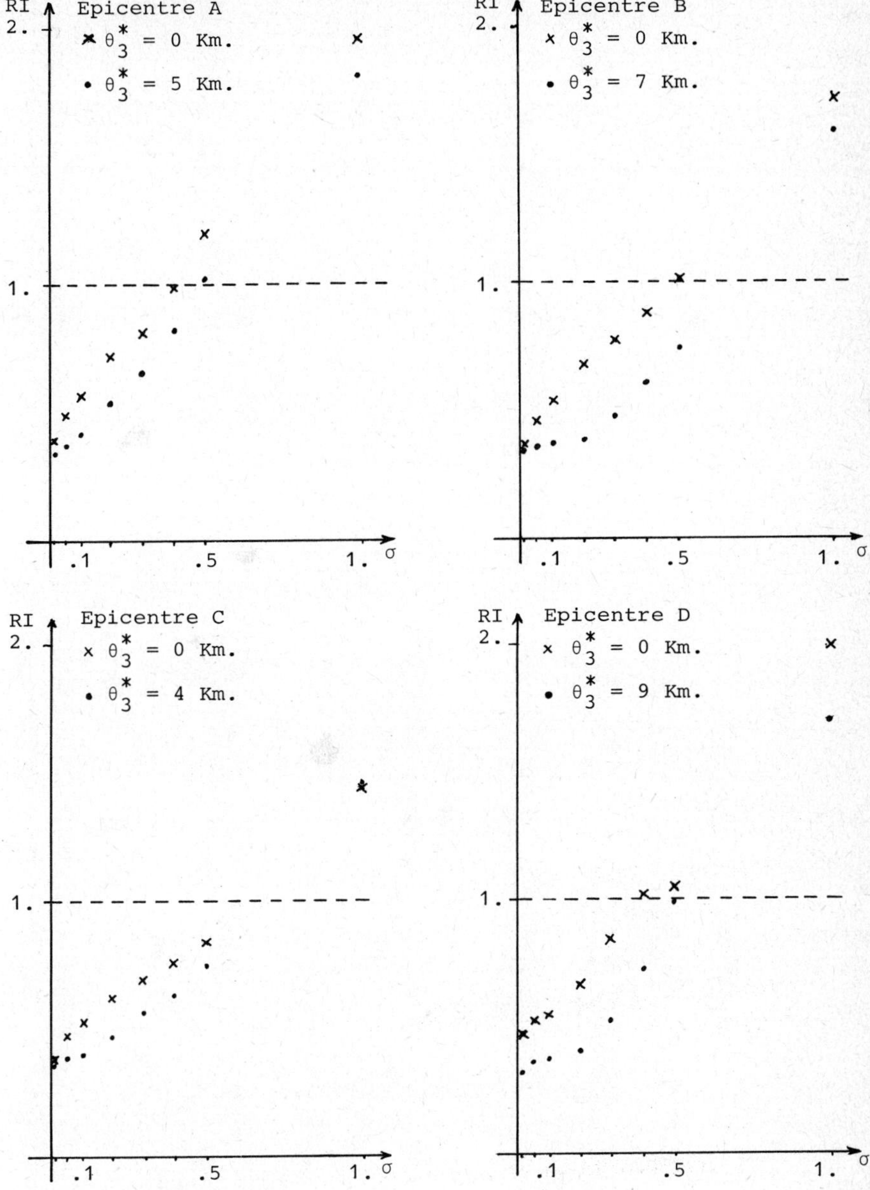

Figure 2. Plots of RI versus σ.

Numerical Techniques for Stochastic Systems
F. Archetti and M. Cugiani (eds.)
© North-Holland Publishing Company, 1980

PARAMETER ESTIMATION AND LEAST SQUARES

E. Spedicato

Istituto di Matematica, Università di Bergamo, Italy

Recent developments in the numerical methods for least squares are reviewed with special reference to important practical problems as the fitting of gaussian and exponential models.

Introduction

The following problem is often encountered in data analysis:

- m values y_i, i=1,2....m, are obtained by measures, recordings..., each one corresponding to a value of an independent or control variable t_i, $y_i=y_i(t_i)$
- it is assumed that a function (model) of the form y=y(t,x) exists, with x a vector of n components, such that

$$y_i = y(t_i,x^+) + e_i \tag{1}$$

  for some vector x^+ and with e_i the error in the data
- it is desired to estimate the unknown vector x^+ using the available information, namely t_i,y_i and the known form of y=y(t,x)

The method of least squares is often used to solve this problem, or as a technique per se, from the times of Gauss, or as a consequence of the maximum likelihood principle in statistical parameter estimation theory. With regard to the latter case we recall that weighted least squares are obtained under the following assumptions:

- no error in the control variables
- the error $e = (e_1,e_2,....e_m)^T$ in the data $y = (y_1,y_2,...y_m)^T$ belongs to a multivariate gaussian distribution f=f(e) with average $E(e_i)=0$ and known variance-covariance matrix W, i.e.

$$f(e) = (2\pi)^{-n/2}/(\det W)^{1/2} \exp(-1/2e^TWe) \tag{2}$$

The weighted least squares method requires minimizing (with some conceptual difficulties for multimodal functions) the object function

$$F = F(x) = \sum_{i=1}^{m} \left\{ (y(t_i,x) - y_i) \sum_{j=1}^{m} W_{ij}(y(t_j,x)-y_j) \right\} \quad (3)$$

or, introducing the vector of the residuals $r=r(x)$, with components $r_i = y(t_i,x)-y_i$:

$$F = F(x) = r^T W r \quad (4)$$

The matrix W usually is not known in advance and often cannot even be estimated bacause of lack of replicated data. In such a case a rather strong but commonly made assumption is the homoscedasticity hypothesis, which states that W is diagonal with identical diagonal elements. This simplifies the object function the the unweighted least squares

$$F = F(x) = r^T r \quad (5)$$

which we shall consider in the following.

Linear least squares

In this important case the model has the form

$$y(t_i,x) = \sum_{j=1}^{n} a_j(t_i)x_j \quad (6)$$

or, in matrix form

$$y(t,x) = A(t)x \quad (7)$$

and we are led to solve, in the least squares sense, the overdetermined linear system $y = A(t)x$, whose solution has the form

$$x^+ = A^+ y \quad (8)$$

where A^+ is the generalized inverse of A. If A is full rank (8) takes the form

$$x^+ = (A^T A)^{-1} A^T y \quad (9)$$

or, equivalently, x^+ is the solution of the classical normal equations

$A^TAx = A^Ty$, which are usually solved by Cholesky decomposition. The normal equation approach however presents two numerical difficulties, namely:

- forming the matrix $B=A^TA$ requires mn^2 multiplications and we note that often $m \gg n$
- as the condition number of B is the square of the condition number of A, which can be large in presence of a poor model or of large errors in the data, the roundoff errors may be very large and make the computed solution senseless

An alternative numerically better approach has been introduced by Golub[1] and has become recently the normal tool in sophisticated libraries for statistical calculations. It goes through the QR decomposition of an m by n matrix A of rank p, which states that orthogonal matrices S and U exist such that

$$R = S^TAU \tag{10}$$

where R is a matrix of the form

$$R = \begin{bmatrix} R_1 & 0 \\ 0 & 0 \end{bmatrix} \tag{11}$$

the sumatrix R_1 being triangular p by p and nonsingular. To construct A and U simple procedures exist using Householder or Givens matrices; the process does not require additional memory, as R can overlap A and Householder matrices are implicitly represented by a vector. To calculate x^+ the procedure is the following:

- evaluate

$$g = S^Ty = (g_1, g_2)^T \tag{12}$$

g_1 being the subvector of the first p components; note that, as S is a product of Householder matrices, it is sufficient to save the intermediate products H_iy

- solve the triangular system $R_1v_1 = g_1$
- calculate $x^+ = Uv$, $v=(v_1,0)$

The features which make the QR method numerically convenient are:

- only $n^2m-n^3/3$ multiplications are required
- perturbations to the solutions are of the form

$$\| dx^+\| = q_1 \| r^+\| \ \mathrm{Cond}(A) + q_2\| r^+\|^2 \ \mathrm{Cond}(A)^2 \tag{13}$$

so that for small residuals (namely good model and small errors in the data) the

computed solution depends only linearly from the condition number and is therefore numerically more stable; for large residuals the QR method may show no advantage over the normal equations.
An even more sophisticated procedure uses the singular value decomposition, which also offers a deeper statistical insight; for details, see Golub[2].

Nonlinear least squares

If the model function is nonlinear, we have a nonlinear minimization problem which can be solved, in principle, by general minimization methods, or, hopefully more efficiently, by methods which exploit the structure of the problem. To put in evidence this structure let us write the gradient of F

$$g = \nabla F = 2J^T r \tag{14}$$

where J = J(x) is the n by m jacobian of r, and the hessian of F

$$G = \nabla^2 F = 2\,(JJ^T + \sum_{i=1}^{m} r_i R_i) \tag{15}$$

where $R_i = R_i(x)$ is the hessian of r_i. Relation (15) shows that the hessian of F is the sum of a term containing only first order information, and of a term containing second order information and which is small if the residuals are small. From this observation one is led to the classical Gauss-Newton iteration

$$x_{k+1} = x_k - a_k (J_k^T J_k)^{-1} J_k^T r_k \tag{16}$$

where a_k, the stepsize, is determined by a line search to approximately minimize F in the direction $s = (J^T J)^{-1} J^T r$; note that s is the solution of the normal equations $J^T J s = J^T r$, which should be solved by QR decomposition. The Gauss-Newton iteration has quadratic rate of convergence only if the residuals are zero at the solution; as this condition is not satisfied usually, slow convergence often results. To improve over the Gauss-Newton method approximations are sought of the neglected second order information. In the Levemberg-Marquardt methods the search vector is defined by

$$s = (J^T J + \mu B)\; J^T r \tag{17}$$

with B a positive definite matrix, often the unit matrix, and μ is a scalar which guarantees s to be a descent direction even if J is singular. Many suggestions have been made on how to select μ and B, giving methods which are robust, but computationally expensive, because system (17) may be solved for possibly different values of μ. Noting that when B=I and μ tends to infinity the search vector tends in direction to the gradient vector, a simple idea is to take s in a set of p easily formed vectors s_j such that s_o is the Gauss-Newton vector and s_p is the gradient vector. Thus we have the dog-leg algorithm of Powell[3], where linear combinations of s_o and s_p are used, and the spiral algorithm of Jones[4], where s_j does not necessarily lay in the space spanned by s_o and s_p. These algorithms are very robust, global convergence to a stationary point following under very mild assumptions, but are less efficient than the methods described below.

Large residuals least squares

When the residuals are "large" (a situation common with poor data or poor models) we expect difficulties with the Gauss-Newton method, like slow convergence and stepsize not tending to one; more precisely, analysis of Ruhe[5], shows that the search directions of the Gauss-Newton method become asymptotically identical with those generated by the steepest descent method on a suitably scaled problem, that they collapse to a two-dimensional subspace and will tend to be pairwise orthogonal. The qualification "large" cannot be meant in absolute value, as a linear scaling of the model changes the residuals but not the Gauss-Newton iterates. Scale invariant measures of "large residuals" have been introduced approximately by Meyer[6] and exactly by McKeown[7] and, in the framework of a clarifying geometrical analysis, by Wedin[8]. These measures relate to the rate of convergence η of the undamped Gauss-Newton method (R-convergence factor), which is given by relation

$$\eta = Q\left\{(J^TJ)^{-1} \sum_{i=1}^{m} r_i R_i\right\} = r^T r/w \qquad (18)$$

where Q(.) indicates the spectral radius, w the smallest curvature radius of the hypersurface r(x) and the quantities are evaluated at x^+. In computational experiments with problems of known η value McKeown[7] has shown that Gauss-Newton and Levemberg-Marquardt methods are more efficient than general purpose (say Quasi-Newton)methods when η is small (even if the rate of convergence is only linear), in agreement with previous results of Bard[9], while for large values of η they are less efficient.

To get benefit from the special structure of least squares in the large residuals situation we need a better approximation of the second order term. As we want to avoid calculation of second derivatives, an obvious idea is to use Quasi-Newton approximations. The first method of this type was proposed by Brown and Dennis[10] in 1971, who suggested approximating each one of the m matrices R_i by Broyden's single rank unsymmetric formula or by its Powell symmetrized (PBS) version. They gave convergence results and limited good computational experience, but the method is unattractive because of the high memory and computational cost. Subsequent research has dealt with approximating the whole term $D = \sum_{i=1}^{m} r_i R_i$. Apart from which Quasi-Newton formula to choose, a question is in the definition of the Quasi-Newton equation. Indeed, as D is the hessian of the function $F = F - x^T J^T Jx$ (with J constant), the natural extension of the usual equation $B_{k+1}(x_{k+1}-x_k) = y_k \equiv \nabla F_{k+1} - \nabla F_k$, would require defining

$$y_k = J_{k+1}^T r_{k+1} - J_k^T r_k - J_{k+1}^T J_{k+1} x_{k+1} + J_k^T J_k x_k \qquad (19)$$

relation simplified by Broyden and Dennis[11] to

$$y_k = J_{k+1}^T r_{k+1} - J_k^T r_k - J_{k+1}^T J_{k+1}(x_{k+1}-x_k) \qquad (20)$$

and by Betts[12] to

$$y_k = J_{k+1}^T r_{k+1} - J_k^T r_k - J_k^T J_k (x_{k+1}-x_k) \qquad (21)$$

An even simpler definition was finally independently considered by Biggs[13] and Dennis,Gay and Welsch[14], namely

$$y_k = J_{k+1}^T r_{k+1} - J_k^T r_{k+1} \qquad (22)$$

and turned out in numerical experiments to be the slightest but clear winner. Considering the choice of the Quasi-Newton formula we first observe that symmetry is a natural requirement while positive definiteness is not. This excludes formulas like the DFP,BFS and self-dual and makes natural candidates the single-rank (which has quadratic termination without exact line searches) or the PBS formula (which corresponds to a minimal correction in Frobenius norm). We would like to have $J^T J + B$ positive definite,what is not guaranteed by those formulas and is still an open problem. Finally, it is important to use self-scaling formulas in the sense of Oren[15], because even if the R_i are constant, full rank changes to D can be made simply by the changes in the residuals. Biggs has used the scaling factor $\gamma = r_{k+1}^T r_k / r_k^T r_k$, while Dennis et al. have used the formula $\gamma = \min \left\{ \left| (x_{k+1}-x_k)^T y_k / (x_{k+1}-x_k)^T B_k (x_{k+1}-x_k) \right| , 1 \right\}$, namely a safeguarded Oren-Spedicato[16] optimally conditioned formula.

The computational experience with the Biggs and Dennis et al. algorithm is very

good, the breakthrough being due to the scaling technique; even on large residuals problems the method is definitely superior to general purpose Quasi-Newton methods like the BFS. Convergence theorems are given by Dennis and Walker[17], who prove a superlinear rate of convergence under mild assumptions.
Quasi-Newton techniques are not the only way of improving the Gauss-Newton method for large residuals problems. Ruhe[5] has proposed a simple conjugate gradient type acceleration procedure which implies the following modification of the Gauss-Newton search vector

$$d_k = J_k^+ r_k$$
$$s_k = d_k - \beta_k s_{k-1} \qquad (23)$$

where $\beta_k=0$ for $k=0,n,\ldots$.(or at restart iterations), otherwise $\beta_k = \delta_k / \delta_{k-1}$ where

$$\delta_k = \| J_k d_k \|_2^2 \qquad (24)$$

Ruhe proves that this procedure has n-step quadratic rate of convergence with asymptotically exact line searches and shows numerically that it is substantially better than Gauss-Newton on large residual problems where it is competitive whith general purpose Quasi-Newton methods. In view of the less memory and computational cost of Ruhe's procedure it will be of great interest to compare it with the Biggs and Dennis et al. modifications.

Separable least squares

Separable least squares are problems where some of the variables appear linearly, some nonlinearly, so that model (7) becomes

$$y(t,x,z) = A(t,z)x \qquad (25)$$

z being the p-dimensional vector of nonlinear variables. Models of this type are very common, for instance sums of exponentials or gaussians.
In dealing with separable models the idea is to solve first for the linear variables, keeping fixed the nonlinear ones. This gives the reduced object function

$$F = (y-A(z)A^+(z)y)^T(y-A(z)A^+(z)y) \qquad (26)$$

which is minimized with respect to z,a problem of dimension less than the original one; finally, when the optimal values z^+ are obtained, the linear parameters are given by $x^+=A^+(z^+)y$. The equivalence of this procedure with the one in the original n+p dimensional space is proved by Golub and Pereyra[18] under the assumption of constant rank of A in an open set containing the solution (this relating to the discontinuous dependence of the generalized inverse from the rank). The implementation of Gauss-Newton type methods on the reduced object function requires heavy use of the derivative DA^+ of the generalized inverse, given by the following formula

$$DA^+ = -A^+DAA^+ + A^+A^{+T}DA^T(I-AA^+) + A^+ADA^TA^{+T}A^+ \tag{27}$$

Moreover, as the generalized inverse is obtained by the QR decomposition, namely (see (10),(11))

$$A^+ = U\begin{bmatrix} R_1^{-1} & 0 \\ 0 & 0 \end{bmatrix} S^T \tag{28}$$

the most natural formulation of separable least squares is done using QR techniques; some simplifications can be done in (27) resulting in various algorithms (see Krogh[19], Kaufman[20], Ruhe and Wedin[21]) for which Wedin[22] has proved the rate of convergence to be better than for the full dimensional formulation. This is confirmed by numerical experiments, see for instance Corradi and Stefanini[23], which also indicate a superior rubustness with respect to the starting guesses for the nonlinear parameters.

A very sophisticated code, named VARPRO, for separable least squares has been developed at Stanford University and has been succesfully used in many applications, even of on line type; a modification of VARPRO, named VARPM, has been developed at CISE,Milano, to deal automatically with pure or mixed exponential/gaussian models and to provide quantities of statistical interest, like variance-covariance matrices of the estimated parameters.

Conclusion

We have reviewed the state of art in unconstrained least squares, showing that highly satisfactory algorithms are now available. We have not dealt with the constrained problem where the state of art is also very good for linear constraints, see for instance Lawson and Hanson[24], but where many problems are of course open for general constraints. Finally we note that least squares play a fundamental role in other parameter estimation techniques, for instance in Ruhe[25]'s method to locate the global minimizer of an exponential model, where the bulk of the calculation is in solving a positively constrained linear least squares problem.

References

1 - Golub G.,Numerical methods for solving linear least squares problems, Numerische Mathematik 7,1965

2 - Golub G., Matrix computations and least squares, in Nonlinear Optimization 1980, L.Dixon,E.Spedicato,G.Szego edts., Springer-Verlag,1980

3 - Powell M.J.D., A hybrid method for nonlinear equations, in Numerical Methods for Nonlinear Equations,P. Rabinowitz ed., Gordon-Breach, 1970

4 - Jones A., Spiral, a new algorithm for nonlinear parameter estimation using least squares, Computer Journal 13,1970

5 - Ruhe A., Accelerated Gauss-Newton algorithm for nonlinear least squares, report UMINF 67/78, University of Umea,1978

6 - Meyer R.R., Theoretical and computational aspects of nonlinear regression, in Nonlinear Programming,A. Rosen, O.Mangasarian, K.Ritter edts., Academic Press, 1970

7 - McKeown J.J., On algorithms for sums of squares problems, in Towards Global Optimization, L.Dixon, G.Szego edts., North Holland, 1975

8 - Wedin P.A., On the Gauss-Newton method for nonlinear least squares, report ITM 24, Lunds Datacentral

9 - Bard Y., Comparison of gradient methods for the solution of nonlinear parameter estimation problems, SIAM J. Numerical Analysis 7,1970

10 - Brown K.M. and Dennis J.E., New computational algorithms for minimizing a sum of squares of nonlinear functions, report CS71/6, Yale University

11 - Dennis J.E., Some computational techniques for the nonlinear least squares problem, in Numerical Solution of Systems of Nonlinear Equations, D.G.Byrne, C.A.Hall eds., Academic Press, 1973

12 - Betts J.T., Solving the nonlinear least squares problem: application of a general method, JOTA 18,1976

13 - Biggs B., The estimation of the hessian matrix in nonlinear least squares problems with nonzero residuals, Mathematical Programming 12,1977

14 - Dennis J.E., Gay D.M., Welsch R.E., An adaptive nonlinear least squares algorithm, report TR 1, Sloan School of Management, 1979

15 - Oren S.S., Self-scaling variable metric algorithms for unconstrained function minimization, PhD Dissertation, Stanford University, 1973

16 - Oren S.S., Spedicato E., Optimal conditioning of self-scaling variable metric algorithms, Mathematical Programming,10,1976

17 - Dennis J.E. and Walker H.F., Convergence theory for least-change secant update methods, report TR 476, Department Mathematical Sciences,Rice University

1979

18 - Golub G.H. and Pereyra V., The differentiation of pseudoinverses and nonlinear least squares whose variables separate, SIAM J. Numerical Analysis 10,1973

19 - Krogh F.T., Efficient implementation of a variable projection algorithm for nonlinear least squares, Communications ACM 17,1974

20 - Kaufman L., A variable projection method for separable least squares problems, BIT 15,1975

21 - Ruhe A., and Wedin P.A., Algorithms for separable least squares, report 434, Computer Science Department, Stanford University

22 - Wedin P.A., On surface dependent properties of methods for separable nonlinear least squares, report ITM 23, Lunds Datacentral

23 - Corradi A. and Stefanini L., Computational experience with algorithms for separable nonlinear least squares, Calcolo 15,1978

24 - Lawson C.L. and Hanson R.J., Solving least squares problems,Prentice-Hall, 1974

25 - Ruhe A., Least squares fitting by positive sums of exponentials, report UMINF 70, University of Umea,1978

Numerical Techniques for Stochastic Systems
F. Archetti and M. Cugiani (eds.)
© North-Holland Publishing Company, 1980

EFFICIENT COMPUTATION OF MAXIMUM LIKELIHOOD ESTIMATES IN ARMA REGRESSION MODELS

Corrado Corradi
Istituto di Scienze Economiche and Istituto di Matematica Applicata, Università di Bologna

Luciano Stefanini
Sogesta s.p.a., Urbino

An efficient computational procedure is described for the maximum likelihood estimation of ARMA regression models based on the linear-nonlinear decomposition of the set of coefficients. The relative merits as compared to the conventional methods are examined, and numerical results are reported showing the effectiveness of the proposed approach.

1. Introduction.

The purpose of this note is to describe an efficient numerical method for computing maximum likelihood estimates in ARMA regression models. We indicate how the problem of interest can be reduced to a separable non linear least squares problem, that is, a least squares problem in which the parameters to be solved for can be partitioned into a nonlinear set - the coefficients of the MA structure - and a linear set - the remaining coefficients. We then outline special purpose iterative procedures that can be used in place of the conventional nonlinear least squares algorithms, taking advantage of the linear-nonlinear decomposition of the parameter set. Numerical results are reported to show the effectiveness of the proposed technique.

2. Reduction to a mixed linear-nonlinear least squares problem. Ad-hoc numerical methods.

Following Hill (1977) we write the model of interest in the form

$$P(\rho)\, y - X\beta = T(\theta)\, e \qquad (1)$$

where y is a m-vector of observations on the "dependent" variable, X is a given non stochastic m by s matrix of observations on the independent variables, e is a m-vector of unobserved disturbances, assumed to be NID $(0,\sigma^2)$ and the matrices P,T are of the form

$$P(\rho)=\begin{bmatrix} 1 & & & & 0 \\ -\rho_1 & \ddots & & & \\ \vdots & \ddots & \ddots & & \\ -\rho_p & & \ddots & \ddots & \\ 0 & -\rho_p & \cdots & -\rho_1 & 1 \end{bmatrix}, \quad T(\theta)=\begin{bmatrix} 1 & & & & 0 \\ -\theta_1 & \ddots & & & \\ \vdots & \ddots & \ddots & & \\ -\theta_q & & \ddots & \ddots & \\ 0 & -\theta_q & \cdots & -\theta_1 & 1 \end{bmatrix}$$

where $\rho=(\rho_1,\dots,\rho_p)^T$, $\theta=(\theta_1,\dots,\theta_q)^T$.

For simplicity, the pre-sample values $y_0,\dots y_{1-p}$, $e_0,\dots,e_{1-q}$ have been set equal to zero and not treated as random variables. Of course, this assumption does not affect the asymptotic properties of the resulting estimates. Under the above assumptions the logarithmic likelihood function has the form

$$L(\rho,\beta,\theta,\sigma^2) = m/2 \ln\sigma^2 - 1/(2\sigma^2)[y-P^{-1}(\rho)X\beta]^T V^{-1}(\rho,\theta) \cdot [y-P^{-1}(\rho)X\beta]$$

except for an irrelevant constant term, where we have denoted

$$V^{-1}(\rho,\theta) = [T^{-1}(\theta)P(\rho)]^T T^{-1}(\theta)P(\rho) .$$

It is then easy to prove the following

LEMMA. The maximum likelihood estimation problem for model (1) is numerically equivalent to the non linear least squares problem

$$(2) \qquad \min_{\eta,\theta} \| b(\theta) - A(\theta)\eta \|^2$$

where $\eta = (\rho^T,\beta^T)^T$, $b(\theta) = T^{-1}(\theta)y$, $A(\theta) = T^{-1}(\theta)\ [Y \vdots X]$,

and

$$Y = \begin{bmatrix} 0 & 0 & \cdots & 0 \\ y_1 & 0 & \cdots & 0 \\ y_2 & y_1 & \cdots & 0 \\ \vdots & \vdots & & \vdots \\ y_{m-1} & y_{m-2} & \cdots & y_{m-p} \end{bmatrix} .$$

The result can be readily proved by eliminating σ^2 to obtain the "concentrated" likelihood function

$$\tilde{L}(\rho,\beta,\theta) = -\, m/2 \ln \{[y-P^{-1}(\rho)X\beta]^T V^{-1}(\rho,\theta)\, [y-P^{-1}(\rho)X\beta]\},$$

using the identity $P(\rho)y = y - Y\rho$.

Least squares problems of the type (2) are asually called "separable", since the parameter set can be separated into two disjoint sets, η, which appears linearly for fixed θ , and θ . The structure of problem (2) can be efficiently exploited in several ways. The general strategy is to employ two-step iterative procedures, in which optimization with respect to the linear parameters is performed first, and corrections to the nonlinear parameters after that.

Now for fixed θ the optimal value for η can be expressed as

(3) $$\hat{\eta} = A^{-}b$$

where A^{-} is a symmetric generalized inverse of A, i.e. $AA^{-}A = A$, $(AA^{-})^T = AA^{-}$. Two basic iterative procedures can then be devised, having in common the computation of $\hat{\eta}$ at each iteration. In the simpler scheme, for any given θ the <u>numerical value</u> $\hat{\eta}$ is inserted into (2), yielding the problem

(4) $$\min_{\theta} \|r_1(\theta)\|^2 = \|b(\theta) - A(\theta)\hat{\eta}\|^2 \quad .$$

This technique has been extensively used in various fields of computational statistics, cf Wold and Lyttkens (1969).

A more sophisticate approach involves the substitution of the <u>analytical expression</u> for $\hat{\eta}$ into (2), yielding the problem

(5) $$\min_{\theta} \|r_2(\theta)\|^2 = \|[I - A(\theta)A^{-}(\theta)]\, b(\theta)\|^2 \quad .$$

This procedure has been investigated recently by Golub and Pereyra (1973) and improved by Kaufman (1975).

It is definitely clear that for both the resulting "purely nonlinear" problems (4),(5) corrections to θ can be computed utilizing any of the usual algorithms for nonlinear least squares problems. For simplicity, we shall refer to the basic Gauss-Newton method

$$\theta \leftarrow \theta + \mu\, h^{(i)} \quad , \quad h^{(i)} = -[\partial r_i/\partial\theta]^{+} r_i$$

$i = 1,2$, and outline an efficient implementation of the main body of the algorithms. As mentioned above, they can be arranged as two-step procedures as follows:

Step 1 (evaluation of $\hat{\eta}$ according to (3)).

Using the Householder transformations (cf e.g.Lawson and Hanson (1974)) we perform an orthogonal trapezoidal decomposition of A :

$$QAS = \begin{bmatrix} U & P \\ O & O \end{bmatrix} ,$$

where Q is an m by m orthogonal matrix, S is a t by t permutation matrix (t = p+s), U is a k by k non singular upper triangular matrix (k = rank(A)), then compute $\hat{\eta}$ as

$$\hat{\eta} = S_1 U^{-1} Q_1 b ,$$

where

$$Q = \begin{bmatrix} Q_1 \\ Q_2 \end{bmatrix} \begin{matrix} k \\ m-k \end{matrix} , \qquad S = [\underset{k}{S_1} \vdots \underset{t-k}{S_2}] .$$

Step 2 (correction to θ).

We compute

$$B = \partial A/\partial\theta \ \hat{\eta} - \partial b/\partial\theta ;$$

then since (cf Kaufman, cit.)

$$- [\partial r_2/\partial\theta]^+ r_2 = (Q_2 B)^+ Q_2 b ,$$

we have

$$\text{(Algorithm 1)} : \quad h = B^+(b - A\hat{\eta})$$

$$\text{(Algorithm 2)} : \quad h = (Q_2 B)^+ Q_2 b$$

It is clear that for the computation of h in both expressions a complete orthogonal decomposition of the relevant matrices can be employed using the Householder transformations again.

In passing, we can note two main computational differences between the Algorithms. First, the dimension of the matrix under $^+$, which is m by q for Algorithm 1 and (m-k) by q for Algorithm 2. Second, Algorithm 1 requires the explicit evaluation of the residual, while Algorithm 2 requires the pre-multiplication of B by the matrix Q of Step 1.

It is hardly necessary to note that in the conventional Gauss-Newton procedure the correction is performed to all parameters simultaneously according to

$$h = [A \vdots B]^{+}(b-A\eta),$$

so that the dimension of the matrix under $^{+}$ is now m by t+q.

Let now summarize the relative merits of the above numerical methods.

(i) Algorithm 1 and Algorithm 2 vs conventional Gauss-Newton.

1. The computational work per iteration is approximately the same, as can be verified by easy but tedious algebra;

2. A reduction in dimension of the parameter space is evidently obtained, and consequently a reduced number of initial guesses has to be provided;

3. The function evaluation is easier for the conventional scheme, however Algorithm 1 and 2 evaluate the optimal η, so probably they give better points at each iteration.

(ii) Algorithm 2 vs Algorithm 1.

It can be shown (cf Ruhe and Wedin (1974)) that Algorithm 2 has the same asymptotic superlinear convergence rate as the conventional Gauss-Newton scheme (in the case of "small" residual and full rank matrices), while Algorithm 1 in general has linear convergence rate. Of course, these are a-priori properties so it is possible to find practical cases where the simpler scheme indeed has a faster convergence rate. Numerical experiments on several test problems confirm that the separation of the parameter set usually improves the rate of convergence, and in some cases a dramatic reduction in the number of iterations can be obtained (cf Corradi and Stefanini (1978)).

Remark 1. By straightforward modifications to Step 2 the corresponding damped schemes, as well as any other method based on the knowledge of the first derivatives only, cf e.g. Dennis and Welsh (1978), can readily be obtained.

Remark 2. In order to compute the vector $b(\theta)$ and the matrix $A(\theta)$ as well as their derivatives there is no need to construct the explicit inverse of matrix $T(\theta)$. Indeed, it is not difficult to obtain the following expressions:

$$b_1 = y_1$$

$$b_k = y_k + \sum_{i=1}^{\min(k-1,q)} \theta_i b_{k-i}\ , \quad k=2,\dots,m$$

$$a_{kj} = \begin{cases} 0 & , \quad k = 1,\dots,j \\ b_{k-j} & , \quad k = j+1,\dots,m \end{cases} \quad , \quad j = 1,\dots,p$$

$$\left.\begin{array}{l} a_{1j} = x_{1,j-p} \\ a_{kj} = x_{k,j-p} + \displaystyle\sum_{i=1}^{\min(k-1,q)} \theta_i a_{k-i,j}\ , \quad k=2,\dots,m \end{array}\right\} \quad j=p+1,\dots,p+s.$$

As a consequence, we obtain the following expressions for the derivatives $\partial b/\partial\theta_n$, $\partial A/\partial\theta_n$, $n = 1,\dots,q$:

$$\frac{\partial b_k}{\partial\theta_n} = \begin{cases} 0 & , \quad k=1,\dots,n \\ b_{k-n} + \displaystyle\sum_{i=1}^{\min(k-1,q)} \theta_i \frac{\partial b_{k-i}}{\partial\theta_n} & , \quad k=n+1,\dots,m \end{cases}$$

$$\frac{\partial a_{kj}}{\partial\theta_n} = \begin{cases} 0 & ,k=1,\dots,j \\ \dfrac{\partial b_{k-j}}{\partial\theta_n} & ,k=j+1,\dots,m \end{cases} \quad , \quad j=1,\dots,p$$

$$\frac{\partial a_{kj}}{\partial\theta_n} = \begin{cases} 0 & , \quad k=1,\dots,n \\ a_{k-n,j} + \displaystyle\sum_{i=1}^{\min(k-1,q)} \theta_i \frac{\partial a_{k-i,j}}{\partial\theta_n} & , \quad k=n+1,\dots,m \end{cases} \quad j=p+1,\dots,p+s.$$

3. Numerical results.

Several test problems have been run in order to make a comparative study of the performances of the various methods, i.e. Algorithm 1, Algorithm 2 and conventional Gauss-Newton (G.-N.), both in undamped and damped form. Complete details of the entire study will not be presented here, however the reported results can be regarded as representative of the whole set of numerical tests.

Here we consider the following three models (we have denoted $B^j z_t = z_{t-j}$):

Problem 1 (Wolfer annual sunspot numbers, cf Anderson (1977), p. 132,

164):

model: $(1-\rho_1 B-\rho_2 B^2)K_t = \beta_1 + (1-\theta_1 B^{11})e_t$;

parameter set: $(\rho_1,\rho_2,\beta_1,\theta_1)$, p=2, s=1, q=1, m=120;

final solution: (1.212, -0.569, 16.630, -0.223);

final residual sum of squares (RSS): 38714;

Anderson's solution: (1.468, -0.591, 6.151, -0.236),
with RSS = 47094;

initial values: (1.217, -0.571, 16.489, -0.2).

Problem 2 (Dow Jones utility index, cf Anderson, cit., p. 107, 132, 165):

model: $(1-\rho_1 B)\Delta L_t = (1-\theta_1 B)e_t$;

parameter set: (ρ_1,θ_1), p=1, s=0, q=1, m=77;

final solution: (0.868, 0.547);

final RSS: 11.053;

Anderson's solution: (0.755, 0.373), with RSS = 11.108;

initial values: (0.711, 0.274).

Problem 3 (Sales data with leading indicator, cf Box and Jenkins (1970), p. 410, 537):

model: $(1-\rho_1 B)\Delta Y_t = \beta_1 + \beta_2 \Delta X_{t-3} + (1-\theta_1 B-\theta_2 B^2)e_t$;

parameter set: $(\rho_1,\beta_1,\beta_2,\theta_1,\theta_2)$, p=1, s=2, q=2, m=149;

final solution: (0.726, 0.005, 4.709, 1.019, -0.298);

final RSS: 9.537;

Box and Jenkins' solution: (0.720, 0.009, 4.820, 1.260, -0.389),
with RSS = 11.250;

initial values: (0.696, 0.012, 4.747, 0.5, -0.5).

In all our experiments the convergence criterion was set to three correct digits in the parameter estimates; the initial values for the linear parameters were simple least squares estimates conditional on the selected values for the nonlinear ones.

Reported in the Tables are the number of iterations, i.e. the number of gradient calls, and the number of function evaluations needed to satisfy the convergence criterion, in the form NI/NF.

In table 1 we compare the performances of the various methods in solving the above problems. The results show that the linear-nonlinear decomposition indeed can yield a considerable improvement in computational efficiency, especially when the damped version is used. It is confirmed that the simpler Algorithm 1 can perform better despite the poor theoretical convergence rate; however it seems to be strongly affected by the position in the starting guess, as it is shown in Table 2, where the performances are compared for various choices of the initial values for Problem 1.

The results reported in Table 3 were obtained using model 2, where the series ΔL was perturbed according to $y_t = \Delta L_t + \varepsilon\zeta_t$, $\zeta_t \sim NID(0,1)$; by varying the value of ε a range of possible data sets could be simulated, allowing to test the robustness of the different methods. It will be seen that Algorithm 1 behaved better when ε was reasonably small, while Algorithm 2 yielded some improvement in iterations even in case of significantly large perturbations.

In summary, we conclude that the experimental results support the theoretical considerations outlined in the previous Section in such a way that Algorithm 2 can generally be suggested as an efficient device for the numerical treatment of the problem under study, since it is more reliable than simpler separation schemes and compares favourably with the conventional procedures even in perturbed cases.

	Undamped Algorithms			Damped Algorithms		
Problem	1	2	G.-N.	1	2	G.-N.
1.	2/2	3/3	3/3	3/3	3/3	>50
2.	8/8	12/12	12/12	9/9	16/16	18/18
3.	failed	4/4	7/7	failed	4/4	7/8

TABLE 1.

	Undamped Algorithms			Damped Algorithms		
Guess for θ_1	1	2	G.-N.	1	2	G.-N.
-0.2	2/2	3/3	3/3	3/3	3/3	>50
-0.4	failed	3/3	5/5	failed	3/3	>50
-0.8	failed	5/5	6/6	failed	5/5	>50

TABLE 2.

	Undamped Algorithms			Damped Algorithms		
ε	1	2	G.-N.	1	2	G.-N.
0.01	9/9	13/13	13/15	9/9	17/17	18/18
0.1	14/14	10/10	11/11	16/16	13/13	14/15
1.	failed	6/6	9/20	failed	6/6	17/17
10.	failed	13/13	14/29	failed	5/5	14/14

TABLE 3.

REFERENCES

(1) Anderson, O.D., 1977, "Time Series Analysis and Forecasting, The Box-Jenkins Approach" (Butterwoths, London).

(2) Box, G.E.P. and Jenkins, G.M., 1970 "Time Series Analysis, Forecasting and Control" (Holden-Day, San Francisco).

(3) Corradi, C. and Stefanini, L., 1978, "Computational Experience with Algorithms for Separable Nonlinear Least Squares Problems", Calcolo XV, 317-330.

(4) Dennis, J.E. and Welsh, R.E., 1978, "Techniques for Non Linear Least Squares and Robust Regression", Communications in Statistics-Simulation and Computation B7, 345-359.

(5) Golub, G.H. and Pereyra, V., 1973, "The Differentiation of Pseudoinverses and Nonlinear Least Squares Problems whose Variables are Separate", SIAM Journal on Numerical Analysis 10, 413-432.

(6) Hill, R., 1977, "Covariance and Estimated Parameters in ARMA Regression Models", Annals of Economic and Social Measurement 6, 109-122.

(7) Kaufman, L., 1975, "A Variable Projection Method for Solving Separable Nonlinear Least Squares Problems", BIT 15, 49-57.

(8) Lawson, C.L. and Hanson, R.J., 1974, "Solving Least Squares Problems" (Prentice-Hall, Englewood Cliffs, N.J.).

(9) Ruhe, A. and Wedin, P.A., 1974, "Algorithms for Separable Nonlinear Least Squares Problems", Stanford Computer Science Technical Report 434.

(10) Wold,A. and Lyttkens, E., 1969, "Nonlinear Iterative Partial Least Squares (NIPALS) Estimation Procedures", Bullettin of the I.S.I. 43, 29-51.

PART TWO

STOCHASTIC DESIGN AND ANALYSIS OF NUMERICAL ALGORITHMS

Numerical Techniques for Stochastic Systems
F. Archetti and M. Cugiani (eds.)
© North-Holland Publishing Company, 1980

A STOCHASTIC ANALYSIS OF ERRORS IN NUMERICAL PROCESSES

Marco Cugiani

Istituto di Matematica - Università di Milano

INTRODUCTION.

The theoretical relevance of a probabilistic analysis of errors in numerical processes is clearly stressed by the fact that even for the simplest numerical problems, for which effective algorithms are available also in classical numerical analysis, the errors cannot be often bounded in a deterministic framework.
It is well known that the error bounds which can be derived by the so called "worst case analysis" are generally so far from the actual behaviour of the numerical process or so cumbersome to apply to be of no use in the every day numerical computation.
Heuristic rules are often used (the so called thumb rules) which enable, on average, a more realistic control of the computations. The validity of such rules cannot be granted in every case, so that their application requires an estimate, that is based on implicit statistical assumptions, of their range of validity.
The stochastic analysis of errors arises quite naturally when we use a Monte Carlo method, as we shall see in the next sections, but can be performed also after a deterministic algorithm has been applied, as we shall see in the section two. Finally in the section three we shall outline a theoretical framework of the implication of stochastic models in the general problem of error evaluation.

Section I. Error analysis in Monte Carlo methods.

In order to show how natural and effective is the probabilistic analysis of errors in Monte Carlo methods we shall consider a typical application, namely the evaluation of a multiple integral.

The problem we shall be concerned with in the most of this section is the numerical evaluation of:

$$J = \int_S f(x)dx$$

where S is the unit-cube in R^n and f(x) a continuous function on S. It is well known that for n > 3 a method frequently used is based on a sample of f(x) in S. We briefly recall some basic facts about how this sample can be performed.

Let Z be a random variable uniformly distributed in S and let $f_i=f(Z_i)$ (for i=1,...,k) be an independent sample of the random variable f(Z). Then an unbiased estimation of J is the mean

$$(1.1) \qquad Y_k = \frac{\sum_{i=1}^{k} f(Z_i)}{k}$$

whose distribution is approximately normal with mean J and whose variance is given by $\frac{\sigma^2}{k}$, where

$$(1.2) \qquad \sigma^2 = \int_S f^2(x)dx - (\int_S f(x)dx)^2;$$

of course this later quantity, whose relevance is crucial in our analysis, cannot be directly calculated after (1.2), but is rather estimated throughout the sample variance

$$\sum_{i=1}^{k} (f_i - Y_k)^2/k$$

or better throughout the unbiased estimator

$$\Sigma(f_i - Y_k)^2/(k-1).$$

A probabilistic bound of the error, in the evaluation of J by means of Y_k, is given by

$$(1.3) \qquad \text{Prob}(|Y_k - J| < \frac{4\sigma}{\sqrt{k}}) > 0.9999$$

by the well known properties of the normal distribution.

It is also known that these bounds can be improved using more sophisticated estimate than Y_k, whose variance can be shown to be smaller than that given by (1.2). We shall not consider in more details this more effective estimates: for the purpose of our paper it is enough to say that the probabilistic bounds thus obtained are good measures of the actual errors in the computation of multiple intergrals for n > 3.

The effectiveness of these methods in bounding the error is such that they often supply the fundamental algorithms in the adaptive processes of integration. We shall briefly recall some basic features of such methods.

We assume that S is split into p subsets S_j so that

$$S_i \cap S_j = \emptyset \quad , \quad \bigcup_{j=1}^{p} S_j = S,$$

and we pose

$$J_i = \int_{S_j} f(x)dx$$

Once J_j (for j=1,2,...,p) have been computed, for example by a Monte Carlo method, we evaluate the bound of the error in each subset S_j and we check in which subsets it exceeds a prefixed accuracy level. These subsets and only these are again subdivided into, say, q subsets for which the whole process is repeated until the overall accuracy level is satisfied in each of the subsets generated by the sequence of subdivisions.

We remark again that, at least in this problem, the probabilistic bound of the error is so good that the very structure of most effective algorithms is determined by the probabilistic analysis of error in each part of the domain of integration.

A "probabilistic control" of the progress of the algorithm can be so effective as to be usually retained even when, for some particular reasons, the approximation of J_j is performed by some classical method of integration (for example the quadrature formulae).

In next section we shall stretch this point even wider extending the probabilistic analysis of errors also to the control of the calculations when these are performed by deterministic methods, starting from a closer look to the problem of multiple integration with an adaptive method when J_i are computed by deterministic techniques.

Section II. Statistical control of deterministic algorithms.

We assume that an approximation of

$$J_j = \int_{S_j} f(x)dx = \int_{S_j} \phi(x)w(x)dx$$

(where w(x) is a suitable weight-function) is given by

$$\sum_{h=1}^{H} A_{h,j}\phi(x_h) = I(j,H) = I_j$$

with a suitable choice of the points x_h in S_j, and of the coefficients $A_{h,j}$.

A statistical control of I_j can be easily obtained by drawing N points x_i from a uniform distribution in S_j, and considering a Monte Carlo estimate

$$Y_N = \frac{V(S_j)}{N} \sum_{i=1}^{N} \phi(x_i) w(x_i)$$

where $V(S_j)$ is the volume of S_j.

If we pose

$$\eta = \frac{V(S_j)}{N} \sum_{i=1}^{N} \phi(x_i) w(x_i) - \sum_{h=1}^{H} A_h \phi(x_h) = Y_N - I_j$$

and $l_j = J_j - I_j$.

We obtain $\eta - l_j = Y_N - J_j$

A relation analogous to (1.3) holds also in this case and we have

$$\text{Prob}\ (|Y_N - J_j| < 4\sigma_o) > 0.9999$$

where σ_o is the standard deviation of the random variable Y_N and can be estimated through the standard deviation of the sample given by the N points X_i.

It follows that with near one probability the error satisfies the relation

$$l_j \leq | \eta | + 4\sigma_o.$$

By methods of this kind it is possible to obtain satisfactory results in the computation of multiple integrals. Regretfully one has to remark that this problem is almost uniquely well tailored for a statistical control of the error.

In what follows of this section we shall briefly consider more general attempts at the statistical handling of the errors in numerical analysis. We start by considering the problem of the summation of N terms (N sufficiently large), all of them affected by errors which can be assumed as random variables independent and identically distributed. In this case, in strict analogy to what has been done in the case of multiple integrals, the central limit theorem enables one to obtain easily a bound of the error in the result of the operation.

Analogous results could be obtained, although less easily, for the multiplication of N numbers α_i, each affected by an error $\delta\alpha_i$. Here we assume that α_i are random variables uniformly distributed in the interval (1, 10) and the $\delta\alpha_i$ are random variables uniformly distributed in the interval $(-0.5\ 10^{-n}, +0.5\ 10^{-n})$, and represent the rounding errors in the data. In this case we are interested mainly in the relative error; this error will be given by two terms, the first representing the effect of inherent errors $\delta\alpha_i/\alpha_i$, and the second representing the rounding errors due to the finite word length of the computer.

Let us consider the product in the form

$$(2.1) \qquad \prod_{i=1}^{N} (\alpha_i + \delta\alpha_i) = \pi\beta_i$$

(in the following we will assume that any product is written in its normalized form $m=\nu 10^r$, where ν (with $1\leq\nu<10$) is rounded to the n-th digit and r is an integer). We'll focus our attention on the relative errors

$$\delta\overset{*}{\alpha}_i = \frac{\delta\alpha_i}{\alpha_i}$$

In a two factors product we can assume that, with close approximation, a relation of the type of the following one holds

$$\delta(\alpha\beta) = (\alpha + \delta\alpha)(\beta + \delta\beta) - \alpha\beta \approx \alpha\delta\beta + \beta\delta\alpha$$

(disregarding the product $\delta\alpha\cdot\delta\beta$) and then

$$\delta^*(\alpha\beta) = \frac{\delta(\alpha\beta)}{\alpha\beta} \approx \frac{\delta\alpha}{\alpha} + \frac{\delta\beta}{\beta} = \delta^*\alpha + \delta^*\beta$$

In a product of two normalized factors α_i, α_j the relative error is given by

$$\delta^*\alpha_i + \delta^*\alpha_j + \eta_{ij}$$

where η_{ij} is the relative truncation error in the product $\alpha_i\alpha_j$.
For a three factors product the relative error turns out

$$(\delta^*\alpha_i + \delta^*\alpha_j + \eta_{ij}) + \delta^*\alpha_k + \eta_{ijk}$$

Generally in a product of, say a hundred normalized factors, the relative error will be given by

$$(2.2) \qquad \sum_{i=1}^{100} \delta^*\alpha_i + \sum_{H=1}^{99} \eta^*_H = \Delta_1 + \Delta_2$$

(here obviously $\eta^*_1=\eta_{12}$, $\eta^*_2=\eta_{123}$ and so on).
The first term has approximately normal distribution, according to the central limit theorem; such term is indeed the sum of a large number of random variables of the type

$$\frac{\delta\alpha_i}{\alpha_i}$$

The distribution $F(x)$ of the variable $\frac{1}{\alpha_i}$ can be easily found, at least in some cases. Under the above assumption we have for example

$$\text{Prob}\left(\frac{1}{x+h} < \alpha_i < \frac{1}{x}\right) = F(x+h) - F(x) = \frac{1}{9}\left(\frac{1}{x} - \frac{1}{x+h}\right) =$$

$$= \frac{h}{9\,x\,(x+h)}$$

and then

$$f(x) = F'(x) = \lim_{h\to o} \frac{1}{9x(x+h)} = \frac{1}{9x^2}$$

Standard computations yield the probability density function of the product

$$\delta\alpha_i \cdot \frac{1}{\alpha_i}$$

which turns out ("omitting" the factor 10^{-n})

$$\begin{array}{ll} f(x) = o & \text{for } x < -\frac{1}{2} \\ f(x) = \frac{1}{18}\left(\frac{1}{4x^2} - 1\right) & \text{for } -\frac{1}{2} \le x < -\frac{1}{20} \\ f(x) = \frac{11}{2} & \text{for } -\frac{1}{20} \le x < \frac{1}{20} \\ f(x) = \frac{1}{18}\left(\frac{1}{4x^2} - 1\right) & \text{for } \frac{1}{20} \le x < \frac{1}{2} \\ f(x) = 0 & \text{for } \frac{1}{2} \le x \end{array}$$

The function $f(x)$ has the following graphic representation

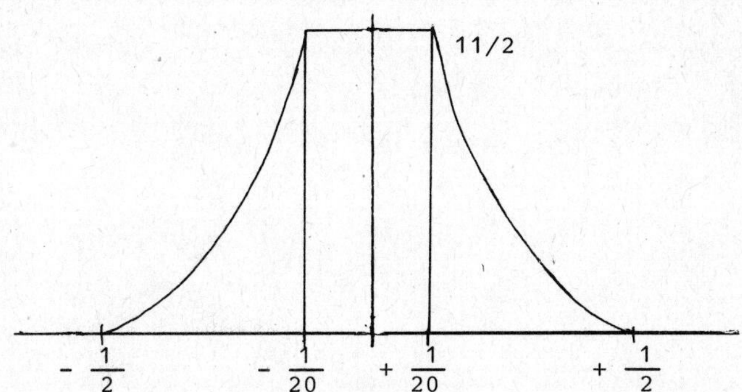

It's very easy now to compute the variance of the random variable $\delta^*\alpha_i\cdot 10^n$ which turns out to be $\frac{1}{120}$

Setting then:

$$\sigma_o^2 = \frac{1}{120}$$

the variance of the random variable $\Delta_1' = \Delta_1 \cdot 10^n$ turns out to be

$$\sigma_1^2 = 100\ \sigma_o^2$$

and hence $\sigma_1 = 10\sigma_o = 10\sqrt{\frac{1}{120}} = \sqrt{0.833}\ \ldots = 0.913\ \ldots$

It follows $P(\Delta_1' < 4\sigma_1) = P(\Delta_1' < 3,652) > 0.9999$ so that (reintroducing the factor 10^{-n}) the inherent relative error results to be not greater than $4\cdot 10^{-n}$ with near one probability.

More difficult is the attempt to give an estimate for the size of the induced error

$$\nabla_2 = \sum_{H=1}^{99} \eta_H^*$$

Any η_H^* is indeed given by a ratio of the type $\frac{\delta\mu_H}{\mu_H}$

Where $\delta\mu_H$ is the error generated in the computation of a certain normalized product μ_H of normalized factors β_i. We can assume, may be, that $\delta\mu_H$ is uniformly distributed; on the contrary the product μ_H cannot any way be considered as uniformly distributed. Indeed it is easily seen that a two factors product has the probability density function

$$f(y) = 0 \qquad \text{for } y < 1 \qquad \text{and for } y > 10$$

$$f(y) = \frac{10 \lg 10 - y \lg y}{81} \qquad \text{for } 1 \leq y < 10$$

When N increases, the N factors normalized product is a random variable with limit distribution

$$\tilde{f}(x) = \frac{1}{x \lg 10} \qquad \text{for } 1 \leq x < 10$$

$$\tilde{f}(x) = 0 \qquad \text{elsewhere}$$

The above considerations suggest that an attempt could be made in order to give an estimate of the variance of Δ_2 (whose distribution can be assumed approximately normal), at least as a first approximation. But we observe, at this point, that the problem is any way not very easy, if we will try to find a solution quite satisfactory.
We can here recall that an analogous approach has been also tried, with the same hardly satisfactory results, to the estimation of rounding errors in numerical solution of ordinary differential equations.
We recall here that Von Neumann and Goldstine [1947],[1951] tried at the very beginning of modern numerical analysis, to settle in a probabilistic framework the evaluation of errors in the solution of a linear system. The results that they obtained are of a relevant theoretical interest but had a negligeable influence on the practice of the numerical computing. The basic features of their work are anyway still worthy to

be remembered in this paper.

Let us consider a linear system

$$AX = B$$

where A is a real matrix $n\times n$, positive definite and X a vector in R^n. The matrix A is assumed to be decomposed as the product

$$A = R^T\cdot D\cdot R$$

where R is an upper triangular matrix and D is a diagonal matrix. Due to rounding errors, which are the only ones considered by the authors, the actual decomposition will be

$$A = R_1^T\cdot D_1\cdot R_1+Z$$

where R_1 and D_1 are approximations respectively of R and D, while Z is an error matrix, whose behaviour we are interested in. Another stage we want to investigate is the inversion of R_1: again due to the rounding errors, we shall obtain a matrix R_2 such that $R_2\cdot R_1=I+U$, where I is the identity and U the error matrix in the inversion of R_1.

In order to characterize Z and U we assume that the elements a_{ij} of A are random variables independent and normally distributed with mean value equal zero and variance σ^2.

Now we observe more closely the matrices Z and U: they are given by sums of products of matrices whose elements are sums of products of random variables: these random variables can be either elements a_{ij} of A (assumed to be normally distributed) or rounding errors (assumed to be uniformly distributed). It is certainly not easy to identify in any single case the distribution of the elements of these matrices, but the task is made somewhat easier by the two following considerations.

First, we are not so much interested in the distribution of the elements of the matrix but rather in the distribution of some suitable function which conveys globally those characteristics of matrices more relevant in the evaluation of the errors. A suitable function is the spherical norm

$$|| A || = \rho (A)$$

where $\rho(A)$ is the spectral radius of A.
If $a_{ij} \sim N(0,\sigma^2)$ it is possible to derive a rather intricated expression for the distribution of $|| A ||$, in which the eigenvalues of A appear. Another idea which can be of some help in this problem is the notion of "dominant variable". A random variable X is said to dominate the r.v. Y if for any convex function ϕ we have

$$E(\phi(X)) \geq E(\phi(Y))$$

More generally it can be proved that if $X_1, \ldots, X_m$ are independent variables, which respectively dominate $Y_1, \ldots, Y_m$ we have

$$E(\phi(X_1,\ldots,X_m)) \geq E(\phi(Y_1,\ldots,Y_m))$$

where again ϕ is any convex function on R^m. Obviously we assume that all before considered expected values exist. The two above remarks advise us to follow, in the handling of the matrices involved, two basic guidelines: to evaluate each matrix according to its norm and to replace a matrix, whenever it is useful for simplifying the computations, by another matrix whose elements dominate those of the former.

The basic results are the following

$$\text{Prob}\ (|| Z || > k_1\ n\ 10^{-r}) < k_2\ 2^{-n}$$

$$\text{Prob}\ (|| U || > k_3\ n\ 10^{-r}) < k_4\ 2^{-n}$$

where k_i are absolute constants, and r is the number of significant digits used in the computation. Some results can also be given considering the effects of rounding errors in the inversion of the matrix A. We represent by

$$V = WA - I$$

(where W is the approximate inverse of A), the error matrix.
The basic result about the inversion is that the numbers n, dimension of the problem, and r, number of significant digits used in computation, must satisfy the relation

$$n < k \cdot 10^{r/3}$$

in order to ensure that with near one probability the error in the approximation of A^{-1} is kept within some bound, on which depends the coefficient k. More recently an analogous approach was tried by C. Blanc [1959] who considered a more effective way (the Cholesky decomposition) in the solution of the linear system.
A completely different approach, which scores some results in dealing with inherent errors, can be found in the book of Rust & Burrus (1972).
Let us consider a linear system $m \times n$ $(m \geq n)$

$$AX = \tilde{B} + D$$

where A is a matrix whose elements a_{ij} $(i=1, \ldots, m;\ j=1, \ldots, n)$ are assumed known without errors, while $\tilde{B}$ and D are random n-dimensional vectors containing the errors in the data of the problem.
The following relations about these vectors can be assumed

$$E(\tilde{B}) = B \qquad E(D) = 0$$

$$\Sigma = E(DD^T) = E((\tilde{B}-B)(\tilde{B}-B)^T)$$

The components of B, $\tilde{B}$ will be indicated by b_i a $\tilde{b}_i$ respectively. $\tilde{b}_i$ are random variables whose distribution, for example, can be postulated to be either normal when $\tilde{B}$ is an output of a measurement process, or uniform in the interval $(b_i - 0.5 \cdot 10^{-r},\ b_i + 0.5 \cdot 10^{-r})$ when $\tilde{B}$ is an approximate value of B rounded to the r-th digit.

We can consider the problem of solving the linear system

$$AX = B$$

whose solution will be indicated by $\bar{X}$.
$\bar{X}$ will be certainly unique if A is of maximum rank, and will certainly exist if moreover $m=n$. We will suppose the existence of $\bar{X}$ also for $m>n$. According to statistical practice we can now approximate the solution mainly in one of two ways. In the first it can be shown that an unbiased estimator of $\bar{X}$ is given by $X=H\tilde{B}$, where H is a suitable matrix, which can be uniquely identified as

$$H = (A^T \Sigma^{-1} A)^{-1} A^T \Sigma^{-1}$$

requiring that random vector $H\tilde{B}$ is a least variance estimator. In the second way we are looking for confidence intervals for the components $\bar{x}_i$ of $\bar{X}$. This can be obtained expressing $\bar{x}_i$ as a scalar function of $\bar{X}$ and defining, in a general form, unbiased estimators of scalar functions of $\bar{X}$. When the distribution of $\tilde{B}$ is known a confidence region for $\bar{X}$ can always be obtained; this process of estimation is particularly straightforward when $\tilde{B}$ follows a normal joint distribution: also the estimators can be shown to be normal so that the confidence region can be easily found by a formula like (1.3).

Section III. Stochastic models in error evaluation

In this section we shall be mainly concerned with the applications of the Wiener process to the probabilistic bounding of the error in some problems of numerical analysis. We remark that the results we are going to outline in this section are only the first steps in the application of stochastic models to numerical problems and their actual computational relevance is still limited to some particular cases. First we shall again consider multiple integration, which, as already remarked, lends itself easily to a stochastic analysis. Let $X(t,w)$

be a Wiener process for $t \in [0, 1]$ such that

$$X(0,\omega) = X_0$$

Let Z be the integral

$$Z = \int_0^1 X(t)\,dt$$

If $X(t)=X(t,\omega)$, Z turns out to be a normal random variable with mean value

$$E(Z) = X_0$$

and variance $\sigma^2/3$, where σ^2 is the variance of the r.v. $X_1=X(1,\omega)$.
We are interested in computing

$$\overline{Z} = \int_0^1 \overline{X}(t)\,dt$$

where $\overline{X}(t)$ will be thought as a sample path of the Wiener process. We assume know m values $\overline{X}_i$ of $\overline{X}$ at the points $t_1, t_2, \ldots t_m$ with $0<t_1<t_2 \ldots <t_m \leq 1$ and we pose of course $\overline{X}_0=X_0$, $t_0=0$.
Now we try to derive the distribution of Z conditioned by $X(t_i)=\overline{X}_i$ (for $i=1, \ldots, m$). The random variables

$$Z,\ X(t_1),\ \ldots\ldots,\ X(t_m)$$

have a joint multidimensional normal distribution, whose expected values form a vector whose components are all equal to X_0, while the covariances are given by

$$\mathrm{cov}\,(X(t_i),\ X(t_j)) = \sigma^2 \min\,(t_i,t_j)$$

$$\mathrm{cov}\,(Z,X(t_i)) = \sigma^2 \int_0^1 \min\,(t_i,t)\,dt = \sigma^2\, t_i\,\left(1 - \frac{t_i}{2}\right)$$

Now we build the covariance matrix

$$\Sigma = \begin{vmatrix} V(Z) & \mathrm{cov}(Z,X(t_1)) & \ldots\ldots & \mathrm{cov}(Z,X(t_m)) \\ \mathrm{cov}\,(Z,X(t_1)) & V(X(t_2)) & \ldots\ldots & \mathrm{cov}(X(t_1),X(t_m)) \\ & & & \\ \mathrm{cov}\,(Z,X(t_m)) & \mathrm{cov}(X(t_1),X(t_m)) & \ldots\ldots & V(X(t_m)) \end{vmatrix}$$

where obviously $V(X(t_i)) = \sigma^2 t_i$.

The matrix $\sum$ can be considered partitioned as

$$\begin{vmatrix} S_{11} & S_{21}^T \\ S_{21} & S_{22} \end{vmatrix}$$

where $S_{11}=V(Z)$; S_{21} is the vector of the covariances of Z and each $X(t_i)$ for $i=1, \ldots, m$ and S_{22} is the covariance matrix of the $X(t_2)$. Now we can compute the conditioned expected value

$$E\,(Z \mid X\,(t_i) = \bar{X}_i,\ i = 1,\ldots,m) =$$

$$= X_o + S_{21}^T \; S_{22}^{-1} \; Y$$

where Y is the m-dimensional column vector

$$Y = (\bar{X}_1 - X_o\,,\ \bar{X}_2 - X_o,\ \ldots,\ \bar{X}_m - X_o),$$

and the conditioned variance

$$V\,(Z \mid X\,(t_i) = \bar{X}_i) = S_{11} - S_{21}^T \; S_{22}^{-1} \; S_{21}$$

We can also easily derive an esplicit expression for S_{22}^{-1}, given by the tridiagonal symmetric matrix (for $m \geq 3$)

$$S_{22}^{-1} = \frac{1}{\sigma^2}\begin{vmatrix} \frac{t_2}{t_1(t_2-t_1)} & \frac{1}{t_1-t_2} & 0 \ldots\ldots\ldots\ldots & 0 \\ \frac{1}{t_1-t_2} & \frac{t_3-t_1}{(t_3-t_2)(t_2-t_1)} & \frac{1}{t_2-t_3} \ldots\ldots & 0 \\ \ldots\ldots\ldots\ldots\ldots\ldots & \ldots\ldots\ldots\ldots & \frac{1}{t_{m-1}-t_m} & \frac{1}{t_m-t_{m-1}} \end{vmatrix}$$

while for $m=2$ and for $m=1$ we have respectively

$$S_{22}^{-1} = \frac{1}{\sigma^2}\begin{vmatrix} \frac{t_2}{t_1(t_2-t_1)} & \frac{1}{t_1-t_2} \\ \frac{1}{t_1-t_2} & \frac{1}{t_2-t_1} \end{vmatrix} \quad \text{and } S_{22}^{-1} = \frac{1}{\sigma^2\, t_1}$$

An approximate value of σ^2 can be given by the formula

$$(3.1)\qquad \sigma^2 = \frac{1}{m} \sum_{i=1}^{m} \frac{(\bar{X}_i - \bar{X}_{i-1})^2}{t_i - t_{i-1}}$$

Indeed the random variables

$$\frac{X(t_i) - X(t_{i-1})}{\sqrt{t_i - t_{i-1}}}$$

are independent, normally distributed with mean 0 and variance σ^2, so that the quantities

$$\frac{\bar{X}_i - \bar{X}_{i-1}}{\sqrt{t_i - t_{i-1}}}$$

can be reasonably considered independent outcomes of a random variable $N(0,\sigma^2)$. Here again we meet the problem that independent samples must be used for obtaining and validating the results of a Monte Carlo process. Thus we can conclude that in this way we can obtain an estimate of the integral $\bar{Z}$ and a probabilistic bound of the error in the approximation. Another problem which lends itself quite naturally to the above analysis is the one-dimension interpolation.

Let again $\bar{X}(t)$ be a sample path of Wiener process. We are looking for an approximation to the value $X^* = \bar{X}(t^*)$, assuming as before the values $\bar{X}(t_i)=\bar{X}_i\,(i=1, \ldots, m)$. Let $t_j < t^* < t_{j+1}\,(0 \le j < m)$ be.

We can easily obtain by the properties of the Wiener process

$$(3.2)\qquad \begin{cases} \mu(t^*) = E\left(\bar{X}(t^*) \mid \bar{X}_i;\ i=1,\ldots,m\right) = \bar{X}_j \dfrac{t_{j+1}-t^*}{t_{j+1}-t_j} + \bar{X}_{j+1} \dfrac{t^*-t_j}{t_{j+1}-t_j} \\[2ex] \sigma^2(t^*) = V\left(X(t^*) \mid \bar{X}_i;\ i=1,\ldots,m\right) = \sigma^2 \dfrac{(t^*-t_j)(t_{j+1}-t^*)}{t_{j+1}-t_j} \end{cases}$$

Therefore $X(t^*) \sim N(\mu(t^*), \sigma^2(t^*))$ and the mean value $\mu(t^*)$ coincides with approximate value given by interpolation, while the error in this approximation can be bounded in probability by the usual formula employing the value σ^2 given by (3.1).

By the very structure of this method, the values $\overline{X}_i$ with $i \neq j$, $i \neq j+1$ influence not the approximated value thus found, but the confidence we can place in it, determined by the value obtained for σ^2 according to (3.1).

Another problem for which the error evaluation can be performed with the methods outlined in this section is the global optimisation of one dimensional functions. The basic result is a theorem about the distribution of the global minimum of the sample path of the Wiener process $X(t,\omega)$.

Let

$$F(z) = \text{Prob}\ (\min_{0 \leq t \leq 1} X(t,\omega) \leq z \mid X(1,\omega) = X_1);$$

it can be shown that:

$$F(z) = \begin{cases} 1 & \text{for } z \geq \min(X_o, X_1) \\ \exp\left(-2 \dfrac{(X_o - z)(X_1 - z)}{\sigma^2}\right) & \text{for } z < \min(X_o, X_1) \end{cases}$$

We assume that we have evaluated the objective function $\overline{X}(t)$, considered as a sample path of the Wiener process, at the points $t_i (i=1, \ldots, m)$ and we pose, as above $\overline{X}(t_i) = \overline{X}_i$.

Let $X^*_m = \min_{1 \leq i \leq m} \overline{X}_i$, by the above theorem we have:

$$\text{Prob}\ (\min_{t_i < t < t_{i+1}} \overline{X}(t) < X^*_m - \gamma \mid \overline{X}(t_j) = X_j,\ j=1,\ldots,m) =$$

$$= P_i = \exp\left\{-2 \frac{(X^*_m - \gamma - \overline{X}_{i+1})(X^*_m - \gamma - \overline{X}_i)}{\sigma^2 (t_{i+1} - t_i)}\right\}$$

where γ is a suitable positive constant.

Therefore

$$\text{Prob}\ (\min_{0<t<1} \overline{X}(t) < X^*_m - \gamma \mid \overline{X}(t_j) = X_j,\ j = 1,\ldots,m) =$$

$$= 1 - \prod_{k=1}^{m-1} (1 - P_k)$$

This probabilistic bound enables to set up an effective stopping criterion for global optimization algorithms which assumes the Wiener process as a stochastic model of the one-dimensional objective function.

Unfortunately, at least to my knowledge, no analogous result to the above theorem has been proved for stochastic processes with multidimensional parameter.

In order to extend the above analysis to the optimisation of multidimensional functions we shall consider a simple statistical model suggested in a recent paper of Zilinskas (1978), based on heuristic arguments rather than on a theoretical analysis of a stochastic process. Let $f(X)$ be a real function mapping a domain $K \subset R^n$ into R, and $f(X_i)=f_i$ its values at the points X_i $(i=1, \ldots, m)$.

In this simple heuristic model we assume for any $X \in K$, $X \neq X_i$, that $f(X)$ is a one-dimensional normal variable with expected value

$$(3.3) \qquad \mu_m(X) = \frac{\sum_i \frac{f_i}{\|X - X_i\|}}{\sum_i \frac{1}{\|X - X_i\|}} \qquad (i = 1, \ldots, n$$

and variance

$$(3.4) \qquad \sigma_m^2(X) = \sigma^2 \min \|X - X_i\|$$

where $\|A\|$ is the euclidean norm of A, and σ^2 is the parameter of the model. A sensible value of σ^2 can be obtained by the formula

$$\bar{\sigma}_2 = \frac{1}{k} \sum_{j=1}^{k} \frac{(f(X_{m+j}) - \mu_m(X_{m+j}))^2}{\min\limits_{1 \leq i \leq m} \|X_{m+j} - X_i\|}$$

Where X_{m+j} for $j=1, \ldots, k$ are random points obtained from a uniform distribution on K (and different from any point X_i with $i=1, \ldots, m$). Even though (3.3) and (3.4) are not derived from a stochastic process, considering the proper correlation functions, but assumed only on an heuristic ground, we shall speak allowing for some improprieties, of conditional probability of the random variable $f(X)$; we mean the probability computed under the normality assumption for $f(X)$ and related through (3.3) and (3.4) to the values of $f(X)$ already observed.

Let $f^* = \min_{X \in K} f(X)$ and consider the sequence $f(X_i) = f_i$, $X_i \in K$ generated by any algorithm.

Let $f^*_m = \min_{1 \leq i \leq m} f(X_i)$ be assumed as an approximated value of f^*.

For any $\varepsilon_1 > 0$ and $X \in K$ we can compute on the basis of the model of the objective function

$$\psi_m(X, \varepsilon_1) = \text{Prob}\{f^*_m - f(X) > \varepsilon_1 \mid f(X_i) = f_i,\ i=1,\dots,m\}$$

and we easily deduce

$$\psi_m(X, \varepsilon_1) = \phi\{f^*_m - \varepsilon_1;\ \mu_m(X),\ \sigma_m(X)\}$$

Where $\phi(x;\ a,\ b)$ is the normal distribution function with mean a, and standard deviation b.

Now let us consider the set

$$E^m_{\varepsilon_1 \varepsilon_2} = \{X \in K\ ;\ \psi_m(X, \varepsilon_1) > \varepsilon_2\}$$

for suitable values of the positive real constants ε_1, ε_2. Let $\mu(A)$ be the Lebesque measure of the set A, normalized so that $\mu(K)=1$. The numerical value $\mu(E^m_{\varepsilon_1 \varepsilon_2})$ can be considered, at least in some instances, as an indicator of a probabilistic bound of the error. The value $\mu(E^m_{\varepsilon_1 \varepsilon_2})$ can be used effectively to build up a stopping rule in global optimisation algorithms. We assume that the sequence $f(X_i)$ has been generated be a Monte Carlo search; X_i are random points from an uniform distribution on K. Given a suitable real $\varepsilon_3 > 0$ we consider as a stopping rule for the sequence $f(X_i)$ the condition

$$\mu(E^m_{\varepsilon_1 \varepsilon_2}) < \varepsilon_3$$

The measure $\mu(E^m_{\varepsilon_1 \varepsilon_2})$ cannot be evaluated analytically; an approximation to it can be obtained by a Monte Carlo method. If Z_i is a random variable uniformly distributed on K and Y_i is a Bernoulli random variable such that:

$$Y_i = 1 \qquad \text{when} \quad \psi_m (Z_i, \varepsilon_1) < \varepsilon_2$$

$$Y_i = 0 \qquad \text{when} \quad \psi_m (Z_i, \varepsilon_1) > \varepsilon_2$$

the expected value $E(Y) = \mu(E^m_{\varepsilon_1 \varepsilon_2})$ is the probability for a point uniformly distributed on K of falling into $E^m_{\varepsilon_1 \varepsilon_2}$.

REFERENCES

[1] Archetti, F., & Betrò B., (1979), "A stopping criterion for global optimization algorithms" Research Report A-61 del Dipartimento di Ricerca Operativa e Scienze Statistiche dell'Università di Pisa.

[2] Blanc, C., (1959), "Sur l'estimation des erreurs d'arrondi, Information Processing, UNESCO.

[3] Cugiani, M., (1980), "Metodi numerico statistici", UTET, Torino.

[4] Goldstine, H.H. & Von Neumann, J.J. (1947), "Numerical inverting of matrices of higher order, I", Bull. Amer Math. Soc., Vol.53, pagg. 1021 - 1099.

[5] Goldstine, H.H. & Von Neumann, J.J. (1951), "Numerical inverting of matrices of higher order II",Proc. Amer. Math. Soc. Vol.2, pagg. 188 - 202.

[6] Mockus, J., Tiesis, V. & Zilinskas, A. (1978), "The application of Bayesian methods for seeking the extremum"in "Towards Global Optimization" Vol. 2, L.C.W. Dixon & G.P. Szegö eds., North-Holland.

[7] Rust B.W. & Burrus W.R., (1972), "Mathematical programming and the numerical solution of linear equations"; Amer. Elsevier Publ. Co. Inc., New Jork.

[8] Tirani, R., (1977), "Un metodo statistico per la stima degli errori nelle formule di integrazione approssimata", Atti Sem. Mat. e Fisica - Università di Modena, Vol. 26, pagg. 205 - 213.

[9] Zilinskas, A., (1978), "On statistical models for multimodal optimization", Math. Operationsforsch Statist., Ser. Statistics, Vol. 9, No 2, pagg. 255 - 266.

Numerical Techniques for Stochastic Systems
F. Archetti and M. Cugiani (eds.)
© North-Holland Publishing Company, 1980

ANALYSIS OF STOCHASTIC STRATEGIES FOR THE NUMERICAL SOLUTION OF THE GLOBAL OPTIMIZATION PROBLEM

Francesco Archetti
Istituto di Matematica dell'Università di Milano

INTRODUCTION.

In this paper we are concerned with the problem of identifying x^* and f^* such that $f^* = f(x^*) \leq f(x)$ for all $x \in K$. where $K \subset R^N$ is a compact set and $f : R^N \to R$.
Many algorithms have been shown to converge to local minimum points: when however the function has many local minima, these algorithms converge to any of them and are likely to fail to find the global minimum.
The numerical problem of locating the global minimum vs. a local one, the global optimization problem, has been the subject of systematic investigation since the early 70's and despite the increasing attention it has been recently attracting, still we don't have a clear - cut picture of the problem, let alone a coherent body of theory and algorithms as in the case of local optimization. It can be safely argued that the global optimization problem is more challenging than what it could be expected, as it is clearly witnessed by the increasing number of papers in this field, the fact that results from a large number of fields of mathematics seem to apply in global optimization studies and the still large "variance" of the numerical methods proposed.
A basic distinction can be drawn between deterministic methods and probabilistic ones: even if the former do not seem at the moment to provide effective numerical algorithms, still the analysis of the computational performance of a class of them gives some insight into the very nature of the global optimization problem and allows us to argue the case for the stochastic approach. (sect.1)
Probabilistic algorithms can be regarded as approximate algorithms,

in which the result has only some probability of being exact: we pay the price of this uncertainty in order to have information about the solution of problems which could otherwise be computationally intractable. A class of these algorithms, namely those based on random sampling (sect. 2), can be now regarded as an effective, even if still partly heuristic, framework for the numerical solution of global optimization problems.

After another approach a stochastic model of the problem, i.e. a probability distribution on its instances, is considered (sect. 3) and an algorithm can be designed on the basis of its "expected performance". (sect. 4).

Despite some difficulties which still restrain the numerical effectiveness of these last algorithms in multidimensonal problems, the development of stochastic models for optimization problems seems a most promising area in global optimization studies: a theoretical framework of these models is given in sect. 5.

Sect. 1: The case for the stochastic approach.

First it can be remarked that the global optimization problem, as stated in the introduction, is not well posed, in the classical sense, even when x^* is unique: indeed the continuous dependence of x^* on the data of the problem cannot be ensured. For example, let $f(x) = \cos(\pi x) - \delta x \quad K = [-2,2]$

If $\delta > o \; x^* \simeq 1$ while for $\delta < o \; x^* \simeq -1$

The global optimization problem in the dependent variable, i.e. the problem of finding $f^* \leq f(x)$ for all $x \varepsilon K$., is well posed if $f(x) \varepsilon C(K)$, as shown by the relation $\| f^* - g^* | \leq \| f - g \|$, where $g(x) \varepsilon C(K)$, g^* is it global minimum and $\|.\|$ the supremum norm in K.

We assume for the moment that f(x) belongs to some class of functions F (K).

Def. 1. For any n, a strategy is a vector valued function mapping F(K) into $K^n = K x \ldots x K$. (n-points strategies will be, in the following of this section, be indicated by S_n and the n-uple they associate to f by $S_n(f) = (x_1(f), x_2(f), \ldots, x_n(f))$.

Def. 2. If the strategy is constant in $F(K)$, i.e. it maps $F(K)$ into a point of K^n, then it is termed passive.

Def. 3. If the strategy depends on $f \in F(K)$ so that $x_j(f)$, $j=2,\dots,n$ depends on x_ℓ and $f(x_\ell)$, $\ell=1,\dots,j-1$, then it is termed sequential.

Def. 4. The accuracy of S_n, $S_n(f) = (x_1(f), x_2(f), \dots, x_n(f)$, for $f \in F(K)$, is $A(f,S_n(f)) = \min_{j=1,\dots,n} f(x_j(f)) - f^*$.

Def. 5. The guaranteed accuracy of S_n in $F(K)$ is

$$A(S_n) = \sup_{f \in F(K)} A(f,S_n(f)). \qquad (1.1)$$

The guaranteed accuracy is now assumed, as a measure of the effectiveness of the strategy S_n, analogously to what is done in the "worst case" analysis of combinatorial algorithms and more generally in the deterministic analysis of errors in numerical processes.

It is now easy to show that with only qualitative assumptions on $F(K)$, $A(S_n)$ can be meaningless, e.g. it can be unbounded for each finite n and all strategies S_n.

Let $f \in C^\infty[0,1]$; then, for any prefixed $\varepsilon>0$, there is a neighborhood U_ε of x^* such that: $f(x)-f^*<\varepsilon$ for $x \in U_\varepsilon \cap [0,1]$.

However if only qualitative assumptions are made about $f(x)$ we don't know how large U_ε is and whether $\min_{j=1,\dots,n} f(x)-f^*<\varepsilon$, whichever are the points x_j, for a finite n.

Indeed it is always possible to define a function $g(x) \in C^\infty[0,1]$ such that $g(x_j)=f(x_j)$, $j=1,\dots,n$, for each finite n, and still for any large finite H: $\min_{j=1,\dots,n} g(x_j)-g^*>H$.

Therefore we must introduce "quantitative" assumptions about $F(K)$: in the sequel of this section we shall assume:

$F(K) = L_{\ell,\rho}(K) = \{f:K\to R; |f(x)-f(y)| \leq \ell\rho(x.y)\}$, $x,y \in K$ for a distance function ρ.

If the objective function $f(x)$ satisfies the above Lipschitz condition, we can introduce the following optimality criteria for n-points strategies:

Def. 6. S_n^* is A-optimal if $A(S_n^*) = \inf_{S_n} A(S_n)$.

Def. 7. Given $\delta>0$ if there exists n^* and S_{n^*} such that $A(S_{n^*}) \leq \delta$ while no strategy S_n exists such that $A(S_n) \leq \delta$, $n<n^*$, then S_{n^*} is n-optimal.

Let $K = [0,1]^N$ and $\rho(x,y) = \max_{i=1,\dots,N} |x^i-y^i|$, $x=(x^1,x^2,\dots,x^N)$,

$y=(y^1,y^2,\dots,y^N)$. Passive optimal strategies can be constructed in the following way.

If $n=\nu^N$, for some natural ν, then the strategy $S_n^*: S_n^*(f)=(x_1^*,x_2^*,\dots,x_n^*)$, where x_j^*, $j=1,\dots,n$ are the centres of the hypercubes with edges of length $1/\nu$ partitioning K, is A-optimal.

If $\nu^N<n<(\nu+1)^N$ then the strategy in which x_j^*, $j=1,\dots,\nu^N$, are chosen as before and x_j^*, $j=\nu^N+1,\dots,n$, are chosen arbitrarily, is A-optimal.

In both cases the guaranteed accuracy is:

$$A(S_n^*) = \frac{\ell}{2\,[\sqrt[N]{n}]} \qquad (1.2)$$

where [x] means the largest integer not larger than x.

The method in which the objective function is evaluated at the points $x_j^* \in S_n^*$, $n=\nu^N$, is usually called the grid search method.

It is now easy to derive n-optimal passive strategies, with the above assumptions about ρ and K.

For any $\delta>0$, let $n^*=\min\{n$ such that $A(S_n^*)\leq\delta\}$: it turns out:

$$n^* = \left(\left[\frac{\ell}{2\delta}\right]'\right)^N, \qquad (1.3)$$

where [x] means the least integer not less than x, and the strategy S_{n^*} is n-optimal.

In general the problem of finding n-optimal strategies can be reduced to the solution of a set covering problem. A discrete model of K, by a square net of mesh-size h, is constructed: a minimum cardinality set covering problem can then be posed for this discrete model.

It can be shown (Th. 7 in Archetti & Betrò (1978a)) that a value $\bar{h}$ exists such that, for $h<\bar{h}$; $n^*(h)$, the cardinality of the discrete optimal covering, is equal to n^*.

The result given by (1.3) is not computationally encouraging in that it states that both A and n-optimal passive strategies require, for a given accuracy δ in the approximation to f^*, a number of function evaluations which increases exponentially in the number of independent variables of the problem.

Still one could think that sequential strategies can perform better than passive ones: this is not the case, at least in terms of guaranteed accuracy, as shown by the following result (Archetti & Betrò (1978a)).

Let $\hat{\Sigma}_n$ and Σ_n be respectively the set of n-points sequential strategies and passive ones.

Let $\alpha_1 = \inf_{S_n \varepsilon \hat{\Sigma}_n} A(S_n)$ and $\alpha_2 = \inf_{S_n \varepsilon \Sigma_n} A(S_n)$. Then $\alpha_1 = \alpha_2$.

For a given $\delta > 0$ let $m_1 = \min\{n$ for which $S_n \varepsilon \hat{\Sigma}_n$ exists such that $A(S_n) \leq \delta\}$ and $n_2 = \min\{n$ for which $S_n \varepsilon \Sigma_n$ exists such that $A(S_n) \leq \delta\}$. Then $n_1 = n_2$.

We remark that the above results are concerned with the guaranteed accuracy in $L_{\ell,\rho}(K)$: for most functions in $L_{\ell,\rho}(K)$, sequential strategies may perform better than passive ones and for this reason some research has been devoted to develope sequential algorithms. (Evtushenko (1971), Shubert (1972) and Strongin (1973)).

Even if the performance of these sequential algorithms can improve over that given by (1.3), still the number of f.e. they require is exponential in N, by the "space covering" architecture of these algorithms, which makes them unpractical for N larger than 2 or 3, when a reasonable accuracy is required.

Two more drawbacks can be pointed out in these algorithms: the value ℓ of the Lipschitz constant must be known a priori and no approximation to the solution is given at the intermediate stages of the algorithm. There are other deterministic approaches of an entirely different nature, who have been widely investigated (e.g. Gomulka (1978a) and Treccani (1978)), but they still seem to have some theoretical and computational difficulties and as yet cannot be regarded as a general computational tool in global optimization problems.

Therefore we could sum up this analysis of the deterministic approach by the term "computational intractability" of the global optimization problem. In order to overcome this computational intractability approximate methods have been developed, namely probabilistic methods in which the result has only some probability of being exact.

In sect. 2 we shall be concerned with a class of probabilistic algorithms, based on random samples of the objective function, whose structure and numerical performance have been widely investigated.

The structure of these algorithms, is basically composed of two parts: a deterministic part, which performs the approximation to the different minima of the objective function by local optimization routines, and a stochastic one, which controls the size of the random sample and the

local optimizations. It can be said that some of these algorithms have proved rather successful even in many variables global optimization problems. (Dixon & Szegö (1978)).
The peculiar structure of these algorithms could be fully exploited in the framework of parallel computing, which can now be regarded as a most promising area in computational global optimization. (McKeown (1980)).
There is another reading of the result of the analysis of deterministic strategies, which introduces us to another class of approximate methods: a main drawback of the "worst case" analysis of algorithms, implied by the criterion of guaranteed accuracy, is that the worst case can happen seldom (perhaps never) in practice. Since in turn the study of the behaviour of algorithms is instrumental in their design it is clear that the "worst case" analysis should be replaced by some more realistic measure of the performance of algorithms. One such measure could be obtained, at least in principle, considering a stochastic model of the problem, i.e. a probability distribution on the elements of the space to which the objective function belongs, and replacing the guaranteed accuracy of a strategy by its expected accuracy.
This approach, which has been recently widely suggested also in connection with combinatorial algorithms (e.g. Karp (1976)), enables to derive a class of global optimization algorithms, first considered by Kushner (1964) and Mockus (1964),(1972),(1975).
The objective function f(x) is considered to be a sample path of a stochastic process $\phi(x,\omega)$, the stochastic model of f(x), where ω is an unknown index belonging to some probability space Ω.
The probability distribution of ω in Ω is given, for any finite n, by the joint probability distribution function:

$$P_{x_1,x_2,\ldots,x_n}(y_1,y_2,\ldots,y_n) = \text{Prob}\ (\omega:\phi(x_1,\omega)<y_1,\ldots,\phi(x_n,\omega)<y_n).$$

The development of global optimization methods based on a stochastic model of the objective function, despite some difficulties which still limit their performance in multidimensional problems (Mockus,Tiesis & Zilinskas (1978)),can be regarded as a most promising area in global optimization studies.
In sect. 3 the most used one dimensional stochastic model, the Wiener

process, and some simplified multidimensional models are considered. In sect.4 the statistical validity of the Wiener model is discussed and an effective one dimensional algorithm is closely analyzed: its structure displays the flexibility allowed in the design of algorithms by the use of stochastic models.

In sect.5 an attempt is given to outlining a general theoretical framework of the use of stochastic models in optimization problems: the conditional expectation of a random function is considered as an approximation to the objective function and the basic properties of random functions are used to analyze the resulting optimization strategies. Two optimality criteria quite naturally arise for such strategies: E-optimality, after which the expected error in the approximation to f^* is minimized, and P-optimality, after which the probability is maximized that this error is kept within a prefixed bound. (Archetti & Betrò (1980b)).

Sect. 2: Methods based on random sampling.

These methods can be framed after the following general pattern:

PHASE A: Draw n points $x_j \varepsilon K$, $j=1,\dots,n$, from a uniform distribution in K and compute $f(x_j)$.

PHASE B: Select the "most promising" points and start from them a local optimization routine obtaining a least value f_c.

PHASE C: Test whether f_c is "probably" the global minimum of f(x) in K.

The above cycle is repeated until the test is satisfied.

In the simplest algorithm that point yielding the least sampled value is considered in PHASE B and the test in PHASE C is satisfied if no improvement over f_c is observed in one (or more) further calls of PHASE A. In this method no more that one local optimization is performed in the region of attraction of a minimum: this anyway at the cost of wasting most of the information contained in the sample.

Effective algorithms require a properly balanced compromise between the conflicting goals of making a good use of all the information contained in the sample and of reducing the risk of converging to local minima already found.

Rather than considering in detail a specific algorithm we shall now outline some considerations about the 3 phases and some suggestions for their implementation.

PHASE A: Let $A \subset K$ such that $\mu(A)/\mu(K) = \alpha > 0$, where $\mu(.)$ denotes the volume. If the sample x_j, $j=1,\dots,n$, is drawn from a uniform distribution in K, it can be shown that with probability 1:

$$\lim_{n\to\infty} \min_{j=1,\dots,n} f(x_j) = f^* \qquad (2.1)$$

and for n sufficiently large

$$\min_{i=1,\dots,n} \mathrm{dist}(A,x_i) = 0 \text{ where } \mathrm{dist}(A,x_i) = \inf_{x\in A} \|x-x_i\|. \qquad (2.2)$$

The probability that at least one point in the sample belongs to A is:

$$P(A,n) = 1 - (1-\alpha)^n.$$

By (2.1), $f^*_n = \min_{j=1,\dots,n} f(x_j)$ could be assumed, in a pure Monte Carlo optimization, as an approximation to f^*: if the objective function $f \in L_{\ell,\rho}(K)$ we can compute the "guaranteed expected accuracy", of uniform random sampling.

Let $T_n = (x_1,x_2,\dots,x_n)$, where x_j, $j=1,\dots,n$, are independent random variables uniformly distributed in K. We define the expected accuracy of uniform random sampling:

$$\tilde{A}(f,T_n) = E_{T_n}(\min_{i=1,\dots,n} f(x_i) - f^*)$$

where E_{T_n} denotes the expected value according to the distribution of T_n and the guaranteed expected accuracy:

$$\tilde{A}(T_n) = \sup_{f\in L_{\ell,\rho}(K)} \tilde{A}(f,T_n).$$

If $\rho(x,y) = \max_{i=1,\dots,N} |x^i - y^i|$ it can be proved that

$$\tilde{A}(T_n) = \ell \frac{\Gamma(1+n)\Gamma(1+1/N)}{\Gamma(1+n+1/N)}.$$

As $n\to\infty$ $\tilde{A}(T_n) \sim \ell \frac{\Gamma(1+1/N)}{\sqrt[N]{n}} \geq \ell \frac{0,8}{\sqrt[N]{n}}$ versus a guaranteed accuracy of $\ell/2[\sqrt[N]{n}]$ for the grid search method.

This result partly disagrees with what is stated in Andersenn & Bloomfield (1975) where, for N>6, uniform sampling is argued to be more effective than the grid search.

The difference can be explained by the different distance functions considered and the assumption, in the above quoted paper, that x^* is far from the boundary of K. Indeed after Th.2 in Archetti-Betrò (1978b), if x^* is the centre of K we have:

$$\tilde{A}(T_n) = \frac{\ell}{2} \frac{\Gamma(1+n)\Gamma(1+1/N)}{\Gamma(1+n+1/N)}.$$

As $n\to\infty$ we have, for N>1, $\tilde{A}(T_n) \sim \frac{\ell}{2} \frac{\Gamma(1+1/N)}{\sqrt[N]{n}} \leq \frac{\ell}{2\sqrt[N]{n}}$ while for N = 1,

$$\tilde{A}(T_n) = \frac{\ell}{2}\,\frac{\Gamma(1+n)\Gamma(1+1/N)}{\Gamma(1+n+1/N)} = \frac{\ell}{2(n+1)} .$$

Quite naturally the question arises whether there are other distributions than the uniform one which might improve the guaranteed expected accuracy.

Let Σ_n be the set of all probability measures in K^n.

An n-points random passive strategy is then defined as an n-uple $(x_1, x_2, \dots, x_n)$ of independent random variables x_j, j=1,...,n, distributed in K with a joint probability $\sigma_n \varepsilon \Sigma_n$ (in the sequel also σ_n is called a random strategy).

If a strategy σ^*_n exists such that $\tilde{A}(\sigma^*_n) = \inf_{\sigma_n \varepsilon \Sigma_n} \tilde{A}(\sigma_n)$ then σ^*_n is optimal.

The analysis of random stretegies, according to their guaranteed expected accuracy, let alone finding optimal ones, is a very difficult task. If $\rho(x,y)=|x-y|$ and $K=[0,1]$, an optimal random passive strategy is exhibited in Sukharev (1971).

Its guaranteed expected accuracy $\ell/2(2n-1)$ is better than for one dimensional grid search.

For N>1 only "asymptotically optimal" strategies can be obtained for $\rho(x,y)=\max_i |x^i-y^i|$ and $K=[0,1]^N$.

Their guaranteed expected accuracy can be shown to be asymptotic to $\frac{N}{N+1}\,\frac{\ell}{2\sqrt[N]{n}}$.

The results here outlined for phase A, even if they can be of some theoretical interest, are not directly relevant to the numerical performance of actual optimization algorithms, where the information contained in the sample is mainly used in phases B and C to control the result of local searches, rather than to provide directly an approximation to f*.

PHASE B: Cluster analysis can be now considered the most effective technique for recognizing the pattern of multimodalities in the objective function.

Cluster analysis is a set of statistical techniques for dividing a large set of data, given a measure of similarity between them, into smaller subsets, so that each data is more similar to the others within its subset than to those outside.

Its application to the global optimization problem, due to Törn (1976), has been widely investigated (Boender (1979)), and successfully implemented by various authors (Gomulka (1978b), de Biase & Frontini (1979), Boender, Rinnooy Kan, Stougie & Timmer (1980)).

The clustering approach in global optimization problems is based on the association of one cluster to each local minimum: one point in each cluster could then be used as a starting point of a local optimization routine.

The clustering procedure is performed iterating a basic cycle structured in the following way:

1) The random sample $\{(x_j)\}$, $j=1,\ldots,n$; $x_j \varepsilon K$ is transformed eliminating a number of points with larger function values and performing some steps of a local optimization routine from the remaining points. This operation transforms the sample into a number of separate groups of points which have come closer together in the process of converging to the same local optimum.

2) The average density $d=n/\mu(K)$ of the points x_j in K is computed: clusters are then characterized as groups of points with a density larger than d and are usually obtained "growing" them around a "seed": the group is enlarged, enclosing new points, as wide as the density of its points is larger than d.

As soon as this density becomes less than d a new seed point is considered and a new cluster is grown around it.

The local minima already found are usually chosen as "seed" points: when clusters have been grown around these local minima and still some points remain to be clustered one can perform a local search from the point with the least function value not yet clustered to obtain a new seed point. When all the points have been clustered a number of points is discarded from each cluster and then the cycle is repeated. The clustering procedure can be terminated when two consecutive cycles give the same number of clusters. If one cluster has been connected to each local minimum the problem is solved: there is of course no guarantee of reaching this ideal situation and the user of the procedure is faced with the problem of setting the proper balance between many factors as the number of initial points in the sample, the rate of reduction of points in a cluster and the number of steps per formed by the local optimization routine between subsequent clusterings.

PHASE C: The control of the result and consequently the development of proper termination criteria is certainly the critical part of the methods based on random sampling.

It is still very difficult to evaluate the probability of the result being exact or more generally that the error in the approximation does not exceed some prefixed value. This implies that the stopping rules incorporated in these algorithms even if usually effective -in this sense also cluster analysis can be regarded as a control of the result- are still basically heuristic.

An approximation, in probability, to the value f^* can be obtained considering the function

$$\Psi(\xi) = \frac{m(E(\xi))}{m(K)}, \quad \xi \in R^1,$$

where $E(\xi)=\{x \in K : f(x) \leq \xi\}$ and $m(.)$ denotes the Lebesgue measure. (Chichinadze (1967), Archetti (1975)).

If $f(x)$ is Lebesgue measurable, then $\Psi(\xi)$ is completely defined because $E(\xi)$ belongs to the σ algebra of Lebesgue sets in K. If $\xi<f^*$, the essential infimum of f in K, then $\Psi(\xi)=0$, $m(f^{-1}(f^*))=0$ implies $\Psi(\xi)=0$ iff $\xi \leq f^*$.

The function $\Psi(\xi)$ apart from some trivial cases cannot be evaluated analytically: a set of independent and unbiased observations of $\Psi(\xi)$, for different values ξ_j, $j=1,\ldots,r$, can be obtained performing, for each ξ_j, a uniform random sample of size q and computing $\tilde{\psi}(\xi_j)=p/q$ where p is the number of points in this sample hitting $E(\xi_j)$.

Thus $\psi(\xi)$ turns out to be an unknown regression function and we are looking for its root ξ^* such that $\psi(\xi^*)=0$ and $\psi(\xi)>0$ for $\xi>\xi^*$.

A method for the uniform approximation of $\psi(\xi)$ based upon a particular spline system, has been developed in Archetti & Betrò (1975) and de Biase & Frontini (1978).

Once an estimate of ξ^* has been obtained, it can be used to control, in a statistical sense, the result of the local searches performed at phase B (de Biase & Frontini (1979)).

An interesting possibility for deriving proper statistical stopping rules is the use of extreme value statistics, suggested by Archetti,

Betrò & Steffè (1975) and de Haan (1979), and recently exploited by Boender, Rinnoy Kan, Stougie & Timmer (1980) in their global optimization algorithm.

Sect. 3: Stochastic models of the objective function.

The stochastic model which is more often used for one dimensional objective functions is the Wiener process $f(x)$, $x \in K=[x_o, \bar{x}]$ for which i) $f(x_o)=\mu$ and ii) $f(x)-f(y) \sim N(0, \sigma^2|x-y|)$; $x,y \in K$.

The choice of the Wiener process, despite some obvious shortcomings -e.g. the a.e. non differentiability of its sample paths- is basically due to the following reasons:

1) it is an acceptable statistical model of the global behaviour of the objective function, whose goodness of fit can be easily tested, as it will be described in the next section.

2) The behaviour of a Wiener sample path, and therefore of the objective function, conditioned by the observations -i.e. function evaluations- already performed can be easily characterized.

The distribution of $f(x)$, conditioned by $z_n=\{f_1=f(x_1), f_2=f(x_2), \ldots, f_n=f(x_n)\}$, $x_i<x_{i+1}$, is normal with expected value $\mu(x)$ and variance $\sigma^2(x)$ given for $x \in \Delta_i=[x_i, x_{i+1}]$, $i=0,\ldots,n-1$, by the following formulas:

$$\mu(x) = E(f(x)|z_n) = f_i \frac{x_{i+1}-x}{x_{i+1}-x_i} + f_{i+1} \frac{x-x_i}{x_{i+1}-x}$$
$$\sigma^2(x) = \text{var}(f(x)|z_n) = \sigma^2 \frac{(x-x_i)(x_{i+1}-x)}{x_{i+1}-x_i} \tag{3.1}$$

and, for $x \in \Delta_n=[x_n, \bar{x}]$, by

$$\mu(x) = E(f(x)|z_n) = f(x_n)$$
$$\sigma^2(x) = \text{var}(f(x)|z_n) = \sigma^2(x-x_n) \tag{3.2}$$

Another characterization of the Wiener process, is given by the distribution: $F(z) = \text{Prob}\{\min_{x \in K} f(x) \leq z | f(\bar{x}) = \bar{f}\}$.

It can be proved, (Archetti & Betrò (1980b)), that:

$$F(z) = \begin{cases} 1 & z \geq \min(\bar{f}, \mu) \\ \exp\left(-2 \frac{(\mu-z)(\bar{f}-z)}{\sigma^2(\bar{x}-x_0)}\right) & z < \min(\bar{f}, \mu) \end{cases} \tag{3.3}$$

The relations (3.1), (3.2) and (3.3) enable to design an effective one dimensional global optimization algorithm, to be analyzed in sect.4, in which the evaluations of $f(x)$ are performed where the conditional probability is maximum of improving over the least function value as yet observed. Moreover (3.3) allows to evaluate the probability that the error in the approximation to f^* does not exceed some prefixed value and thus to derive an effective stopping criterion for the algorithm. These results do not admit a straightforward extension to multidimensional problems both for the excess of overhead computations connected with their implementation and the mathematical intricacies of stochastic processes with a multidimensional parameter.

For this reason some simplified multidimensional models have been proposed, which are based on heuristic arguments rather than derived from a stochastic process, considering the proper correlation functions. One such model has been proposed by J. Mockus (1980); another model, which we shall briefly discuss below, has been suggested by Zilinskas (1978): who assumes that f(x) is a normal variable with expected value:

$$\mu_n(x) = \frac{\sum_{i=1}^{n} f(x_i)/\|x-x_1\|}{\sum_{i=1}^{n} 1/\|x-x_i\|} \tag{3.4}$$

and variance $\sigma_n^2 = \sigma^2 \min_{i=1,\dots,n} (x-x_i)$ (3.5)

where $\|\,.\,\|$ is the euclidean norm and σ a parameter of the model. Even if (3.4) and (3.5) are assumed only on heuristic grounds, we shall speak, allowing for some improperty, of conditional probability of the random variable f(x), meaning the probability, computed under the normality assumption for f(x), and related by (3.4) and (3.5) to the values of f(x) already observed. A possible use of this simplified model has been suggested in Archetti & Betrò (1979), in order to derive a probabilistic stopping rule for global optimization in R^N, $N>1$, where no result like (3.3) is as yet available.

Sect. 4: The structure of a probabilistic algorithm.

In order to display, in a simple mathematical framework, the possibilities allowed, in the design of a global optimization algorithm, by the use of a stochastic model of the objective function, we shall describe a 1-dimensional algorithm based on the Wiener process. (Archetti & Betrò (1980a)).

First we shall consider whether the Wiener process can be accepted as a "good" stochastic model of the objective function f(x), $x \in K=[x_0,\bar{x}]$: this can be accomplished testing the hypothesis that the observed increments $q_i = f(x_i) - f(x_{i-1})$, where x_i are equidistributed in K, are drawn from a normal distribution. The test is performed in two steps:

a) The randomness of the sample $Q_n = q_i$, $i=1,\dots,n$ can be checked by the sample correlation coefficient:

$$R = \frac{\frac{1}{n}\sum_{i=1}^{n} iq_i - \bar{q}_n \bar{I}}{S_q S_I}$$

where $\bar{q}_n$ is the mean of Q_n, S_q^2 is the standard deviation of Q_n, while $\bar{I}$ and S_I are some coefficients depending on n.

If the r.v. q_i, $i=1,\dots,n$, are independent on i, the r.v.

$$T_n = R[(n-2)(1-R^2)^{-1}]^{\frac{1}{2}}$$

follows a t-distribution with $\nu=n-2$ degrees of freedom.

Thus, from a tabulated t-distribution, one can reject, for different sample sizes and significance levels, the hypothesis that the increments q_i, $i=1,\dots,n$, are stochastically independent.

b) The normality of a r.v., whose mean and variance are unknown, can be rejected, when the quantity $D_n = \max_{i=1,\dots,n} \delta_i$ exceeds a critical value $D_{\alpha,n}$, tabulated for different significance levels α.

Here $\delta_i = \max\{F_o(q_{(i)};\bar{q},S_q) - \frac{i-1}{n}, \frac{i}{n} - F_o(q_{(i)};\bar{q},S_q)\}$ where $q_{(i)}$ are the values q_i, arranged in increasing order and $F_o(x;\bar{x},s) = \int_{-\infty}^{\frac{x-\bar{x}}{s}} \frac{1}{\sqrt{2\pi}} \exp(-\frac{u^2}{2})du$.

The tests a) and b) have been performed for some 1-dimensional functions and result in the rejection of the hypothesis that Q_n is a random sample drawn from a normal distribution when n exceeds some value $\bar{n}$, depending on the particular function being considered and the significance level of the tests.

This fact allows for a simple explanation: as n increases the distribution of the r.v. q_i is increasingly influenced by the "local" features of f(x), which, for f(x) reasonably smooth, cannot be modelled by the Wiener process, due to the a.e. nondifferentiability of its sample paths.

After these introductory remarks about the validation of the model, we consider, in some detail, the structure of the algorithm. We remind that before starting the algorithm some evaluation of f(x) are spent in order to derive a first estimate of the parameters μ and σ of the Wiener process, which is updated during the progress of the algorithm.

1) <u>The selection of the observation points.</u>

We assume that n-1 evaluations of f(x) have been performed, at x_i i=1,...,n-1, arranged so that $x_o \le x_1 \le \ldots x_{n-1}$: let $f^*_{n-1} = \min(\mu, f_1, f_2, \ldots \ldots, f_{n-1})$ and γ some positive value.

First we define the interval Δ_ℓ by the condition:

$$\text{Prob}\{\min_{x\in\Delta_\ell} f(x) < f^*_{n-1} - \gamma | z_{n-1}\} = \max_i \text{Prob}\{\min_{x\in\Delta_i} f(x) < f^*_{n-1} - \gamma | z_{n-1}\}.$$

The latter probability is equal, by the Markov property, to

$$\text{Prob}\{\min_{x\in\Delta_i} f(x) < f^*_{n-1} - \gamma \mid f(x_i) = f_i,\ f(x_{i+1}) = f_{i+1}\}$$

and can be computed, by (3.3), as: $\exp\left(-2\,\dfrac{(f^*_{n-1}-\gamma-f_{i+1})(f^*_{n-1}-\gamma-f_i)}{\sigma^2(x_{i+1}-x_i)}\right)$

Let, for $x\in\Delta_\ell$: $P_\ell(x) = \text{Prob}\{f(x) < f^*_{n-1} - \gamma | f(x_\ell) = f_\ell,\ f(x_{\ell+1}) = f_{\ell+1}\} =$

$$= \int_{-\infty}^{f^*_{n-1}-\gamma} \frac{1}{\sqrt{2\pi}\,\sigma(x)} \exp\left[-\frac{1}{2}\,\frac{(z-\mu(x))^2}{\sigma^2(x)}\right] dz$$

where $\mu(x)$ and $\sigma(x)$ are given by the formulas (3.1) and (3.2); x_n is defined by the condition $P_\ell(x_n) = \max_{x\in\Delta_\ell} P_\ell(x)$, and can be easily computed as

$$x_\ell + \frac{(f^*_{n-1}-\gamma-f_\ell)\ (x_{\ell+1}-x_\ell)}{2(f^*_{n-1}-\gamma)-f_\ell-f_{\ell+1}}\ .$$

2) <u>A stopping criterion of the algorithm.</u>

$$\text{Let } P_k = \text{Prob}\{\min_{x\in[x_k,x_{k+1}]} f(x) < f^*_n - \varepsilon | f(x_k) = f_k,\ f(x_{k+1}) = f_{k+1}\} \tag{4.1}$$

where $\varepsilon > o$ is a prefixed accuracy level

$$\text{By (3.3) } P_k = \exp\left[-2\,\frac{(f^*_n-\varepsilon-f_k)\ (f^*_n-\varepsilon-f_{k+1})}{\sigma^2(x_{k+1}-x_k)}\right]$$

The algorithm terminates when $\text{Prob}\{\min_{x\varepsilon k} f(x) > f^*_n - \varepsilon | z_n\} = \prod_{k=0}^{n-1}(1-P_k)$ is larger than a prefixed probability level.

Given the poor "local" fit of the Wiener process, the numerical performance of the algorithm can be improved using a local optimization routine in those intervals where f(x) can be reasonably assumed unimodal. After a simple effective rule, one such interval $L_h = [x_{h-3}, x_{h+3}]$, is characterized by the relation:

$$f(x_{h-3}) \ge f(x_{h-2}) \ge f(x_{h-1}) \ge f(x_h) \le f(x_{h+1}) \le f(x_{h+2}) \le f(x_{h+3})$$

If $[x_k, x_{k+1}] \varepsilon L_h$, for some h, the corresponding P_k in (4.1) is set to zero, in view of the local optimization already performed in L_h. The value of γ can be controlled according to the progress of the algorithm: a sensible sequence $\{\gamma_j\}$ can be obtained on the basis of (3.3). Let $P_{\gamma_j} = \text{Prob}\{\min_{x \varepsilon k} f(x) > f_n^* - \gamma_j | z_n\}$:
γ_j is kept constant as long $P_{\gamma_j} \leq \bar{P}$, where $\bar{P}$ is a prefixed probability level and halved as $P_{\gamma_j} > \bar{P}$.

Sect. 5. Towards a theory of stochastic models in optimization problems.

Let (Ω, σ, P) be a probability space and (U, B) a measurable space. Given a set X we consider the mapping $\xi_w(x) : X \to U$, $w \varepsilon \Omega$ and the set $\Omega_{x,b} = \{w : \xi_w(x) \varepsilon b\}$.
The mapping $\xi_w(x)$ is a random function iff, for any $x \varepsilon X$ and $b \varepsilon B$, $\Omega_{x,b} \varepsilon \sigma$ i.e. for arbitrary fixed x, $\xi_w(x)$ is a measurable mapping of (Ω, σ) into (U, B). The subscript w in $\xi_w(x)$ will be later dropped unless strictly necessary.
A useful characterization of a random function is by its marginal distribution: given any n points in X (n arbitrary integer) we consider in (U^n, B^n) the random element $\{\xi(x_1), \xi(x_2), \ldots, \xi(x_n)\}$.
Its probability distribution in B^n

$$P_{x_1, x_2, \ldots, x_n}(b) = P\{w : \xi_w(x_1), \xi_w(x_2), \ldots, \xi_w(x_n) \varepsilon b\} \; b \varepsilon B^n$$

is called the marginal distribution of the random function $\xi(x)$.
Conversely Kolmogorov's theorem states that, given a family of distributions

$$\{P_{x_1, x_2, \ldots, x_n}(b); \; n=1,2,\ldots; x_k \varepsilon X, \; k=1,2,\ldots,n; \; b \varepsilon B^n$$

under the so called compatibility conditions there exists a random function for which the given family of distributions is the family of marginal distributions.
Let's consider a random function $\xi_w(x)$ such that, for each $x \varepsilon X$, we have:

$$\int_\Omega |\xi_w(x)|^2 P(dw) < +\infty.$$

For any such function, assuming the values ξ_j at $x_j \varepsilon X$, $j=1,\ldots,n$ the following relation can be shown to hold for any $x \varepsilon X$:

$$E_w\{E_n(\xi_w(x) | \xi(x_1)=\xi_1, \xi(x_2)=\xi_2, \ldots, \xi(x_n)=\xi_n) - \xi_w(x)\}^2 =$$
$$\min_{\eta \varepsilon H} E_w\{\eta(x; \xi_1, \xi_2, \ldots, \xi_n) - \xi_w(x)\}^2$$

where E_w denotes the expected value with respect to P, E_n the conditional expectation and H is the family of all measurable functions with respect to $\xi_1, \xi_2, \ldots, \xi_n$.

A real random function $\xi(x)$, $x \in X$, is called gaussian if its marginal distributions are normal.

We recall that if $a(x)$ and $b(x,y)$, $x,y \in X$, are such that for any n and any n-uple $x_i \in X$, $i=1,\ldots,n$ the matrix with entries $b(x_i,x_j)$ $i,j=1,\ldots,n$ is symmetric positive definite, there exists a gaussian random function $\xi(x)$, $x \in X$, for which the expected value $E(\xi(x))=a(x)$ and $E(\xi(x)-a(x))(\xi(y)-a(y)) = b(x,y)$.

Setting $u_i = \xi(x_i)$, $\underline{u} = (u_1,u_2,\ldots,u_n)$, $\quad a = (a(x_1), a(x_2),\ldots,a(x_n))$,

$$B = [b_{ij}],\ b_{ij} = b(x_i,x_j)\quad i,j = 1,\ldots,n$$

the joint p.d.f. of $\underline{u}$ is: $p(\underline{u};\underline{a},B) = \dfrac{1}{(2\pi)^{n/2}|B|^{1/2}} \exp\{-\frac{1}{2}(\underline{u}-\underline{a})^T B^{-1}(\underline{u}-\underline{a})\}$

The gaussian random function $\xi(t)$, $t \in [0;T]$ for which $a(t)=\mu$ and $b(t,s)=\sigma^2\min(t,s)$, is the Wiener process $\xi(t)$ such that $\xi(0)=\mu$ and $\xi(t)-\xi(s) \sim N(0,\sigma^2|t-s|)$.

The marginal distribution of a random gaussian function $\xi(x)$, conditioned by any number n of observed values, is normal: thus the expected value and the covariance of the random vector $\underline{\alpha}_1=(\xi(y_1),\xi(y_2),\ldots,\xi(y_k))$ $y_i \in X$, $i=1,\ldots,k$, conditioned by the observations already performed $\underline{\alpha}_2 = (\xi(x_1),\xi(x_2),\ldots,\xi(x_n))$, can be computed.

Let $\underline{a}_1=(a(y_1),a(y_2),\ldots,a(y_k))$, $\underline{a}_2=(a(x_1),a(x_2),\ldots,a(x_n))$, and

$$B=\left[\begin{array}{c|c} B^{11} & B^{12} \\ \hline B^{21} & B^{22} \end{array}\right] \qquad \begin{array}{ll} b^{11}_{i,j}=b(y_i,y_j) & i,j=1,\ldots,k \\ b^{12}_{i,j}=b^{21}_{j,i}=b(y_i,x_j) & i=1,\ldots,k;j=1,\ldots,n \\ b^{22}_{i,j}=b(x_i,x_j) & i,j=1,\ldots,n \end{array}$$

It turns out:

$$E(\underline{\alpha}_1|\underline{\alpha}_2) = \underline{a}_1 - B^{12}(B^{22})^{-1}(\underline{\alpha}_2-\underline{a}_2)$$

$$Cov(\underline{\alpha}_1|\underline{\alpha}_2) = B^{11}-B^{12}(B^{22})^{-1}B^{21}$$

Let X be a measurable space: an n-steps strategy is a vector $d=(d_0,d_1,\ldots,d_{n-1})$ of measurable functions d_i, $i=0,\ldots,n-1$, mapping $(X\times R)^i$ into X.

An n-steps strategy d applied to a random function $\xi(x)$ produces an n-uple x_o^d, $x_1^d = d_1(x_o,\xi(x^d)),\ldots,x_{n-1}^d = d_{n-1}(x_o,\xi(x_o^d)),\ldots,x_{n-2},\xi(x_{n-2}^d))$.

In order to characterize an n-steps optimization strategy we introduce the functions:

u_{n-1}: $(x_o,\xi_o,\ldots,x_{n-2},\xi_{n-2}) \to \inf_{x\varepsilon X} E\{\xi(x)\mid\xi(x_o)=\xi_o,\ldots,\xi(x_{n-2})=\xi_{n-2}\}$

d^*_{n-1}: $(x_o,\xi_o,\ldots,x_{n-2},\xi_{n-2}) \to x_{n-1}$, the point where the above infimum is attained.

The function u_{n-2} and d_{n-2} can be subsequently defined by:

u_{n-2}: $(x_o,\xi_o,\ldots,x_{n-3},\xi_{n-3}) \to$

$\to \inf_{x\varepsilon X} E\{u_{n-1}(x_o,\xi_o,\ldots,x_{n-3},\xi_{n-3},x,\xi(x)\mid\xi(x_o)=\xi_o,\ldots,\xi(x_{n-3})=\xi_{n-3}\}$

d^*_{n-2}: $(x_o,\xi_o,\ldots,x_{n-3},\xi_{n-3}) \to x_{n-2}$

Further steps lead in an analogous way to $u_{n-3}, d^*_{n-3},\ldots$, and finally to:

$u_1:(x_o,\xi_o)\to\inf_{x\varepsilon X} E\{u_2(x_o,\xi_o,x,\xi(x))\mid\xi(x_o)=\xi_o\}$ and $d_1^*:(x_o,\xi_o)\to x_1$.

d_o^*, the first step in the strategy, selects x_o from the condition:

$$E\{u_1(x_o,\xi(x_o))\} = \inf_{x\varepsilon X} E\{u_1(x, \xi(x))\}.$$

It is easy to verify that $d^* = (d_o^*, d_1^*,\ldots,d_{n-1}^*)$ is such that:

$E\{\xi(x_{n-1}^{d^*})\} = \inf_{d\varepsilon\Delta} E\{\xi(x_{n-1}^d)\}$, where Δ is the set of all n-steps strategies.

The strategy d^* is termed E-optimal (optimal with respect to the expected value).

As far as the convergence of $\xi(x_n^{d^*})$ to $\xi^* = \inf_{x\varepsilon X}\xi(x)$ is concerned, under the following conditions:

i) $X \subset R^n$ ii) $E\{\sup_{x\varepsilon X}\xi(x) - \xi^*\} < +\infty$

iii) $P\{w:\xi_w(X)\,\varepsilon\, C(X)\} = 1$

it can be shown (Archetti-Betrò (1980 b)) that:

a) $E\{\xi(x_n^{d^*}) - \xi^*\}\to 0$ as $n\to\infty$.

b) $P\{\xi(x_n^{d^*}) - \xi^* > \varepsilon\}\to 0$ as $n\to\infty$, for $\varepsilon > 0$.

Part b of this result leads naturally to another optimality criterion related to the speed of convergence in probability of $\xi(x_n^d)$ to ξ^*.

A strategy $\bar{d}$ defined by: $P\{\xi(x_n^{\bar{d}}) - \xi^* > \varepsilon\} = \min_{d\varepsilon\Lambda} P\{\xi(x_n^d) - \xi^* > \varepsilon\}$ is termed P-optimal (optimal in probability).

P-optimal strategies can be computed by an analogous recursive construction to that described for E-optimal ones. The convergence in probability of $\xi(x_n^{\bar{d}})$ to ξ^* follows from the relation

$P\{\xi(x_n^{\bar{d}}) - \xi^* > \varepsilon\} \leq P\{\xi(x_n^{d^*}) - \xi^* > \varepsilon\}$ and part b of the above result.

REFERENCES

[1] Andersenn, R.S. & Bloomfield, (1975), "Properties of the Random Search in Global Optimization". Journal Optimization Theory & Applications 16, n. 5/6.

[2] Archetti, F., (1975),"A sampling technique for global optimization" in"Towards Global Optimization" L.C.W. Dixon & G.P. Szegö eds., North Holland (1975).

[3] Archetti, F. & Betrò, B., (1975),"Recursive Stochastic evaluation of the level set measure in global optimization problems". Research Report A-21 del Dipartimento di Ricerca Operativa e Scienze Statistiche dell'Università di Pisa.

[4] Archetti, F., Betrò, B. & Steffé, S., (1975) "A theoretical framework for global optimization via random sampling". Research Report A-25 del Dipartimento di Ricerca Operativa e Scienze Statistiche dell'Università di Pisa.

[5] Archetti, F. & Betrò, B., (1978a) ,"A priori analysis of deterministic strategies for global optimization problems" in "Towards Global Optimization", Vol.2, L.C.W. Dixon & G.P. Szegö eds. North Holland (1978).

[6] Archetti, F. & Betrò, B., (1978b) ,"On the effectiveness of uniform random sampling in global optimization problems" Research Report A-51 del Dipartimento di Ricerca Operativa e Scienze Statistiche dell'Università di Pisa.

[7] Archetti, F. & Betrò, B.,(1979),"A stopping criterion for global optimization algorithms" Research Report A-61 del Dipartimento di Ricerca Operativa e Scienze Statistiche dell'Università di Pisa.

[8] Archetti, F. & Betrò, B., (1980a), "A probabilistic Algorithm for global optimization" to appear in "Calcolo".

[9] Archetti, F. & Betrò, B.,(1980b), "Stochastic Models and Optimization" to appear in "Bollettino dell'Unione Matematica Italiana.

[10] Boender, G., (1979), "Cluster Analysis in Global Optimization" M. Sc. Thesis, Erasmus University, Rotterdam.

[11] Boender, G., Rinnoy Kan A.H., Stougie, L. & Timmer, T. "Global Optimization: a stochastic approach" in this volume.

[12] Chichinadze, V.K., (1967),"Random Search to determine the extremum of functions of several variables" Engineering Cybernetics 1967, 1.

[13] de Biase, L. & Frontini, F., (1978), "A stochastic method for global optimization: its structure and numerical performance" in "Towards Global Optimization" Vol. 2, L.C.W. Dixon & G.P.

Szegö eds., North-Holland.

[14] de Biase, L. & Frontini, F., (1979), "A stochastic method for global optimization" in Proceedings COMPSTAT 1978.

[15] de Haan, L., (1979),"Estimation of the minimum of a function using order statistics" Report 7902/S, Econometric Institute, Erasmus University, Rotterdam.

[16] Dixon, L.C.W. & Szegö,G.P., (1978), "The global optimization problem: an introduction" in "Towards Global Optimization" Vol. 2, L.C.W. Dixon & G.P. Szegö eds., North-Holland.

[17] Evtushenko, Yu. G., (1971), "Numerical methods for finding global extrema (case of a non-uniform mesh),U.S.S.R. Computational Math. & Math. Physics, 11,6.

[18] Gomulka, J., (1978a), "Two Implementations of Branin's method: numerical evidence" in "Towards Global Optimization" Vol. 2 L.C.W. Dixon & G.P. Szegö eds., North-Holland.

[19] Gomulka, J., (1978b), "A user's experience with Torn's clustering algorithm" in "Towards Global Optimization" Vol. 2, L.C.W. Dixon & G.P. Szegö eds., North-Holland.

[20] Karp,R.M.,(1976), "The probabilistic analysis of some combinatorial search algorithms" in "Algorithms and Complexity: New directions and recent results "J.F. Traub ed. Academic Press.

[21] Kushner, M.J., (1964), "A new method for locating the maximum point of on arbitrary multipeak curse in presence of noise Journal of Basic Engineering - pp. 97-106.

[22] Mc Keown, J.J., (1980), "Aspects of parallel computation in numerical optimization" in this volume.

[23] Mockus, J., (1964), "On a method for allocation of observations for the solution of extremal problems" U.S.S.R. Computational Math. & Math. Physics, n.2.

[24] Mockus, J., (1972), "On Bayesian methods for seeking the extremum" Automatics and Computers, n.3.

[25] Mockus, J., (1975), "On Bayesian methods of optimization" in "Towards Global Optimization", L.C.W. Dixon & G.P. Szegö eds., North-Holland.

[26] Mockus, J., Tiesis, V. & Zilinskas, A. (1978), "The application of Bayesian methods for seeking the extremum in "Towards Global Optimization" Vol. 2, L.C.W. Dixon & G.P. Szegö eds., North-Holland.

[27] Mockus, J., (1980), "The simple Bayesian algorithm for multidimensional global optimization" in this volume.

[28] Shubert, B.O., (1972), "A sequential method seeking the global maximum of a function" SIAM Journal on Numerical Analysis 9,3,

[29] Strongin, R.G., (1973), "On the convergence of an algorithm for finding a global extremum" Engineering Cybernetics, 4.

[30] Sukharev, A.G., (1971), "Optimal Strategies of the search for an extremum" U.S.S.R. Computational Math. & Math. Physics 11,4.

[31] Törn, A., (1976), "Cluster analysis using seed points and density determined hyperspheres with an application to global optimization" Proceedings of the third international Joint conference an Pattern Recognition, IEEE Transactions an Systems, Man & Cybernetics.

[32] Treccani, G., (1978), "A global descent optimization strategy" in "Towards Global Optimization", Vol. 2, L.C.W. Dixon & G.P. Szegö eds.,North-Holland.

[33] Zilinskas, A., (1978), "On statistical Models for multimodal Optimization" Math. Operations forsch. - stat. Ser.Stat., Vol. 9.

Numerical Techniques for Stochastic Systems
F. Archetti and M. Cugiani (eds.)
© North-Holland Publishing Company, 1980

ASPECTS OF PARALLEL COMPUTATION IN NUMERICAL OPTIMIZATION

J.J. McKeown

Numerical Optimization Centre, Hatfield Polytechnic

INTRODUCTION

Numerical optimization addresses the problem of minimising a function of n variables, possibly subject to constraints. At the present time it is firmly established as a branch of numerical analysis. The techniques of local optimization which have been developed during the last fifteen years or so are now well represented in program libraries and used in many different fields of application; the principles underlying them are taught at undergraduate and in some cases even at school level. In such circumstances one may well ask why it is necessary to introduce a new idea such as that of parallel processing into such a well established field; why, for example, this topic may not safely be left to the programmer and computer specialist. The first part of this paper will be devoted to an attempt to answer this reasonable question; in fact, to establish a case for parallel computation. Later sections will be used to expand upon some of the issues involved, and to suggest approaches to solving the problems arising in realising the full potential of parallel methods.

THE CASE FOR PARALLEL PROCESSING

Although it is probable that few would question the remarks made in the introduction about the respect enjoyed by the various methods of numerical optimization now available, it is necessary to make some qualifying comments about their power and range of validity. In various fields of application they are continually placed in the position of justifying their use in the face of competition from more

---

The author is now with International Computer Ltd., Oil Industries Project, Computer House, 322, Euston Road, London. Some of the work described in this paper was carried out while the author was CNR visiting Professor, University of Bergamo, during 1978.

intuitive or specialized methods. A good illustrative example of a case in which strict methods of numerical optimization have actually lost ground is provided by the optimization of aircraft structural design. During the early 1960's optimization algorithms were developed for the solution of the large nonlinear constrained problems involved, notably by workers such as Gallagher and Gellatly [1]. Their method was a strict optimization algorithm in that it sought to minimize directly a defined objective function subject to nonlinear constraints. However, such were the problems encountered in developing these methods to the level at which the advantages they provided were not overwhelmed by the cost of the computational effort involved that, by the beginning of the 1970's, they had been largely abandoned in favour of more intuitive techniques known as 'Optimality Criterion' methods, e.g. [2]. These algorithms differ from optimization methods in that they do not aim to minimise a function directly, but to satisfy some criterion which is associated, however loosely, with an optimal design. In the context of structures, these criteria might be uniform stress or strain-energy density. Such methods are in effect little more than automated versions of traditional design techniques. It is known that they need not produce truly optimal designs, nor even converge. However, they are simple, credible to non specialists in optimization, and in many cases able to produce acceptable designs with much lower computational cost than the strict optimization required. Not until recently has any success been gained in relating such methods to the mathematical framework of numerical optimization. Clearly, rigorous methods can, in practice, be prohibitively costly.

A second weakness of local optimization techniques - and all the well-established algorithms fall into this category - is that they rarely solve the problem explicitly posed. It is probably quite rare to find an application in which a local optimum, as such, is required. Indeed it seems likely that most users of optimization techniques, if offered a choice between a locally optimal solution and another non-stationary but of lower function value, would choose the latter. Their problems are not local but global, and if we turn to global optimization we find a much less satisfactory situation.

Although many methods for solving global optimization problems have been proposed, few if any are yet well established in general use.

The reason for this is simply that, if local optimization may on occasion be too time-consuming to be useful, this is almost always the case with global optimization methods. Typically these require perhaps an order of magnitude more effort, on small problems, than would be needed to establish a local minimum - and the solution produced is still not verifiable as such. Unfortunately, it is clear that Global Optimization must always be a computationally costly endeavour. The certainty with which the global minimum of a function may be determined or located must be a function of the amount of exploration carried out. This, in turn, requires the repeated evaluation of the objective function which is often a non-trivial operation and indeed is conventionally taken as the main component of cost involved in numerical optimization. Given the fact of an irreducible minimum in the amount of computation required, it is natural to turn to the latest developments in computer technology to find an answer to the problem.

Parallel processing systems of various kinds are now becoming increasingly common. This is due in part to the steady development of computer architecture and software over the years, but there is no doubt that the most significant factor leading to the popularity of such systems is the availability of microcomputers. The word in fact covers a wide range of processors of varying levels of sophistication; but all have in common the feature of low cost per unit of processing power, in whatever terms this may be expressed. It is natural to enquire how such low-cost units may be combined to form larger systems, and how algorithms and the programs which implement them may be designed to use such systems to best advantage. Clearly, any problem which is to be solved efficiently must posess, or must be capable of being formulated so as to posess, a high degree of parallelism in its structure. The global optimization problem happens to have this property inherently, for the same reason that the effort required to solve it cannot be reduced below a certain level. The need for exploration which leads to the latter condition can clearly be satisfied by carrying out numerous nearly independent - and therefore parallelisable - operations. It seems, therefore, that given the availability of parallel computers the basic difficulty of global optimization suggests its own solution. It is this coincidence of need and suitability which makes this problem such an attractive subject for the application of parallel methods, and is the reason why global optimization will be taken as the key theme of the present paper. However, it must be stressed in passing that other aspects of numerical optimization, such as integer programming, provide equally promising targets for these techniques.

At this point it might be conceded that a case has been made for using the most powerful of available computers for solving problems in optimization, but not that this requires any particular effort at the level of algorithm design. It is true that in discussing parallel computation it is easy to fall into a habit of thinking in what might be called 'computer science' terms, and to argue that in the long run advances in the speed of computers, arising from whatever source, will overcome computational difficulties without special intervention from algorithm designers. Indeed, although it is true that parallel computation is simply a way of increasing the throughput of an automatic computer system, it should be remembered that the aim of all research in numerical analysis is to increase the efficiency and accuracy which computations may be carried out. It would be unusual, however, to hear it suggested that the recent advances in the computation of eigensystems, in solving linear equations or indeed in sequential numerical optimization ought to be neglected simply because the increasing effectiveness of computer systems will soon render the older methods quite acceptable. The achievements in these areas, and in many others, mark a real improvement in the understanding of numerical processes, and the benefits which flow from them are additional to any improvements stemming from simple increases in speed of computation. In this sense, research in numerical analysis is worthwhile for its own sake. The question at issue is whether parallel computation is of importance on this level, and it seems that the answer must be in the affirmative. To see why, let us glance once again at traditional methods. The assertion just made as to their independence of computer systems must be qualified in one respect: they embody a principle of sequentiality which is in practice a restriction which allows their implementation on standard computer systems. This restriction is so natural, perhaps because the mode of conscious thought is serial, that we are barely aware of it. Once the idea of parallel automatic computation becomes feasible, however, it is clear that sequential methods are a subset of a potentially more general class of numerical techniques. If this assertion is admitted, then the idea of parallel computation in general, and of course parallel optimization in particular, become worth pursuing independently of advances in computer technology on the same basis as other ideas in numerical analysis. However, although the justification for such research is thus established, it would be foolish not to recognise that such new freedom does depend on a recognition of the fact that algorithms arising from it will be of limited interest unless they can be implemented; practical limitations must therefore be kept in view.

Summing up the case for parallel computational methods in optimization, then, it is claimed that research in this area is justified on three counts. Firstly, there is a case to be made on severely practical grounds for developing algorithms to take advantage of the parallel systems rapidly becoming available; the opportunities they provide ought not to be ignored. Secondly there is a strong methodological motive for such work to be done. The situation sketched in the preceding paragraphs, in which strictly mathematical methods are hard-pressed to provide solutions to practical problems, is clearly unsatisfactory. Agreeable though it may be to find that non-rigorous methods give good results in particular cases, such fragmentation calls into question the idea of an optimization methodology. If such a methodology, with all its advantages, is to be established on a firm foundation, there must exist methods which are general, rigorous and practical. Parallel techniques should help provide them. Thirdly, and on the level of numerical analysis, it is attractive in the long term to explore the possibilities which arise from the relaxation of restrictions imposed by the requirements of sequentiality upon the design of numerical algorithms.

Having, it is hoped, established a motivation for studying parallel computation, it must be said that a considerable amount of exploratory work is required before the real limitations and advantages of such methods in the context of optimization become clear. The field is a wide one, cutting across the usual boundaries between numerical analysis and computer science. The aim of the present paper is to do no more than to indicate one or two of the directions in which such exploratory probes are being made. The starting point will be a simple classification of parallel systems into two main groups; algorithms appropriate to each of these will then be briefly discussed.

TYPES OF PARALLEL PROCESSING SYSTEMS

There is a bewildering variety of parallel processing systems that can be constructed, and even among those already built the range is very wide. For example, Thurber and Wald [3] list five classification schemes which have been proposed. For the purposes of the present paper, a simple classification based on that of Flynn [4] will be used. This involves two groups defined as follows:

A) Single Instruction, Multiple Data (SIMD) Machines

These consist of sets of processors connected together, each with its own memory but all controlled by a single master controller. Each processor therefore carries out the same sequence of instructions as the others but on its own specific data set. Perhaps the best known of such systems is the Burroughs ILLIAC IV; later systems have appeared, such as the ICL Distributed Array Processor [5]. Because of the simplicity of the architecture, such systems can consist of large numbers of processors; the ICL machine, for example, has 4096 processors in a 64x64 grid arrangement. These processors are simple compared with the central processor of a mainframe machine; they carry out operations on one bit at a time and are therefore significantly slower, for most operations, than a mainframe processor. In addition, the architecture obviously imposes limitations on flexibility of application. However, such machines are clearly ideally suited to carrying out numerous simple, independent or quasi-independent tasks. The applications for which these machines were originally designed - data processing, numerical solution of Partial Differential equations etc. - are such that the problems are easily formulated so as to take advantage of whatever degree of parallelism is available in computation. That this is also a feature of Global Optimization will be shown below; it stands in contrast to problems, for example the computation of the inner product of vectors of length n, in which the degree of parallelism is fixed by the nature of the problem itself.

B) Multiple Instruction, Multiple Data (MIMD) Machines

These are machines in which the processors may be programmed individually; examples are CMMP [6] and CM*[7]. They are clearly quite different from SIMD machines; the components are usually computers in their own right, capable of operating independently.

> The possibilities of increased flexibility are obvious, but the penalties in terms of memory congestion and operating system complexity can be such that the maximum effective number of processors is much more limited than in the SIMD case - the upper limit may be no more than a few dozen, although this depends upon many factors including the particular application and is still a matter of research. The questions which arise in regard to the design of algorithms for the most effective exploitation of such systems are many and varied, and will certainly not be resolved in the near future.

The differences between SIMD and MIMD machines are so evident that it goes without saying that each requires a quite distinctive approach to algorithm design. The single instruction machine will demand a problem formulation which allows solution in terms of multiple, simple tasks; furthermore, the number of such parallel tasks must not be fixed for any particular problem or algorithm but must be variable to match the particular system being used. Other requirements will become clearer when an algorithm is described. The MIMD machine allows us to dispense with the requirement that tasks should be identical and synchronous; however, other factors such as memory congestion, program indeterminacy and convergence problems must be faced if synchronism is dispensed with, while waiting losses must be considered if synchronism is imposed.

In the remaining sections of this paper some illustrative algorithms will be discussed. The first is well suited to the requirements of SIMD machines; it is based upon well established methods of Grid Search. The second and third methods are from the class of stochastic methods which are now the most often used algorithms for Global Optimization. The aim in this case will be to suggest, not a parallel algorithm as such, but an approach which may be applied to the parallelisation of existing algorithms or to the design of new ones.

GRID METHODS FOR SIMD MACHINES

Any attempt to design a grid search technique for SIMD machines must take, as its starting point, the present state of the art in the design of such methods for sequential machines. These techniques are well represented by that of Evtushenko [8]. This begins by making certain assumptions about the function whose global minimum is to be found. Such assumptions are essential; if, for example, there exists no upper limit on the magnitude of the derivatives of such a function, then the minimum could be determined only by evaluating it at every conceivable point in the feasible set. We will briefly describe the algorithm proposed by Evtushenko for the case in which the objective function satisfies a Lipschitz condition. Of such functions $f(x)$ we can say that, given any $\underline{x}_1$ and $\underline{x}_2$ in the feasible set, a number C can be found s.t.

$$|f(x_1) - f(x_2)| \leq C \, ||\underline{x}_1 - \underline{x}_2||_2 \tag{1}$$

This is in effect a bound upon the first derivative of such functions. For example, let us consider the one-dimensional function $f(x)$ satisfying (1) for some known value of C, and let $f(x)$ be evaluated at points x_1 and x_2 (Fig. 1)

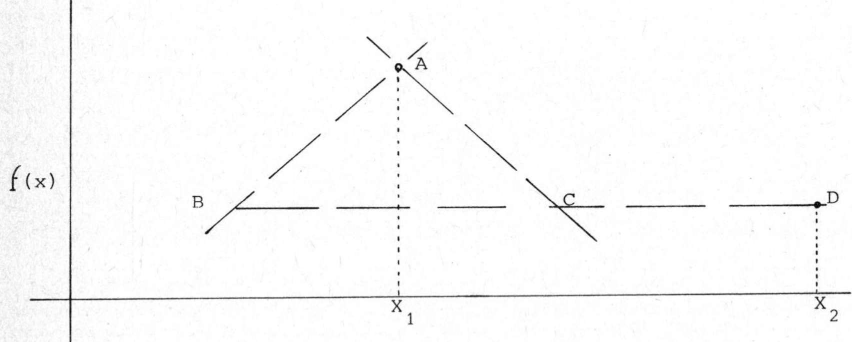

Fig. 1

Let $f(x_2)$ be less than $f(x_1)$ as shown. Let lines AB and AC be

drawn as shown, with slope $\pm$ C. It is clear that the function cannot have values lying below such lines and, in particular, that on the interval BC the function values may not be less than $f(x_2)$. Hence, if we are seeking the global minimum of $f(x)$, we way exclude this interval from further discussion; its length is clearly equal to $2(f(x_1) - f(x_2))/C$.
If the minimum value of f(x) is required to an accuracy ϵ, then clearly AB can be extended at each end by an amount ϵ/C. These observations form the basis of the Evtushnko algorithm, which, for an n-dimensional function, is as follows: Let the feasible set be defined by the limits $a_i \leq x_i \leq b_i$, $a_i < b_i$, $i = 1,2,\ldots,n$.
The first step is to set each element of $\underline{x}$ to $a_i + \frac{\varepsilon}{C\sqrt{n}}$ and evaluate f(x). Call these values x^1 and f^1. Thereafter, x_1 is increased by an amount equal to:

$$\frac{\varepsilon}{C\sqrt{n}} + \frac{(\varepsilon + (f^i - F^J))}{C\sqrt{n}} \quad \text{on the j'th step}$$

where F^J is the lowest function value recorded up to that point. Clearly, as the value of F^j decreases, this interval will become larger in those regions where f^i is greater; the step taken thus adjusts itself to the nature of the function.
When x_1 eventually exceeds its upper limit, it is reset to $a_1 + \frac{\varepsilon}{c\sqrt{n}}$, while x_2 is increased by an amount equal to the smallest step so far taken, i.e.

$$x_2 = a_2 + \frac{2\varepsilon}{C\sqrt{n}} + \underset{k}{\text{Min}} \frac{(\varepsilon + (f^k - F^k))}{C\sqrt{n}} .$$

It is the held at this value while x_1 goes through the same process as before; all other variables are held at their original values. This is repeated until x_2 exceeds its upper limit, at which point x_3 is increased and so on. The process will eventually cover the feasible set and determine the minimum value of f(x) in that set to within an accuracy of ε.
The most important feature of the Evtushenko algorithm (which can, with little modification, make use of other assumptions on f(x) such as bounded second derivative) is the fact that the intervals between points at which the function is evaluated are continually improved in the light of previous observations. Its drawbacks are fundamental to the nature of grid search methods, and perhaps to the Global Optimisation problem as such.
Firstly, if the number of function evaluations on each linear search

is of order N_S, then the total number of evaluations required is of order $(N_S)^n$. Thus, although the algorithm should result in a value for N_S which approaches the minimum possible, it cannot avoid the effects of dimensionality. Secondly, although the minimum function value f^* will be found to accuracy ε, it cannot be guaranteed that the position of the point, x^*, at which this minimum occurs, will be found with a similar accuracy. Arg $f_{min}(\underline{x})$ is therefore not an approximation to x^* in the sense in which f_{min} is an approxximation to f^*.

When we come to consider how the Evtushenko algorithm might be modified to take advantage of parallel computation with P processors, we immediately discover one of the fundamental problems of such computation, namely the reduction in 'learning'. In the present case, the obvious way to employ parallel processors would be by using them to evaluate the function at points along a line. However, we must now preassign a step length for P steps of the line search; and the maximum possible value of this quantity is $2\varepsilon/C\sqrt{n}$. Let us assume that this results in a number of nodes less than or equal to the number of processors. We then evaluate the function at all the nodes along the line simultaneously. Let N_P be the number of such nodes. Then:

$$N_s \leq N_p.$$

The speed-up ratio, that is, the ratio of the time required by a single processor to that needed by the parallel system, is then not P but $P(N_s/N_p) \leq P$. At first glance this might seem even a dangerous situation. It might appear that the number of function evaluations will increase from $0(N_s{}^n)$ to $0(N_p)^n$. The speed up would then be $P(N_s/N_p)^n$. This would mean (since $N_p \geq N_s$) that for n sufficiently large, the sequential algorithm must be faster than the parallel one, no matter how many processing elements there are, since only P function evaluations can be made at a time. However, this situation is avoided by noting that actual line searches are carried out by Evtushenko only along x^1, while the steps along other axial directions are conditioned by the minimum step found on such searches. This situation will be little altered in the parallel case, since it will not be difficult to compute at the end of each line search, without additional function evaluations, a good estimate of the minimum step length which <u>would</u> have been taken by the

Evtushenko method. Steps along directions other than x^1 will therefore be similar in both cases. Hence the total number of line searches would be more or less equivalent for the sequential and parallel algorithms, and the overall speed-up ratio would be of the order of the average of the speed-ups for such individual searches. However, it does show that the unthinking use of parallel processing might give less improvement than might be expected.

Let us consider another way of conducting a grid search which might prove more appropriate to an array processor. Assume once again that the set in R^n in which the search for a global minimum is to be carried out is defined by simple bounds on the x_i; let this domain be covered by a regular mesh whose grid-size is determined by the number of processors available - for example, is such that the number of nodes is equal to the number of processing elements in the parallel system, or to some simple multiple of it. The function, and perhaps the gradient, are computed at such nodes. Let these values be denoted by $\underline{F}_i$, $\nabla \underline{F}_i$ on the nodes of the i'th cell of the mesh, and let the length of the sides of the cell be denoted by the vector $\underline{\delta}_i$. Then, consistent with the assumptions made about the objective function, we can compute a value $\emptyset_1(\underline{\delta}_i, F_i, \nabla F_i)$ which is the minimum possible function value which can occur in, or on the boundary of, this cell. Let $f_{i,min}$ be the minimum value in $\underline{F}_i$. Let $f_{min,min}$ be the minimum value of f over all the nodes of the mesh and let, as before, ε be the required accuracy with which f, the globally minimum function value, is to be determined. Then $f_{min,min}$ is an acceptable approximation to f if the following relation is satisfied:

$$f_{min,min} \leq \min_i \emptyset_i(\underline{\delta}_i \underline{F}_i, \nabla F_i) + \varepsilon \equiv \emptyset_{min} + \varepsilon. \qquad (2)$$

In general, this will not of course be so on the first step. Next, we select one or more cells for further investigation; clearly these must belong to the set J for which:

$$\emptyset_i(\underline{\delta}_i \underline{F}_i, \nabla \underline{F}_i) + \varepsilon \leq f_{min,min}, \quad i \varepsilon J .$$

The number of such cells selected on each iteration depends on the number of processors and the minimum required mesh size (corresponding to $2\varepsilon/C\sqrt{n}$ in the Evtushenko algorithm). The aim of the selection is to use as many of the processors as possible by covering the cells with a new, finer mesh. When all candidate cells have thus been investigated, we have a new value of $f_{min,min}$ and $\emptyset_{min}$. Denoting them by $f^{+}_{min,min}$ and $\emptyset^{+}_{min}$, we have:

$$f^{+}_{min,min} \leq f_{min,min}$$
$$\emptyset^{+}_{min} \geq \emptyset_{min}.$$

We can expect, therefore, that some of the candidate cells will have been eliminated, while others justify more detailed treatment; the process continues until (2) is satisfied.

This algorithm might be more effective than the straightforward parallelisation of Evtushenko because it avoids using the finest possible mesh size until a good deal of information has been gathered about the nature of $f(\underline{x})$; it therefore allows some of the learning behaviour which is the best feature of sequential Evtushenko to become a feature of the parallel algorithm also. However, there are other possible features which could be incorporated into either of the two parallel algorithms proposed here. The most obvious is to introduce the possibility of local minimization starting from selected points; this is already a part of the practical Evtushenko algorithm, and could be carried out by an array processor. However, one of the weaknesses of SIMD systems is their inability to deal efficiently with branch instructions. Suppose that each processor is carrying out a sequence of instructions; since the data are different for each one, it follows that, on occasions, some processors will branch differently from others. This means that these processors will require to carry out a stream of different instructions, which is impossible to do on a SIMD system in parallel. Some processors must therefore wait, imposing a delay upon the system. This Characteristic of SIMD machines therefore imposes a further penalty on them: parallel tasks, as well as being identical, must be logically simple. Of course, there is no absolute requirement to eliminate branching entirely; in some contexts the inefficiency induced may be acceptable. Such questions can only be resolved by a combination of theoretical and empirical study. Such an investigation will be far from trivial, given the known difficulty of designing rational test procedures for Global Optimization algorithms, and analysing their results, even in the sequential case.

Before leaving this section on methods suitable for array processors, it is worth mentioning the possibility of designing interactive algorithms for the determination of global optima by using a parallel processor as a rapid means of computing function values as before, but by presenting this information in a graphical form to a human operator. This would provide the possibility of scanning any two-dimensional subspace of a space of higher dimension; the question of how such information should be presented and manipulated is a fascinating one involving aspects of psychology, mathematics and computer science. It is likely to be useful mainly for the optimization of functions which are cheap to evaluate; but it allows the qualities of human perception to be brought to bear upon a problem which involves fundamental difficulties from a purely mathematical point of view.

Array processors would seem to be particularly suited to this approach because of their fast response time in such an application, which allows their parallel capabilities to be fully exploited. Not only the topology of the array but also the parallel modus operandi seem to match very well the information - perceiving requirements of interactive operations; one could think of the array of processors as mapping directly onto a display screen. The resulting output could also be made to exploit the kind of a-priori knowledge of the function already discussed in connection with the grid algorithm.

METHODS FOR MIMD MACHINES

The flexibility of MIMD machines allows parallelism to be exploited in algorithms which would be quite unsuitable for use with a single instruction machine. However, when a number of processors are carrying out different tasks in parallel, an obvious problem of synchronisation occurs; one cannot expect all such tasks to be completed at the same instant. Such synchronisation can of course be organised by suitable software; the more fundamental difficulty is the time wasted by processors which must wait for others to complete a task before the overall computation can proceed. Because processors in a multiple instruction machine are less closely coupled than in the single instruction case, such waiting may occur even when identical calculations are being carried out in parallel; fluctuations in processor speed may occur in a random way, and the expected time for a set of such calculations to be completed is equal to the expected maximum value of a group of random variables.

Although there are several other sources of inefficiency inseparable from MIMD systems, the synchronisation one is so fundamental that it has led to a new approach to algorithm design, perhaps mainly associated with the Department of Computer Science at Carnegie Mellon University, Pittsburg. This aims to avoid the problems of synchronisation by designing parallel algorithms in which the individual tasks are sufficiently complete in themselves to be allowed to carry out more then one cycle even if other tasks have not finished. The best known example of this is their Chaotic Relaxation algorithm [9] for solving sets of linear equations, based upon the original work of Chazan and Miranker [10]. Each processor repeatedly solves a subset of equations, using the most recent information available from the other processors, and contributing its own results to a pool as soon as they are ready. Excellent results have been achieved by this approach on large sparse sets of equations, with the time needed to find a solution being reduced by a factor approaching the maximum possible of $1/P$, where P is the number of processors in use.

Asynchronous algorithms have a number of special features. Most significantly, unlike synchronous algorithms, they produce quite different results from sequential ones. Clearly, by synchronising parallel tasks so that no computation is carried out until all the information needed has been provided, we can ensure that a parallel synchronised algorithm produced exactly the same results as its sequential counterpart. If, however, synchronisation is abandoned, the sequence of iterations produced by, say, an optimisation algorithm will be quite different from the same algorithm in a synchronised or sequential mode. The result is that the convergence properties of such an algorithm must be re-analysed, and the rate of convergence in particular becomes a function of the relative speeds of the various processors. Thus, the deterministic framework within which such analysis has been carried out in the sequential case must be replaced by an explicitly statistical one for asynchronous parallel algorithms. This requirement is strengthened by the existence of other characteristics of Multiple Instruction systems - notably the requirement for processors to queue in order to gain access to memory - whose effects are statistical in nature.

In order to illustrate and clarify some of the problems which may arise during the design of Global Optimisation algorithms for use on MIMD machines, let us consider one of the most effective of the current sequential algorithms, namely that of Törn [11]. This can be summarised as follows.

TÖRN'S ALGORITHM

Step 0: Select at random an arbitrary number of points, say M, in the region to be searched for a global minimum. The number M is significantly greater than N, the dimension of the problem.

Step 1: Using each of the M points as a starting point, run a few iterations of a local minimisation routine. The number of iterations is small, but otherwise arbitrary. The result of this step is that the original set of points is replaced by an 'improved' set, the two sets being, in general, in one-to-one correspondence.

Step 2: By the technique of 'cluster analysis', separate the improved set into a number of groups or 'clusters'; the number of such clusters will depend upon arbitrary parameters. Each cluster consists of points which, in position and function value, are closer to one another than to members of other clusters.

Stop if termination criterion met.

Step 3: Reduce the number of points in each cluster according to some pre-arranged strategy. For example, the worst half of the set, that is, the half with the highest function values, may be discarded. Repeat from Step 1 with the reduced value of M.

The aim of Törn's algorithm is to locate the position, as well as to determine the value, of the global minimum; in this it differs from the method previously described. In addition, it should find more than one local minimum if any exist; each cluster is intended to converge to such a point. The number of iterations of the local minimizer is kept small during each iteration of the overall algorithm so that during the early stages, when M is large, no effort is wasted. A notable feature of the algorithm is its degree of arbitrariness - the number of points, the number of local iterations per major iteration, the clustering parameters, the reduction strategy. It is intended that these degrees of freedom be used to tune the algorithm to particular cases. The procedure terminates when, for example, the set of points remains sufficiently unchanged for a number of major iterations.

It is clear that parallelism exists in this algorithm on a number of levels; for example, the local minimisation algorithm used may be a gradient method, involving matrix operations. These are certainly amenable to parallel treatment, even using an SIMD machine. There may also be parallelism at a higher level within the local algorithm. However, a discussion of these various levels of parallelism and their treatment is beyond the scope of the present paper. Instead, we shall focus on parallelism at the highest level of organisazion of the algorithm, that is, parallelism which is directly apparent from the description given above. On this level, it is clear that most of the parallelism inherent in the method is contained in step 1. The partial local minimisations can clearly be carried out in parallel. The fact that MIMD processors are being used means that the branch instructions which certainly occur within the local minimizers cause no inherent difficulty. The parallel algorithm is shown in figure 2.

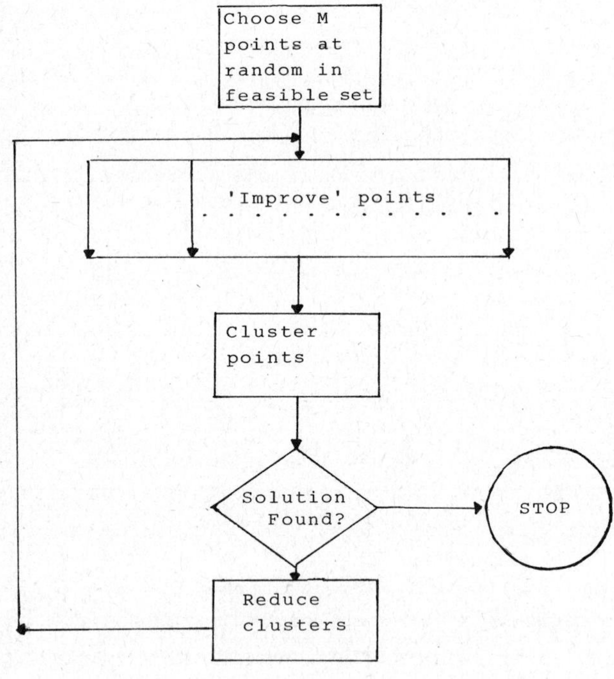

Fig. 2

Before Step 2 can proceed, that is, before the improved set of points can be clustered, all the local minimizers must complete their iterations. This involves a synchronisation cost - the faster processes must wait for the slower to finish. The inefficiency involved could be serious, since the time taken to complete a fixed number of iterations of a local minimization routine is certainly a variable quantity; it is the degree of variability which determines the synchronisation loss. However, this loss can be minimised by noting that the number of iterations is not a fixed and sensitive quantity; since it is arbitrary there can be no fundamental objection to allowing a particularly lengthy iteration to be stopped, if other processors have completed their tasks and are waiting; the result would simply be that more iterations would be carried out from some starting points than from others. Such 'trimming' would, however, not cure the synchronisation problem completely.

Another drawback to the scheme arises from the fact that M, and therefore the number of parallel tasks, decreases progressively as the iterations proceed. If $M = P$, the number of processors (and there would be little point in choosing M to be much smaller), then there would be $(1-\alpha)P$ processors idle on the second major iteration, if M is reduced to αM. Clearly, the speed-up induced by the parallelisation will tend to resemble LogP rather than P in this case. If, on the other hand, $P \ll M$, so that the tasks must be distributed between processors and all the processors are busy until M becomes very small, we might expect a cost due to the frequent re-assignments necessary.

Whatever the reduction in the time required to carry out step 1 of the algorithm, only one processor is needed to carry out the other steps. Almost certainly these steps will in any case require much less time than step 1; the more serious problem is that if a breakdown of the processor carrying out these steps should occur, the whole algorithm might be blocked. Indeed, unless some system of detecting and discarding 'dead' processors is used, and reassigning their tasks to other processors, such blockage could easily occur at any stage.

This very brief analysis seems to indicate that the obvious parallelisation of Törn's algorithm has definite drawbacks. It is likely to require a fairly sophisticated operating system to handle the task assignments involved; its speed-up is likely to be of order LogP rather than P; it will suffer from inefficiency due to synchronisation, and it may be prone to blocking from non-functioning processors. In the next section a general approach to designing parallel algorithms will be proposed which, when applied to the Törn algorithms will be shown to be less subject to these difficulties.

HOLISTIC ALGORITHMS

The parallel algorithm sketched above was derived by exploiting the parallelism inherent in one of the steps of Törn sequential algorithm. This approach to the parallelisation of sequential algorithms is a natural one, but the disadvantages that may arise from it are well illustrated by the present case. In this section a general technique will be proposed which can be used in the design of parallel algorithms based upon sequential predecessors, or indeed may be used to design such algorithms ab-initio. It may be viewed as turning the previous approach inside out, or as defining the highest level of parallelism possible; it consists in modifying the complete sequential algorithm so that it may be applied as a whole as one of many identical parallel tasks. The term 'Holistic' is tentatively applied to this approach because the whole of the algorithm is expressed by, and inherent in, each of the parallel tasks which comprise it. Note however, that the resulting algorithm differs from those which may be used on an SIMD machine because no synchronisation is involved; at any instant individual processors may be carrying out quite distinct instructions. However, since the stream of instructions is the same for all processors, we may talk of the algorithm as being defined by an 'archetypal task'. Since the tasks must not interfere with one another, but must nevertheless contribute to the solution, it is possible to state some evident properties which the archetypal task must posess.

(1) It must be iterative

(2) all tasks operate on the same global data set

(3) the archetypal task is complete, that is, it can solve the problem when run as a single task, i.e. as a sequential algorithm.

(4) a P-fold implementation will reach a solution more quickly than a (P-1) fold, at least for P less than some non trivial upper limit

(5) the archetypal task never changes the form of the global data set, although its contents must change.

It follows from (3) that the archetypal task may not contain any instruction whose execution depends upon the existence of other parallel tasks. Hence, a breakdown in up to (P-1) processors of a P-fold implementation will not prevent the eventual solution of the problem. It is of course envisaged that each processor will be dedicated to a task; the very simple implementation implied is shown in Figure 3.

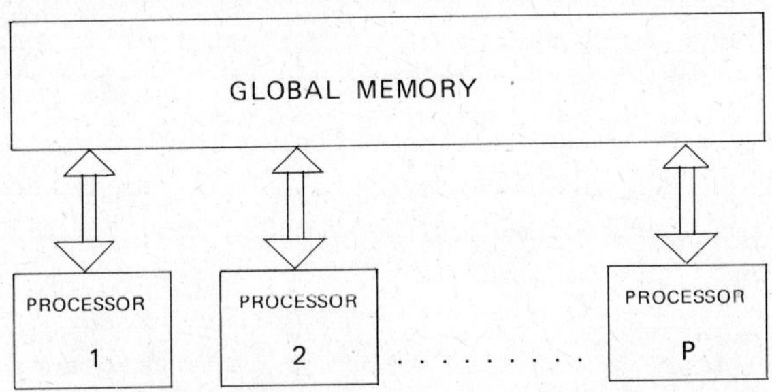

Fig. 3

Because the processors are dedicated, no task-assignment software is required, and the only significant function of the operating system is to ensure that clashes do not occur between various processors as they access global memory. For example, no processor may try to change an item of data being read by another. This requirement may be met by the use of 'critical sections' (e.g. dijkstra [12]; the details are not important in the present context, althought it should be noted that a time delay is implied.

Let us consider how a Holistic version of the Törn sequential algorithm may be designed. The first task is the choice of global data set;the obvious and only candidate is the pool of (initially random) points generated in step 0. Each processor can then act upon this data as though it were alone; it selects points, improves them by running a local optimizer within its private memory, and replaces them. An obvious difficulty arises when it attempts to reduce the clusters which it eventually generates. This process might seem to violate requirement 5, in that the number of elements of the data set would be reduced; clearly such reduction could not be allowed to be carried out by all processors willy-nilly. However, if one processor, say, is given sole rights to carry out the cluster-and-reduce function the structure of the algorithm would be lost, since tasks would no longer be identical.

This dilemma is resolved by having each processor carry in its private memory a list of the index numbers of the points in the global data set, as it sees them. Clustering involves grouping these indices and reduction means eliminating some of them. The result is that, as time passes, each processor will see the global data as a subset of the set which actually exists, and the subsets seen by each processor may be different. The Holistic Törn algorithm is as follows:

Step 0 : Choose the M random points defining the initial pool in the same way as for the sequential algorithm. This can be done by a single processor, and the process may be seen as that of initiallising the global memory. The following steps are carried out by all processors.
Step 1: Set up a list vector L of length M, the values of L(1), L(2)..., being 1, 2, ..., M.
Step 2: Select an element L(I) from L corresponding to an element in the global data set (i.e. a point from the pool) which has its flag unset. Set its flag.
Step 3: Perform several iterations of the local minimizer from the point labelled L(I), and replace the old point in global memory by the new. Unset the flag. If M distinct points have been improved in this way since step 4 was last performed, proceed to step 4; otherwise repeat from step 2.
Step 4: Transfer the contents of the global data set from global to private memory, and carry out a cluster analysis. The clusters are then reduced as in the sequential algorithm, the discarded points having their indices removed from L and the value of M (in private memory) being reduced accordingly. If the termination criterion is satisfied, stop; otherwise repeat from step 2.

The use of the flag in steps 2 and 3 is of course intended to ensure that no two processors attempt to improve the same point, thus uselessly duplicating effort. The setting and unsetting of this flag must be protected using standard logic (Dijkstra [12]). This is the only way in which the processors are, on the algorithm level, aware of each other's existence.

The analysis of this algorithm will be discussed in a later section. for the moment, however, its plausibility may be indicated by considering an implementation with a small number of processors, say P=2, and a value of M>>2. As each processor selects and improves points from the global data set, the risk of interference between them is small; if they should stay in step and finish their improvements at the same time, then one would have to wait while the other transferred the contents of the global set to its private memory for clustering.
If they were out of step, this wait would be reduced and indeed, since a partial local minimization is likely to take more time than a memory transfer, there would often be no need to wait at all. The main source of interference would be due to the occasions when both processors tried to capture the same point for improvement; this would not be likely to cause major delays. Hence, the speed-up in this case wuold probably approach the value 2. Adding more processors would increase the interference rate, and so the returns would, as is normal, diminish. However, it is plain that condition 4 is likely to be satisfied by this algorithm.

A holistic parallel version of the Price algorithm

Another sequential algorithm which lends itself very naturally to parallelisation by the holistic approach is that of Price [13]. This is a simple method intended for use on minicomputers, and may be summarised as follows.

Step 0: Select M points at random in the region of search. Evaluate the function at each, and store the coordinates of all points together with associated function values. Call this data set the 'pool'.

Step 1: Choose the point with the worst (greatest) function value.

Step 2: Choose (N+1) distinct points at random from among the set available and determine the position, $\underline{G}$, of the centroid of the first N. The next trial point is: $2\underline{G} - \underline{R}_{N+1}$, where $\underline{R}_{N+1}$ is the positions of the (N+1)'th point. This point is accepted if it is better than worst point in the pool. If not, repeat from Step 1. If, however, this repetition occurs too often (if the success rate drops below 50%, say) then carry out internal reflections $(\underline{G} + \underline{R}_{N+1})/2$ instead of external.

Step 3: Replace the worst point in the pool by the new point obtained in step 2.

Step 4: If termination criterion is satisfied, stop; otherwise repeat from step 0.

The details of the performance of this method, and a discussion of aspects such as the choice of stop criterion, will be found in Ref. 13.

The price algorithm may be translated into Holistic parallel form virtually without change. Tne natural choice of Global Variable set is once again the pool. Each processor has access to this, choosing points from the current set and updating with new points, all without direct reference to the other processors. In principal the only safeguard which must be applied is to ensure that no processor attemps to read a vector which another processor is trying to change. When this happens then one or more processors must wait;the frequency of its occurence will depend upon the number of processors and the value of M, the total number of points in the pool. A holistic parallel Price algorithm can be summarised as follows.

Step 0: As sequential algorithm. The pool is stored in global memory. As with the Törn algorithm, this loading process may be carried out by a single processor

Step 1: Choose (N+1) points from the pool, as in step 2 of the sequential algorithm, and transfer to local memory

Step 2: Generate a new point, as in step 2 of the sequential algorithm

Step 3: Replace the worst point in global memory by the new point

Step 4: As step 4 of the sequential algorithm.

It can be seen that the parallel algorithm differs hardly at all in description from the sequential one. Of course, there are differences in operation. As each processor tries to implement the instructions listed above, it must compete with others for access to memory. Also, in implementing the various steps, it must be remembered that the pool is constantly changing. Thus, when finding the worst point in step 3, it is understood that some process such as the following takes place: Let A be the (N+1) x M matrix defining the points in the pool and their function values ((N+1'th row). Then , to find the "worst point':

```
     FW     =-10^ℓ  (ℓ is sufficiently large for 10^ℓ to represent infi
               nity).
     DO 1 J = 1, M
     D      = A(N+1,J)
     IF (D.LT.FW) GO TO 1
     FW     = D
1 CONTINUE
```

This is a typical procedure for finding the greatest element of a set. However, in the present case, the set may have been changed by the time the search is completed; the index number found to be associated with the worst point might now be associated with a point of much lower function value. Hence, to implement step 3 as stated would imply either on iterative algorithm or a locking of the global data ser while the search was being performed. Neither of these operations in intended; the aim is to carry out exactly the same instructions as would be performed in the sequential case and to accept any deterioration of performance as one of the numerical costs of parallelisation.

Comparing this algorithm with our definition of a holistic process, we see that:

(i) It is iterative

(ii) all tasks are identical

(iii) all processors operate on the same data set

(iv) the archetypal task is complete; in fact, in the case of a one processor system the holistic algorithm in the sequential algorithm

(v) increasing the number of processors from one to two can certainly be expected to reduce the time taken to reach a solution, since for N and M reasonably large, the interference between the processors should be negligible. The upper limit on the number of processors, above which no improvement can be expected, will depend upun various factors discussed below

(iv) no processor operates on the data in such a way as to obstruct others. Since no processor reduces the size of the pool or alters its form, this property is satisfied. It must be noted, however, that any system involving multiple access to global data must involve some protection devices which will require processors to wait on occasion.

Analysis and testing of Holistic algorithms

As with alloptimisation algorithms, the final measure of effectiveness of those described above can only be gained from numerical testing and experimentation. However, some aspects of their performance are open to analysis and simulation, and these will now be briefly discussed.

The performance of holistic algorithms is affected by two different sets of parameters. On the one hand there are what may loosely be termed 'systems effects', such as memory congestion; on the other there are the numerical characteristics of the algorithm. While it is difficult to state in general that these are truly independent, it seems reasonable to try to treat them separately as a way of gaining insight into the behaviour of the algorithm. The Holistic Price algorithm will be used as an illustration of the way in which this can be done.

System analysis

Because of the nature of holistic algorithms - specifically, because each processor, on average, does the same as any other - a system with P processors can easily be analysed in terms of the states in which any processor may exist. All processors share a common pool of data, and update them without reference to the others. Any processor must therefore be in one of four distinct states at any instant:

State 1: The processor is operating on (N+1) points in its private memory to produce a new point for the pool

State 2: The processor is updating common memory by replacing the current worst point with a new one computed while in state 1. This may involve queueing for the available channels between processor and shared memory.

State 3: The processor is operating in private, choosing (N+1) random integers in the range in the range [1,M] where M in the number of points in the pool (fixed).
State 4: The processor is transferring (N+1) vectors from global to private memory.

The flow from one state to another in shown in fig. 4.

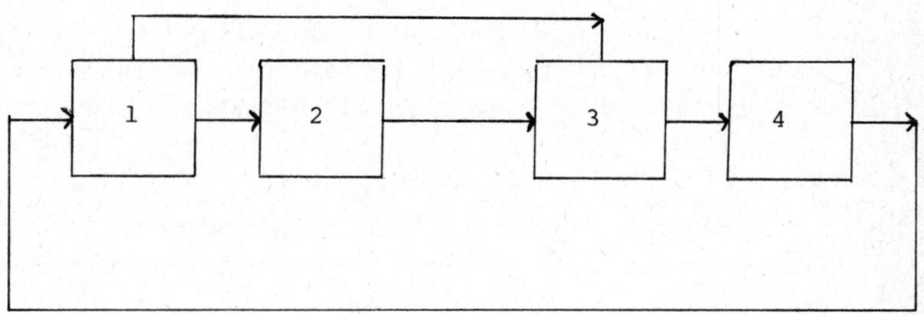

Fig. 4

The direct flow from state 1 to 3 occurs when a failure to find an acceptable point has occurred and a new set of (N+1) index points must be found.

For the purpose of the analysis in hand, all that need be known about these states is the length of time spent in each. These are as follows.
State 1: The time taken to reflect the (N+1)'th point through the centroid of the remainder of the simplex. This can be taken to be a random variable, whose mean and standard deviation are parameters of the system (some distribution being assumed). If a realistic simulation is required, then the likelihood of repeated attempts having to be made to generate an acceptable point be allowed for by specifying the probability, P_1, that an internal reflection will have to be tried. The probability, P_2, that a new set of (N+1) points must be chosen must also be specified.
State 2: The time spent in this state is two parts.

(i) The time spent waiting for access to global memory. This will depend on P, the number of processors, and C, the number of channels to the memory. On any particular iteration it will also depend upon the queueing convention used - for example, whether strict queueing occurs, with new processors joining a queue at the end (first in, first out system); or whether all waiting processors have an equal chance of service independently of arrival time. This distinction is unlikely to have much effect on long term queueing time, which is our main concern.

(ii) The time spent in transferring (N+1) words to global memory. Again, the choice of strategy will affect the short-term; are all (N+1) words passed in one move once a channel has been claimed, or is only one word passed at a time, with the channel relinquished in the interim? The basic unit of time is, in any case, that taken to write a word to memory; this must be specified as the mean and standard deviation of a random variable. This is the fundamental 'service time' which determines how long each processor must queue.

State 3: The time taken to choose (N+1) random points must be specified as to mean and standard deviation.
State 4: As for the write operation in state 2, the time for reading a word from memory must be specified. Note that the channels used for transferring data to and from global memory are the same, so that the queues implied by states 2 and 4 are the same; each 'customer', however, must be assigned a different service time depending upon whether he requires to write (N+1) vectors (state 2) or read (N+1) words (state 4).

The fundamental parameters of the system are, therefore, the following: the number of access channels between the processors and the memory; the parameters of the distributions defining the times spent in the various basic operations described above; and the probabilities associated with the branches in state 1.

Given this set of data, it is a simple matter to write a digital computer program to simulate the system. A theoretical analysis is more complex, and has not yet been carried out. In any case, the simulation approach has the advantage both of simplicity and of allowing the effect of various strategies and other modifications to the system to be easily and quickly investigated.

Such a simulator has been written for the Price algorithm and is being used to determine the effect of the values of the various parameters on such characteristics as the proportion of time wasted in queueing for access to memory. The results of these investigations are described in ref.[14]. An example is shown in figure 5 which illustrates the typical effect of P, the number of processors, on the effective number of processors in the system, P_e. The effective number is defined to be P multiplied by the proportion or time spent in doing useful work, that is, not waiting. The times assumed for the various operations were based upon the values measured on the DEC-10 sequential computer.

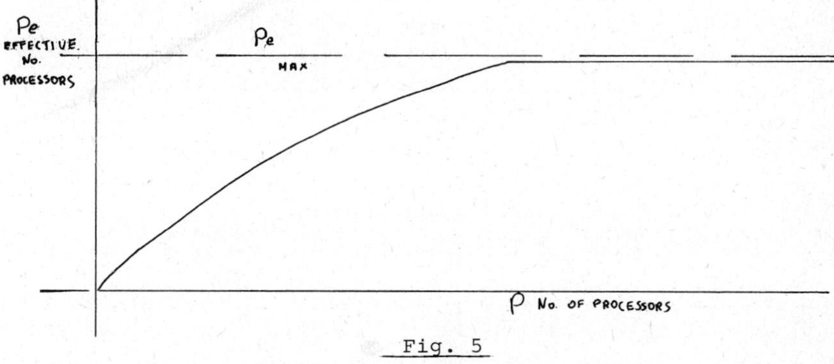

Fig. 5

It can be seen that under the assumed conditions, the effective number of processors rapidly tends towards an asymptotic limit; this limit will depend upon the assumed values of the various parameters involved. Clearly, the effect of alternative architectural and algorithmic features can be easily and quickly evaluated using such methods.

Numerical Analysis

The analysis descibed above will provide a good deal of insight into the performance of a given algorithm as a queuing system. However, it provides no information whatsoever about those aspects of its performance of most interest to numerical analysis, namely its convergence properties. In any case there exists relatively little theoretical background for algorithms such as those of Price and Törn, even in the sequential case. Such convergence results as there are are usually based upon numerical experiments [15], and so the most that can be expected from any theoretical analysis of parallel Holistic algorithms in some insight into the way in which parallelisation is likely to effect suche empirical results. This way be approached in a number of ways. One simple method would be to exploit once again the Holistic nature of the algorithm by focussing on the behaviour of one typical processor. During one complete iteration of such a processor, the pool is 'enriched' by the work of the processor in question and also by the other P - 1 members of the set processors. Ideally, then, one iteration of the typical processor would in effect be P times more 'effective' than it would be if P = 1. In this ideal case the number of iterations of each processor would be 1/P times the number required by the sequential case and so the numerical speed-up ratio would be P. (This would of course be reduced by the factor due to queueing loss). In practice if it is assumed that the speed-up is not NP but α NP, $\alpha < 1$, then the value of α can be predicted statistically and the result compared with empirical measurements. In this way a simple model can be constructed of the various effects tending to degrade performance. Of course, there are problems associated with the difficulties of establishing reliable results for the performance of global optimisation methods, even in the sequential case; but it is hoped that the research currently underway will lead to such results being established.

CONSLUSION

The aim of the foregoing paper has been to examine some aspects of the ways is which parallel computation may be applied to numerical optimisation, and in particular to global optimisation. It must be stressed that a considerable selection of information has had to be made in order to present a clear picture within the confines of a short paper. For example, the classification of parallel systems into SIMD and MIMD, although commonly used and significant, is too simple to give more than a rough idea of the range of available or potential parallel systems; Ref. [3] may help to fill this gap. In particular, the whole area of data flow techniques has been omitted, although these may well turn out to provide a general approach which will help to introduce real order into the present rather ill-defined situation. Ref. [16] provides an introduction to this topic. Another significant area which has barely been touched upon is the exploitation of parallelism at low levels in an algorithm, for example, in the various linear algebraic operations involved in a typical local minimisation technique such as might be used as a subprogram in the Törn technique. Heller's review, Ref. [17], will provide information to balance this omission.

In spite of such unavoidable limitations, it is hoped that the paper ha performed one or two useful functions. In the first place, I have tried to show that parallel computation requires and justifies a more throughgoing approach to algorithm design than might have been expected, if its potential is to be realised to the full. Secondly, at least some of factors which must be taken into account in the design of parallel algorithms for global minimisation have been outlined, and a general approach to such designs, and the analysis of the resulting algorithms, has been suggested. Finally, it is hoped that a framework has been provided which will prove useful as a means to initial access to this rapidly developing field.

REFERENCES

[1] Gellatly, R.A., Gallagher, R.H., and Lubracki, W.A.: Development of a Procedure for automated synthesis of minimum weight structures, AFFDL-TR-64-141, 1964.

[2] Gellatly, R.A., Berke, L.: Optimal Structural Design., AFFDL-TR -70-165, 1971.

[3] Thurber, K.J., Wald, L.D.: Associative and Parallel Processors. A.C.M. Computing Survey, Vol. 7, No. 4, Dec. 1975.

[4] Flynn, M.J.: Some computer organizations and their effectiveness. IEEE Transactions on Computers, Vol. C-21, No. 9, Sept. 1972.

[5] Parkinson, D.: An introduction to array processors. Systems International, Nov. 1977.

[6] Wulf, W.A., Bell, C.G.: C.MMP-a multi-mini-Processor. AFIPS Proc. 1972, FJCC Vol. 41, AFIPS Press, pp 765-777.

[7] Swan, R.J., Fuller, S.H., Siewiorek, D.P.: Cm -a modular, multi-microprocessor, AFIPS Proc. Vol. 46, pp 637-644, 1977.

[8] Evtushenko, Yu.G.: Numerical Methods for finding global extrema (case of a non-uniform Mesh), Zh. Vychisl. Mat. mat. Fiz., 11, 6, pp 1390-1403, 1971.

[9] Baudet, G.M.: The design and analysis of algorithms for asynchronous multiprocessors. Carnegie-Mellon Tech. Rep. CMU-CS-78 -116, 1978.

[10] Chazan, D., Miranker, W.L.: Chaotic relaxation, Linear Algebra and Appl., 2(1969), pp 199-222.

[11] Törn A.A.: A search-clustering approach to global optimization. In: Towards Global Optimization 2 (Dixon and Szegö Eds.) North Holland 1978.

[12] Dijkstra, E.W.: Solution of a problem in concurrent programming control. Comm. ACM 8, 9(Sept. 1965) 569.

[13] Price, W.L.: A controlled random search procedure for global optimization. In: Towards Global Optimization 2 (Dixon and Szegö Eds.) North Holland 1978.

[14] MCKeown, J.J.: Simulated behaviour of parallel global optimization algorithms. Numerical Optimization Centre, Hatfield Polytechnic. T.R. (to appear)

[15] Gomulka, J.: A user's experience with Törn's clustering algorithm. In: Towards Global Optimization 2 (Dixon and Szegö Eds.) North Holland 1978.

[16] Dennis, J.B.: First version of a data flow procedure language. Lecture notes in computer science, 19, Springer Verlag, 1974.

[17] Heller, O.: A survey of parallel algorithms in numerical linear algebra. SIAM review, Vol. 20, No. 4, October 1978.

Numerical Technqieus for Stochastic Systems
F. Archetti and M. Cugiani (eds.)
© North-Holland Publishing Company, 1980

EXPERIMENT DESIGN FOR NONLINEAR REGRESSION

Bruno Betrò
Istituto Matematico Università di Milano - Milano (Italy)

The problem is considered of the optimal design for nonlinear regression models. The choice of suitable optimality criteria is discussed under various circumstances, stressing the limits of criteria deriving from linear regression theory. A criterion is introduced, based on simulation of the estimation process, aiming at overcoming such limits. As an application, the optimal design of a seismographic network is considered.

INTRODUCTION

A relevant amount of statistical literature is devoted to regression problems. Only a small fraction of it, however, considers the case of nonlinear dependence of observations on the unknown parameters, although nonlinear models are fitted in an increasing number of applications. The fact is that linear regression problems admit an elegant analytical treatment and statistical properties of the solutions are known exactly. In particular the smallest confidence regions for the unknown parameters have the simple form of ellipsoids.

The nonlinear case is usually treated assuming as applicable to a finite number of observations, asymptotic properties of Maximum Likelihood estimators, or equivalently linearization of the regression functions in a neighbourhood of the unknown parameters. When the number of available observations is small, as typically in case of costly observations, the goodness of the approximation yielded by such asymptotic properties is doubtful.

In the framework of regression problems, in recent years the study was introduced of the so called optimal design problem, that is the problem of optimally distributing the points where observations are to be taken. A certain number of optimality criteria for designs and algorithms converging to optimal designs have been proposed until today. Most of the results are concerned with the case of linear regression.

The need of more carefully considering the optimal design problem when the regression functions cannot be assumed as linear or asymptotic considerations cannot be applied, was suggested to the author by the problem of the optimal design of a network of seismographic stations; in such problem the distribution of a certain number of stations has to be optimized in order to reduce the error in the location of a seismic source. The regression model describing the observation taken at a station is nonlinear and the number of stations is usually low because of the costs connected with installation and maintenance.

The present paper aims at discussing the main difficulties connected with the formulation of a framework for optimal design in nonlinear regression, as comprehensive as the one available in the linear case. An attempt is made to overcome some of such difficulties by the introduction of a new optimality criterion not related to asymptotic results or approximations. Its evaluations require a simulation of the process of parameter estimation and its optimization can benefit by an increasing number of stochastic optimization techniques.
The advantage of such criterion over the approximate one deriving from D-optimal design theory for linear regression is shown in the case of the optimal design of a seismographic network.

1. Formulation of the problem

Let us consider the regression model

$$y_x = f(x,\Theta) + \varepsilon_x \tag{1}$$

where x is a control variable in some compact set $X \subset R^n$, θ is a vector of unknown parameters in $\Theta \subset R^k$, y_x is a scalar observation of θ which can be viewed as taken at the point x, ε_x the observation noise. Observations will be assumed independent random variables and the noise at x distributed according to $N(0,\sigma_x^2)$.
A collection of points in X, not necessarily distinct

$$\{x_1,\ldots,x_m\}$$

will be called a design of m observation points for the regression model (1), or briefly design. A design can be viewed as a policy for selecting points where observations are to be taken in order to estimate the vector of unknown parameters θ. As some points may coincide, the possibility of multiple observations at the same point is not ruled out. Given a design $\{x_1, \ldots, x_m\}$ and taken the m observations $y_{x_1}, \ldots, y_{x_m}$, an estimate $\hat{\theta}$ of the unknown Θ can be gained by the least squares method:

$$\min_{\theta \in H} S(\theta), \; S(\theta) = \sum_{i=1}^{m} (y_{x_i} - f(x_i, \theta))^2 / \sigma^2_{x_i}.$$

Suppose that it is of interest to estimate "in the best possible way" some components, say the first $r \leq k$, of the vector θ (in the following the first r components of θ will be indicated by $\theta^{(1)}$ and by $\hat{\theta}^{(1)}$ the first r components of $\hat{\theta}$). Then a suitable scalar criterion should be introduced to measure the effectiveness of designs $\{x_1, \ldots, x_m\}$ in estimating $\theta^{(1)}$. Its minimization should correspond to the "best possible" estimation of $\theta^{(1)}$ in some sense. As the distribution of $\hat{\theta}$ depends, for given $f(.,.)$ and σ^2_x, on $x_1, \ldots, x_m$ and θ, such criterion will be in the form

$$A = A(x_1, \ldots, x_m, \theta), \tag{2}$$

A mapping R^{mn+k} onto $[0, +\infty]$. Criteria of the form (2) will be in general, of course, only of theoretical interest, because they involve the unknown θ. For actual evaluation of the effectiveness of the design $x_1, \ldots, x_m$, we will assume that a probability measure $P(d\theta)$ is given on the parameter space θ. Such a measure may reflect a priori information available about θ or relative relevance assigned to accurate identification of different possible values of θ. Averaging (2) with respect to P, we obtain

$$A(x_1, \ldots, x_m) = \int_{\Theta} A(x_1, \ldots, x_m, \theta) \; P(d\theta). \tag{3}$$

A criterion in the form (3) allows to define an optimal design $\{x_1^*, \ldots, x_m^*\}$ by the equation

$$A(x_1^*, \ldots, x_m^*) = \inf_{\substack{x_i \in X \\ i=1, \ldots, m}} A(x_1, \ldots, x_m). \tag{4}$$

An optimal design yields the points at which observations are to be taken in order to estimate $\theta^{(1)}$ in an optimal way, in the sense of definitions (2), (3), (4).

The crucial problem is the following: how to give a suitable form to $A(x_1, \ldots, x_m, \theta)$ so that the solution of (4) corresponds to an actual

sensible gain in the knowledge about $\theta^{(1)}$?
The next three sections are devoted to investigate about possible answers to such question.

2. Optimal designs for linear regression

As mentioned in the introduction, the problem of giving optimal designs for linear regression has been widely studied during the last twenty years. In the linear case indeed the problem allows for a comprehensive analytical treatment and satisfactory answers can be given. We are going to outline the main results as yet available.
Let's consider the linear regression

$$y_x = \theta^T f(x) + \varepsilon_x$$

where $f^T(x) = (f_1(x), \ldots, f_k(x))$.

Given a design $\{x_1, \ldots, x_m\}$ the L.S. estimator of θ turns out to be

$$\hat{\theta} = (F^T \Sigma^{-1} F)^{-1} F^T \Sigma^{-1} y ,$$

$$F^T = [f(x_1) | \ldots | f(x_m)] ,$$

$$\Sigma = \begin{bmatrix} \sigma^2_{x_1} & & 0 \\ & \ddots & \\ 0 & & \sigma^2_{x_m} \end{bmatrix}, \quad y = \begin{bmatrix} Y_{x_1} \\ \vdots \\ Y_{x_m} \end{bmatrix}.$$

$\hat{\theta}$ is an unbiased estimator and its covariance is

$$M^{-1} = (F^T \Sigma^{-1} F)^{-1} . \tag{5}$$

In particular $\hat{\theta}^{(1)}$ is an unbiased estimator of $\theta^{(1)}$ with covariance

$$(M^{(1)})^{-1} = (M_{11} - M_{12} M_{22}^{-1} M_{21})^{-1},$$

where

$$M = \begin{bmatrix} M_{11} & M_{12} \\ M_{21} & M_{22} \end{bmatrix}, \qquad M_{11} \quad r \times r \text{ matrix.}$$

It is natural to give to criterion A the form of some functional Φ of $M^{(1)}$. The most common examples of such criteria are (Kiefer (1974)

$$\Phi_o(M^{(1)}) = \det (M^{(1)})^{-1} \tag{6}$$

$$\Phi_1(M^{(2)}) = \operatorname{tr} (M^{(1)})^{-1} \tag{7}$$

$$\Phi_\infty(M^{(1)}) = \text{maximum eigenvalue of } (M^{(1)})^{-1} = \lambda_{max}((M^{(1)})^{-1}) \tag{8}$$

The last criterion is more seldom used as its degree of regularity may be lower than the one of the coefficients of $M^{(1)}$.
We remark that the covariance matrix doesn't depend on θ, so that the criterion based on such matrix will not need averaging as in (4). Designs minimizing Φ_o, Φ_1, Φ_∞ are called D-optimal, A-optimal and

E optimal respectively (Kiefer (1958), Kiefer and Wolfowitz (1959)).

An attractive geometrical interpretation of Φ_0, Φ_1, Φ_∞ as optimality criteria is related to confidence regions for $\Theta^{(1)}$.

Indeed a confidence region at level $1-\alpha$ for $\Theta^{(1)}$ has the following ellipsoidal form

$$E \equiv \{\Theta^{(1)}: (\Theta^{(1)} - \hat{\Theta}^{(1)})^T M^{(1)} (\Theta^{(1)} - \hat{\Theta}^{(1)}) \leq \chi^2(x, 1-\alpha)\} \qquad (9)$$

where $\chi(M, 1-\alpha)$ is the $(1-\alpha)$ quantile of the χ^2 distribution with r degrees of freedom, and it is

$$\text{vol (E)} \propto \sqrt{\det(M^{(1)})^{-1}};$$

$$\text{sum of squared semiaxes of E} \propto \text{tr} ((M^{(1)})^{-1});$$

$$\text{maximum semiaxis} \propto \sqrt{\lambda_{max}((M^{(1)})^{-1})}.$$

From a computational point of view, optimization of criteria (6), (7), (8) involves the handling of mxn constrained variables. Thus, when the product mxn is large, solving the resulting constrained optimization problem may result in a very difficult task.

For large m, the complexity of the problem can be sensibly reduced considering in place of a collection of m points, a probability measure ξ on X. Such measures are called approximate designs: when ξ is concentrated on s points $x_1, \ldots, x_s$, $s \leq m$ and $\xi(x_i) = r_i/m$, r_i positive integers such that $\Sigma r_i = m$, then such new definition of designs includes the previous one given in sect. 1, otherwise it can be viewed as an approximation to those designs. The question has been examined in Kiefer (1961).

Given an approximate design ξ, the matrix $M(\xi)$ can be defined, extending (5), as

$$M(\xi) = \int_X \frac{1}{\sigma_x^2} f(x) f^T(x) \xi(dx).$$

Optimality criteria for approximate designs are stated in a straightforward way, for instance using functionals (6), (7), (8).

A fundamental result is that any design ξ can be considered, as far as $M(\xi)$ is concerned, as concentrated on a finite number of points. Indeed the following lemma has been proven (Karlin and Studden (1966)).

Lemma: For any design ξ_1 there is always a design ξ_2 concentrated on at most $\frac{1}{2}$ k(+1)+1 points such that $M(\xi_1) = M(\xi_2)$.

In other words, it is necessary to consider only approximate designs of the form

$$\xi = \sum_{i=1}^{k(k+1)/2+1} p_i \, \delta_{x_i}$$

where δ_{x_i} is the δ function concentrated at x_i, $\sum_{i=1}^{k(k+1)/2+1} p_i = 1, p_i \geq 0$

and the computation of optimal designs is reduced to an optimization problem in

$$\ell = (\frac{k}{2}(k+1)+1)n + \frac{k}{2}(k+1)$$

variables.

Further simplifications to such optimization problem have been proposed in the literature. Usually the ℓ variables problem is further reduced to a sequence of n variables problems, giving a sequence of measures of the type

$$\xi_{i+1} = (1 - \alpha_i)\xi_i + \alpha_i \, \delta_{x_i} \, , \quad \alpha_i > 0 \tag{10}$$

converging to optimal approximate designs, both in the sense of A-optimality and in the sense of D-optimality and recently also for more general forms of the functional Φ (Atwood (1973),Fedorov (1972), Wynn (1972), Wu and Wynn (1978), Wu (1978)).

A basic tool for the construction of sequences (10) is the general equivalence theorem which can be stated assuming for simplicity $\sigma_x^2 = \sigma^2$ in the following form:

<u>General Equivalence Theorem</u> (Kiefer (1974)) Let M be the collection of matrices $M(\xi)$ for all probability measures ξ on X. Assume that a functional Φ is defined on M and that a design ξ^* is Φ-optimal, that is

$$\Phi(M(\xi^*)) = \inf \{\Phi(M(\xi)) : \; M(\xi) \in M\}.$$

If Φ is continuously differentiable in a neighbourhood of $M(\xi^*)$ and $d(x,\xi) = - x^T \nabla\Phi(M(\xi))x$, $\bar{d}(\xi) = \max_{x \in X} d(x,\xi)$, $d^{\ddagger}(\xi) = - \operatorname{tr}(\nabla\Phi(M(\xi))M(\xi))$,

$(\nabla\Phi(M))_{ij} = \partial\Phi(M)/\partial M_{ij}$, then

a) $\quad \bar{d}(\xi^*) = d^{\ddagger}(\xi^*)$;

b) if there exists a strictly increasing function G on $\Phi(M)$ continuously differentiable at $\Phi(M(\xi^*))$ such that $G \circ \Phi$ is convex on M, then a) is sufficient for Φ- optimality;

c) $\bar{d}(\xi^*) = d^{\ddagger}(\xi^*) \iff \xi^*\{x : d(x,\xi^*) = \bar{d}(\xi^*)\} = 1$.

The theorem gives actual conditions for verifying Φ-optimality. If for example $\Phi(M) = -\log \det M$, then $\nabla\Phi(M) = -M^{-1}$ and $d\ddagger(\xi) = \mathrm{tr}(M^{-1}M) = k$ so that D-optimality of ξ^* can be checked by the condition that ξ^* must be concentrated on the set

$$\{x : x^T M^{-1}(\xi^*)x = k\} .$$

3. The problem of estimates accuracy in nonlinear regression

The development of a theoretical framework for designs in nonlinear situations as comprehensive as the one outlined in the preceding section for the linear case is severely restricted by the lack of a general theory for the estimates accuracy in nonlinear regression, and more generally in nonlinear models, for a finite number of observations. In nonlinear models, the most popular procedure for gaining confidence regions for the unknown Φ is based on linearization of the model in a neighbourhood of Φ, relying on the asymptotic theory of maximum likelihood estimators. For a finite number of observations, a certain number of assumptions, which the user is not always aware of, are required to ensure that linearization yields good results. For nonlinear regression, such assumptions are as follows:

a) the vector valued function $f(\Theta) = (f(x_1, \Theta), \ldots, f(x_m, \Theta))$ is adequately fitted in a neighbourhood U of the true value Θ^* by the first order Taylor expansion

$$f(\Theta) = f(\Theta^*) + J(\Theta^*)(\Theta - \Theta^*) \qquad \Theta \in U$$

$$J_{ij}(\Theta^*) = (\partial f(x_i, \Theta)/\partial\Theta_j)_{(\Theta=\Theta^*)} ;$$

b) the probability that $\hat{\Theta}$ falls outside U is close to zero, so that $\hat{\Theta}$ can be considered as the L.S. solution of the equation

$y = f(\Theta) + J(\Theta^*)(\Theta - \Theta^*)$

and hence normally distributed with mean Θ^* and covariance matrix

$V(\Theta^*) = (J^T(\Theta^*)\Sigma^{-1}J(\Theta^*))^{-1}$.

b) implies that confidence regions for Θ^* (or $\Theta^{*(1)}$) can be built just like in the linear case by means of the matrix $V(\Theta^*)$. As Θ^* is unknown, this statement may appear useless, but a) implies $J(\hat{\Theta}) \simeq J(\Theta^*)$ with probability close to 1 and hence $V(\hat{\Theta})$ is a close approximation to $V(\Theta^*)$.

On the other hand it is hard to predict in actual situations if linearization is valid, mainly when the number of observations is small.

Exact confidence regions have been given only in some particular cases, for example in some partially linear models (El-Shaarawi and Shah (1978)). In more general situations, a criterion for the construction of confidence regions relies on the likelihood ratio test (Cox and Hinkley (1974)). Such regions are bounded by contours of constant likelihood so that parameters lying inside are more likely than those outside.

The generalized likelihood-ratio principle (Mood et al. (1963)) dictates that the hypothesis

$$H_o : \Theta_1 = \Theta_1^*, \ldots, \Theta_r = \Theta_r^*$$

is to be rejected for small values of the ratio

$$\lambda_n = \sup_{\theta \in \Theta_o} L(\theta; y_{x_1}, \ldots, y_{x_m}) / \sup_{H} L(\theta; y_{x_1}, \ldots, y_{x_m})$$

where Θ_o is the section of Θ formed by the hyperplanes

$$\Theta_1 = \Theta_1^*, \ldots, \Theta_r = \Theta_r^*.$$

A precise form of the principle is possible only when the distribution of λ_n, or of some function of it, is known; for large n , $-2\log\lambda_n$ is distributed as a chi-square distribution with 2 degrees of freedom, but such approximation is in most cases a good approximation even for small n.

Once the distribution of λ_n is available, either in an exact or approximate way , confidence regions can be obtained in a standard way using the quantiles of such distribution. Regardless the closedness of the approximation, such regions are anyway exact level sets of the likelihood. Under such approach, confidence regions for nonlinear regression (1) at an approximate level 100 $(1-\alpha)$% are, for $r=k$ and $\sigma_x^2 = \sigma^2$ in the form (Draper and Smith (1965))

$$S(\Theta) \leq S(\hat{\Theta}) + \chi^2(k, 1-\alpha) \tag{11a}$$

$$S(\Theta) \leq S(\hat{\Theta}) \{1 + \frac{k}{n-k} F(k, n-k, 1-\alpha)\}, \tag{11b}$$

where $F(k, n-k, 1-\alpha)$ is the $1-\alpha$ quantile of the F distribution with k and n-k degrees of freedom.

Regions (11) are not as easily analitically manageable as ellipsoidal ones and this prevents their wide use in the applications. However, such aspect should be not so important today, because of the increasing availability and powerfulness of modern computers and computational techniques.

As suggested by Ross (1978), mathematical programming techniques may result in valuable help in gaining confidence bounds for the unknown parameters or functions of them in regions (11).

4. Optimal design for nonlinear regression

According to the analysis developed in the preceding section, we'll distinguish two cases: a) linearization is applicable for any $\theta \in \Theta$; b) linearization results in poor approximation. Let's consider the two cases in some details:

a) In such case the usual criteria for linear regression admit a very natural extension considering in place of the matrix $M(\xi)$ the matrix

$$M(\xi, \theta) = \int_X \frac{1}{\sigma_x^2} \frac{\partial f(x,\theta)}{\partial_x} \left(\frac{\partial f(x,\theta)}{\partial_x}\right)^T \xi(dx)$$

and averaging functionals $\Phi(M^{(1)}(\xi,\theta))$ ((6) - (8), e.g.) with respect to θ as in (3); thus optimal designs, exact or approximate, can be defined according to (4) as minimizing functionals of the type

$$\Phi(\xi) = \int_\Theta \Phi(\xi,\theta)\, P(d\theta).$$

Here ξ is of the form $\sum_{i=1}^{n} \delta_{x_i}$ if n is small or in the form of a generic probability measure on X if n is large.

Optimality of designs under linearization has been considered by Box and Lucas (1959), Box and Hunter (1965) Draper and Hunter (1967), Atkinson and Hunter (1968), Box (1970), White (1973). All these papers are concerned with the case of prior probability $P(d\Theta)$ concentrated at a"most probable point" or in the form of a multivariate normal distribution and also consider the possibility of repeating observations when needed. Thus all cases in which experiments cannot be repeated and prior knowledge or interest about Φ is not in the form of a distribution "centred" at a most probable value are excluded by the analysis.

An intriguing task would be to investigate about the optimum properties of functionals $\bar{\Phi}(\xi)$ for more general forms of $P(d\Theta)$, extending the results reported in sect. 2 for functionals of $M^{(1)}(\xi)$.

b) When linearization results in poor approximation, as typical in the case of few observations, the introduction of criteria not related to asymptotic properties becomes necessary.
Linearization is rather attractive as the deriving optimality criteria can be stated in close analytical form. We saw in the proceding section that in the case of confidence regions the lack of such analytical form can be overcome by the use of efficient numerical techniques. We think that such approach could result fruitful also for optimal design in nonlinear regression.
In what follows we'll show how an exact criterion, that is a criterion not based on approximate assumption, can be obtained in a purely numeric way, suitable for computer processing, through Montecarlo simulation.
Let $\hat{\theta}$ be the L.S. estimate, or in principle any other estimate of θ, derived from observations taken at $x_1,\ldots, x_m$. As in actual situations is, let us assume that a numeric approximation $\hat{\theta}_c$ to $\hat{\theta}$ is gained through a suitable algorithm. Then the error in the estimation process can be defined

$$\hat{e}(\theta;\ x_1,\ldots,x_n) = \mathrm{err}(\hat{\theta}_c,\theta), \tag{12}$$

where err $(\cdot,\cdot)$ is some"error" function, that is a scalar nonnegative function which vanishes when the arguments coincide. $\hat{e}(\theta;\ x_1,\ldots,x_m)$ is a r.v., depending on the noise in the observations y_{x_i}, with mean

$$e(\theta;\ x_1,\ldots,x_m) = E\ \{\hat{e}(\theta;x_1,\ldots,x_m)\}.$$

In presence of a prior distribution $P(d\theta)$ on θ $e(\theta;x_1,\ldots x_m)$ can be further averaged obtaining

$$\bar{e}(x_1,\ldots,x_m) = \int_\Theta e(\theta;x_1,\ldots,x_m)\ P(d\theta). \tag{13}$$

Then we can define a design $x_1^*,\ldots,x_m^*$ optimal if

$$\bar{e}(x_1^*,\ldots,x_m^*) = \inf_{\substack{x_1,\ldots,x_m \\ x_i \in X}} \bar{e}(x_1,\ldots,x_m). \tag{14}$$

Of course, the distribution of $\hat{e}(\theta;x_1,\ldots,x_m)$ cannot be given analitically, but nevertheless the Montecarlo method furnishes the tool for evaluating $e\ (\theta\ ;\ x_1,\ldots,x_m)$ for any $\theta, x_1,\ldots,x_m$, simulating independent outcomes of $y_{x_1},\ldots, y_{x_m}$ according to $N(0,\sigma^2_{x_i})$.

If P is in the simple form of a finite sum $\Sigma p_i \delta_{x_i}$, then $\bar{e}$ $(x_1,\ldots,x_m)$ is in such a way easily evaluated, otherwise also Θ^i has to be randomly generated according to P.
In this way unbiased estimates of $\bar{e}$ $(x_1,\ldots,x_m)$ are obtainable for any design $\{x_1,\ldots,x_n\}$.
The optimization problem (14) is hence in the form of a constrained stochastic optimization problem; the increasing attention devoted in the literature to such problems (see e.g. Kushner and Lakshmivarahan (1977) and Zielinski (1979)) legitimizes the outlined approach as a suitable tool for the actual optimization of designs.

5. Optimal design of a seismographic network

The problem of the optimal design of a seismographic network was treated in Kijko (1978), Archetti and Betrò (1979), Betrò (1979), as a problem of optimal allocation of the observation points, in a nonlinear regression scheme.
Features of the problem are the small number of observations which can be taken (networks for the study of the seismicity of a restricted area are usually formed by $6 \div 10$ stations and obviously observations cannot be repeated) and the fact that seismic sources need not to be concentrated around a single point.
In the above quoted papers the problem was solved by the application of the linearization procedure. The restricted number of observation points makes suspect the validity of linearization. In what follows we'll show how the use of a criterion of the type (13) may result in a better allocation of the sesmic stations.
The arrival times to a station of the seismic waves propagating through the earth crust are described by the model

$$t_i = t_o + f(x_i,x_o) + \varepsilon_i$$

where the function f is the so called travel time function from a seismic source localized at the point x_o to the station placed at the point x_i; t_o is the origin time of the seismic event.
The simplest form of f is when the waves propagate in an homogeneous and isotropic medium with constant velocity v. In this case

$$t_i = t_o + \frac{1}{v} \|x_i - x_o\| + \varepsilon_i . \tag{15}$$

Of the four unknown parameters, only the three components of the vector x_o are of practical interest. This leads very naturally to give (12) the form

$$\hat{e}(x_o;\ x_1,\ldots,x_m) = \|\hat{x}_o - x_o\| \ , \tag{16}$$

where $\hat{x}_o$ is the L.S. estimate of x_o.

As in Betrò (1979), we are going to consider optimization of the seismic network existing around the town of Ancona, on the Adriatic coast of Italy (fig. 1). The network consists of six stations (black dots in the figure), all placed at null depth. In the region earthquakes occur mainly under the sea and hence the coast represents a serious constraint for the allocation of the seismographic stations. The simple model (15) is assumed with v= 5 km/sec. and σ_{x_i}=0.1. Two different points are assumed to represent equally possible seismic sources: referring to the coordinate system of the map completed with the depth expressed in kilometers, their coordinates are

$$x_o^1 = (25,\ 25,\ 5); \qquad x_o^2 = (27,\ 17,\ 1).$$

We'll consider here the simple case, particularly attractive for the possibilities of graphical displaying, in which a better allocation for one station is sought. The five stations held fixed are in fig. 1 indicated by MB, PL, MS, MT, CF. The region in which the best allocation for the sixth station x_6 is sought is the one bounded by the coast and the borders of the figure.
Fig. 2 displays some level contours of the function

$$\bar{V}(x_6) = \frac{1}{2}(V_1(x_6) + V_2(x_6)) \tag{17}$$

where $V_i(x_6)$ is the volume of the confidence ellipsoid for x_o^i, i=1,2, considered as a function of x_6. The details for the evaluation of such volumes are given in Betrò (1979).
The level contours reveal the presence of the constrained global minimum x_6' of $\bar{V}(x_6)$ near the projection of x_o^2 onto the x-y plane with a value less than 180.
In fig. 3 are represented some level contours of the function

$$\bar{e}(x_6) = \frac{1}{2}(E\{\hat{e}(x_o^1;\ x_6)\} + E\{\hat{e}(x_o^2;x_6)\}) \tag{18}$$

where $\hat{e}(x_o^i;x_6)$ is defined by (16) for n=6 and fixed $x_1,\ldots,x_5$.
$\bar{e}$ (x_6) was evaluated by the Montecarlo method up to an accuracy of one decimal digit. The L.S. estimates involved in the computations were obtained by OPLS routine of the Numerical Optimization Centre

Library, a particularly efficient routine for solving this type of problems (Bacchelli and Di Natale (1979)).
The figure clearly supports the fact that near the global minimum of $\bar{V}(x_6)$ there is only a local minimum for $\bar{e}(x_6)$. The constrained global minimum x''_6 of $\bar{e}$ (x_6) is near the upper left corner of the figure, where the coast intersects the upper border.
This implies that, if the sixth station is placed at x''_6, then, on the average, the location of seismic events originating at x_o^1 and x_o^2 will be better than if the station were placed at x'_6.
The sensible reduction in the average error (about 0.4 km) should stress the advantage of the exact criterion (18) based on a natural evaluation of the error, over the approximate one (17) based on the linearization procedure.

References

[1] Atkinson A.C. and Hunter W.G.(1968). The design of experiments for parameter estimation. Technometrics 10, 271-289.

[2] Archetti F. and Betrò B. (1979). Optimization problems arising in the design and exploitation of seismographic networks. In "Numerical Methos for Dynamical Systems", Dixon L.C.W. and Szegö G.P. eds. North Holland.

[3] Atwood C.L. (1973). Sequences converging to D-optimal designs of experiments. Ann. Statist. 1, 342-352.

[4] Betrò B.(1979). Optimal allocation of a seismographic network by nonlinear programming. Proceedings IX FIP Conference on Optimization Tecniques.

[5] Bacchelli B. and Di Natale M. (1979). General purpose vs. specialized algorithms for an ill-conditioned estimation problem. This volume.

[6] Box G.E.P. and Lucas H.L. (1959). Design of experiments in nonlinear situations. Biometrika 46, 77-90.

[7] Box G.E.P. and Hunter W.G. (1965). Sequential design of experiments for nonlinear models. IBM Scientific Computing Symposium in Statist., 113-137.

[8] Box M.J. (1970). Some experiences with a nonlinear experimental design criterion, Technometrics 12, 569-589.

[9] Cox D.R. and Hinkley D.V. (1974). Theoretical statistics. Chapman and Hall, London.

[10] Draper N.R. and Smith H. (1965). Applied regression analysis. J.Wiley & Sons.

[11] Draper N.R. and Hunter W.G. (1967). The use of prior distributions in the design of experiments for parameter estimation in nonlinear situations. Biometrika 54, 147-153.

[12] El Shaarawi A. and Shah K.R. (1978). Interval estimation in non linear models. T.R. STAT -78-06. Faculty of Mathematics, Waterloo University.

[13] Fedorov V.V. (1972). Theory of Optimal Experiments. Academic Press, New York.

[14] Kiefer J. (1974). General equivalence theory for optimum design (approximate theory). Ann. Math. Statist. 2, 849-879.

[15] Kiefer J. (1958). On the nonrandomized optimality and randomized nonoptimality of symmetrical designs. Ann. Math. Statist. 29, 675-699.

[16] Kiefer J. and Wolfowitz J. (1959). Optimum design in regression problems. Ann. Math. Statist. 30, 271-294.

[17] Kiefer J. (1961). Optimum designs in regression problems, II. Ann. Math. Statist. 32, 298-325.

[18] Karlin S. and Studden W.J. (1966). Optimal experimental designs. Ann. Math. Statist. 37, 783-815.

[19] Kijko A. (1977). Methods of the optimal planning of regional seismic networks. Publ. Inst. Geophys. Pol. Acad. Sc. A-7(119).

[20] Kushner H.J. and Lakshmivarahan S. (1977). Numerical studies of Stochastic Approximation Procedures for Constrained Problems. IEEE Trans. Automat. Contr. Ac 22, 428-439.

[21] Mood A.M., Graybill F.A. and Boes D.C. (1963). Introduction to the theory of statistics. McGraw-Hill.

[22] Ross G.J.S. (1978). Exact and approximate confidence regions for functions of parameters in nonlinear models, in COMPSTAT 1978, Proceedings in Computational Statistics. Physica-Verlag, Wien.

[23] White L.V. (1973). An extension of the general equivalence theorem to nonlinear models. Biometrika 60, 345-348.

[24] Wu C.F. (1978). Some algorithmic aspects of the theory of optimal designs. Ann. Statist. 6, 1273-1285.

[25] Wu C.F. and Wynn H.P. (1978). The convergence of general steplength algorithms for regular optimum design criteria. Ann. Statist. 6, 1273-1285.

[26] Wynn H.P. (1970). The sequential generation of D-optimum experimental designs. Ann. Math. Statist. 41, 1655-1664.

[27] Wynn H.P. (1972). Results in the theory and construction of D-optimum experimental designs. J. Roy. Statist. Soc. Ser. B 34, 133-147.

[28] Zielinski R. (1979). Global stochastic optimization: a review of results and some open problems. This volume.

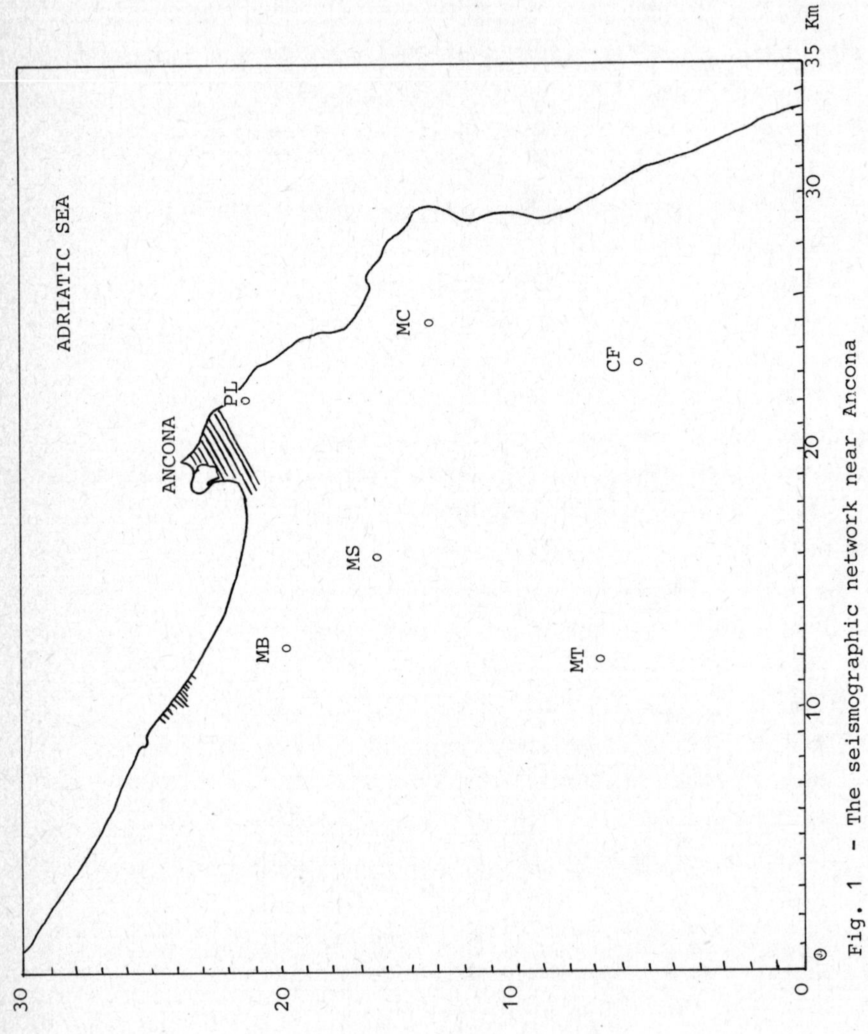

Fig. 1 - The seismographic network near Ancona

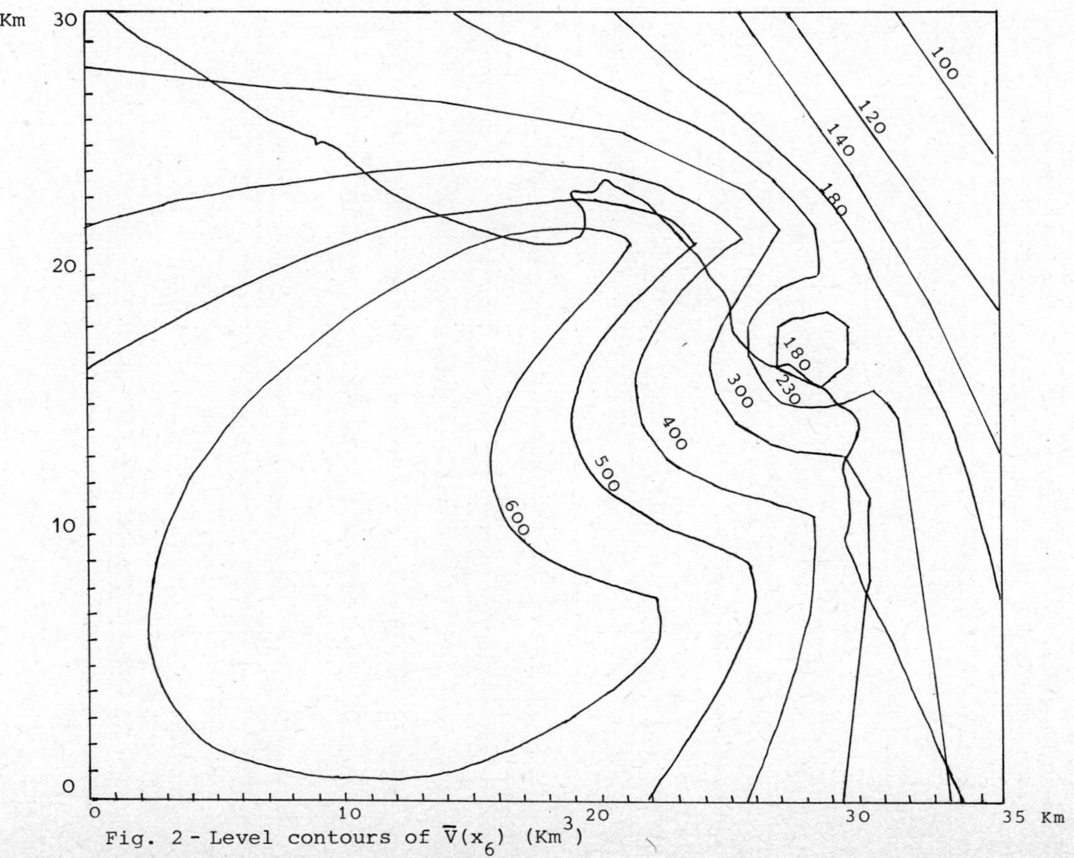

Fig. 2 - Level contours of $\overline{V}(x_6)$ (Km^3)

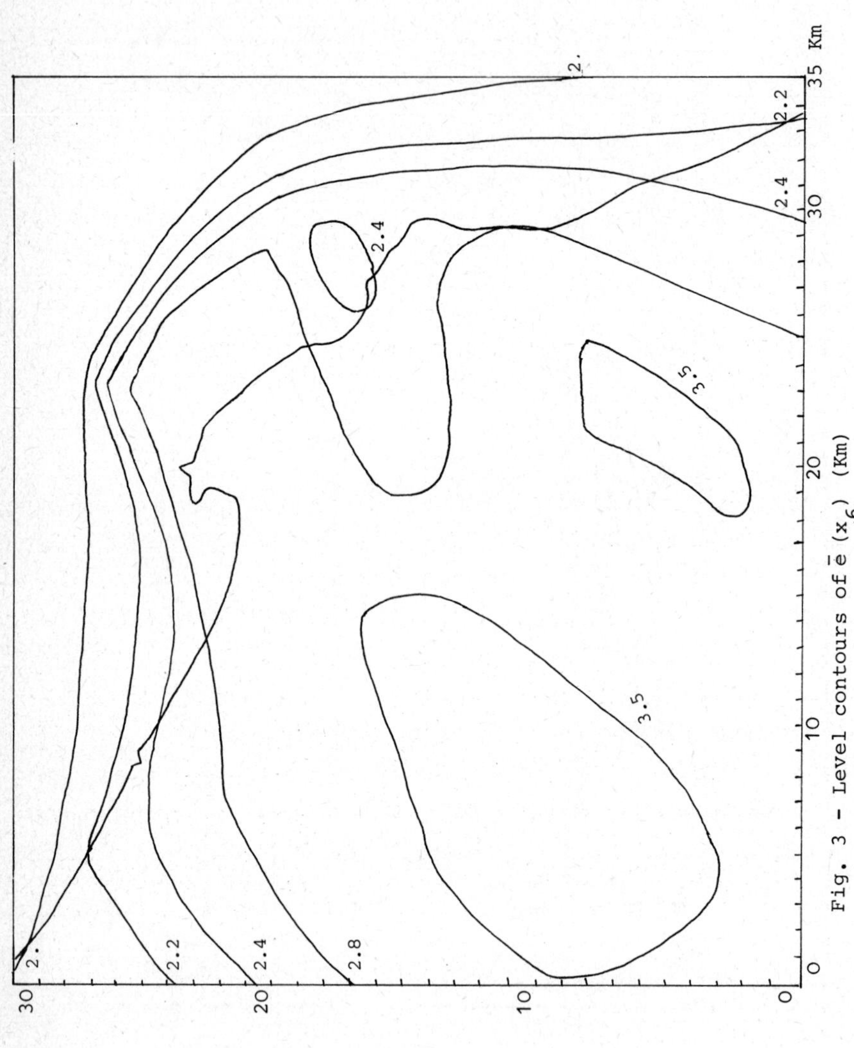

Fig. 3 - Level contours of $\bar{e}(x_6)$ (Km)

Numerical Techniques for Stochastic Systems
F. Archetti and M. Cugiani (eds.)
© North-Holland Publishing Company, 1980

OUTLINE OF A PROBABILISTIC FRAMEWORK FOR COMBINATORIAL OPTIMIZATION

Gianfranco d'Atri

Dipartimento di Matematica - Università della Calabria

INTRODUCTION.

1. Due to Operations Research applications, optimization problems have gained importance in combinatorial analysis, and now, a relevant theory exists about Combinatorial Optimization [Lawler (1976)]. It is mainly devoted to the study of parameters measuring the "time" required by algorithms to solve problems and the "goodness" of solutions found by them.

Referring to [Garey, Johnson (1979), Lawler (1976)] for precise definitions and results about "time complexity" and "approximation", here we remark that for most algorithms we only know the worst-case behaviour, that is, an upper bound to the number of steps required or to the approximation of the solution found [(1)].

This fact creates a gap between theory and practical situations in which the worst-case (mathematically speaking) rarely occurs: as an example, the Simplex Method is efficiently and reliably used to solve large Linear Programming problems (thousands of variables); yet, it has been shown to require exponential time (number of steps growing exponentially with the number of vars.) for families of problems.

2. From a practical point of view, the "average" behaviour should be a more significant measurement of algorithmic performance; thus, expected time of sorting subroutines may be studied under the assumption that all n! permutations of the n numbers to be sorted are equally likely.

[(1)] The approximation may be measured by the ratio (value of solution)/(optimal value).

In fact, a first probabilistic approach to algorithmic combinatorics has been this kind of analysis for sorting algorithms, as collected in Knuth's book (1973).
His approach suffers from the drawback that uniform distribution (as any postulated one) of the input is not necessarly realistic; however, it provides a correct way of analyzing situations in which the probabilistic assumption is reasonable.

3. Pioneering papers in probabilistic combinatorics (not algorithmic) are to be considered those of Erdös and Renyi (1959) on "random graphs". Though their approach is similar to Knuth's one -all graphs on n vertices with m edges are supposed equally likely- , the motivation is quite different.
Their results are the natural development of enumeration techniques on graphs and contain asymptotic expressions, rather than precise or approximate evaluations.
The possible application of random graphs analysis to the study of algorithms for Optimization problems was proposed by Karp (1976) and, since then, the field has been rapidly developing.
Now, we cannot distinguish between algorithmic and non-algorithmic combinatorics, under a probabilistic assumption about input data; however the (Knuth - Erdös - Karp's) approach will be called "Hungarian", for simplicity's sake.

4. Another way of using probability theory in combinatorial optimization was proposed by Rabin (1976) who suggested the use of "random" steps in the algorithms: while input data are considered fixed, the behaviour of the procedure is, in some points, non deterministic but controlled by a known stochastic process.
Surprisingly, in some cases these algorithms are more powerful than non-random ones.
However, this approach (we call it Cricket approach) is successful for a limited number of problems in which the combinatorial structure is unknown or too difficult to explore.

5. As we show in the paper, the seeming opposition of the two tech-

niques [2] -Hungarian and Criket - is subsumed by unified approach (we call it Fully Randomized) which allows random steps in the algorithm and assumes a random distribution of the input.
Such point of view is not only "unifying theory", but it naturally arises in developing Hungarian or Cricket analysis of some problems.

6. Lastly, apart from their historical theoretical background, let us motivate the two approaches by their practical Operations Research counterpart.

a) The usual technique for comparing or analyzing algorithmic performance in practice is to write down computer codes, run them on test problems and collect experimental data on time, accuracy and s.o. In this procedure, test-problems are obtained by (pseudo)-random generation of data.
The Hungarian approach is just the probabilistic formalization of such statistical analysis.

b) Any algorithm has, in some point, to break ties, i.e. to decide among equivalent choices.
For example, in the Simplex Method, at each iteration, a negative reduced cost column has to be selected, but many such columns may exist in the matrix. Usually, some criterion is added to avoid ties, or, more simply, a random choice is suggested by instructions as "take one column of negative reduced cost".
The Cricket approach is just the probabilistic formalization of those seemingly non-random algorithms, containing tie-break steps.

PART 1

Section 1. Probabilistic approach.

1. For a general discussion on probabilistic approaches to Combinatorial Optimization, let us introduce a simplified version of a well-known game: the "naval battle".
The "sea" is a strip of n sectors and m < n ships are placed on the

---

(2) As Rabin says:"This approach (Hungarian) suffers from the drawback...we are on shaky ground in assuming a particular distribution...In this paper we present a different approach".

sea by one player, the second player has to choose a sequence of k sectors to explore in order to find out a ship and destroy it.
In our simplified form, the game stops when the first ship is found, and each sector contains one ship at most. We are interested in the number of steps required to terminate.

A natural formalization of the study is the following: given a "problem instance" $v \in \{0,1\}^n$, we solve it, i.e. we find a j such that $v_j=1$, by an algorithm $s \in \{1,2,...,n\}^n$, i.e. a sequence s(j) of indices specifying the components of v to be tested ($v_{s(j)}=1$ or not?). For a given algorithm, we consider its time complexity for problem v: $t_n(v) = \min \{j/v_{s(j)}=1\}$ (set $t_n = \infty$ when $V_{s(j)} = o$ (for all j).

Fig. [1 | 2 X | X | X | n] X ships; 1 2 ... n sectors

A deterministic (as opposed to probabilistic) analysis of time-complexity consists in bounding $t_n(v)$ by a function B_n^s of n, not depending on v, that is $B_n^s \geq \max_v t_n(v)$.
If we consider an algorithm s^o which is a permutation of $\{1,...,n\}$ (to ensure termination of the game) and we take as set of feasible problem instances $V_n = \{v/ \sum_1^n v_j = m\}$ -simply called the problem-, we obtain

$$B_n^{s^o} = \max \{t_n(v)/v \in V_n\} = n-m+1$$

2. When we are interested in the "mean" time or in a bound for "almost all cases", i.e. almost all instances, we define a probability measure on the set V_n, thus obtaining a random problem.
The function $t_n : V_n \to N$ is, now, a random integer variable and we may compute its average value or assign a probability to statements of the type $\{t_n \leq k\}$.
It is easy to verify that for an algorithm-permutation, under the assumption of uniform distribution of all $\binom{n}{m}$ possible allocations of m ships, we have

$$\text{Prob } (t_n \leq k) = 1-(n-k)!(n-m)!/(n-k-m)!$$

indeed, $\text{Prob}(t_n \leq k)$ = 1-Prob (k different sectors contain no ships)=

$= 1-\binom{n-k}{m}/\binom{n}{m}$.

In particular, when $m/n \to p$ as $n \to \infty$ $(p \neq 0,1)$:

$$\text{Prob}(t_n \leq \lg n) \to 1 \quad ^{(1)}$$

Remark, that for $m/n \to p \neq 0,1$, the worst-case bound is

$$B_n^{s^o} \approx (1-p)n.$$

3. While in the previous Hungarian approach to the naval battle, we analyzed a given algorithm s^o, now we turn our attention to the time spent by algorithms of a given class on a <u>fixed</u> problem instance v^o.

Let $t_n^*(s) = t_n(v^o)$ for algorithm s, take as set of feasible algorithms $S_n = \{1,2,\ldots,n\}^n$, i.e. <u>all</u> possible n-sequences, and define a probability measure on S_n, then $t_n^*: S_n \to N$ is a random variable which may be analyzed in the same way as t_n. The randomized space S_n is a <u>random algorithm</u>. For example, let us consider the uniform distribution on S_n and vector v^o with exactly m 1's, then

$$\text{Prob}(t_n^* \leq k) = 1-(1-\frac{m}{n})^k$$

Indeed $\text{Prob}(t_n^* > k)$=Prob(the first components of s avoid m sectors)= $=n^{n-k}(n-m)^k/n^n$.

In particular, when $m/n \to p$ as $n \to \infty$ $(p \neq 0,1)$:

$$\text{Prob}\ (t^* \leq \lg n) \to 1 \quad ^{(1)}$$

Further, the same result is obtained for any particular input instance $v \in V_n$, so we derive the following statement:

"Random algorithm S_n solves problem V_n within lg n steps with probability approaching one".

Such a statement is a typical conclusion of a probabilistic analysis according to the Cricket approach.

The same result is obtained, more easily and in a more fascinating way, by stating the algorithm as a sequence of steps, some of which is random, as follows

step 1: $\ell \leftarrow 1$
step 2: $s(l)$ is randomly chosen in $\{1,\ldots,n\}$
step 3: $\ell \leftarrow \ell+1$ and go to 2

(1) lg n should be replaced by any function approaching infinity.

then, $\text{Prob}(t_n^*>k)=\text{Prob}(s(j)=0 \text{ for } j=1,\ldots,k)=[1-\text{Prob}(s(j)=1)]^k$. Obviously, this last model of random algorithm is equivalent to the previous one.

Section 2. Deterministic versus Probabilistic Approach.

1. If we make no difference between probability approaching 1 and probability equal to 1, i.e. certainty in the discrete, probabilistic results for the naval battle may be considered better than deterministic ones.

However, this doesn't qualify the probabilistic approach as superior, as it is showed by analyzing the game under modified probabilistic assumptions.

First, suppose $m/n=O(1/n)$, that is m bounded by a constant factor, then $\text{Prob}(t_n \leq \lg n)$ approaches a constant and only for $k=O(n)$ we have $\text{Prob}(t_n \leq k) \to 1$. So, when the number of ships is almost constant, the worst-case analysis provides a result similar to that of the Hungarian approach.

Now, consider the behaviour of the random algorithm of section 1; we have showed that, when $m \approx p\cdot n$, $\lg n$ steps suffice to assure with large probability that a ship is found, and this for any problem instance v such that $\sum_1^n v_j = m$.

But, suppose that the m ships are uniformly spaced on the sea as showed in the figure

Fig.

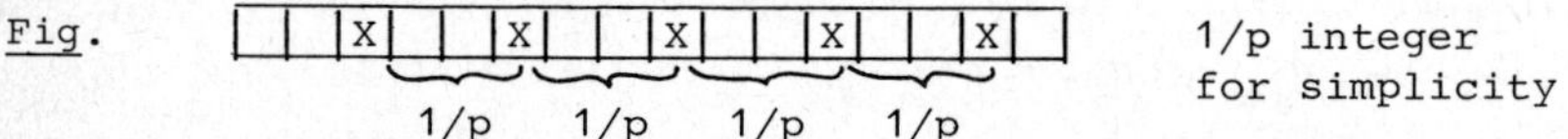

then the random algorithm behaviour is unchanged (because it ignores the special structure of the problem), while any algorithm s which performs a sequential search ($s(j+1)=s(j)+1 \bmod. n$) has an assured worst-case performance of $1/p$ steps (constant).

2. Apart from the "quality" of the results obtained, which type of analysis should be preferred in practice? For example is the probabilistic one more adherent to reality?

No unique reply may be given.

Let us consider the travel time from a point X to a point Y along a given route, by taking into account the time spent at lights when they are in the "red" state.

First case: we go daily from X to Y for delivering newspapers

Second case: we go from X to Y = airport for the last flight to Peking, where we must attend an important meeting.

In the first case, the mean travel time is an appropriate measure of the "goodness" of the route, while in the second one the worst-case (all lights red) time is more useful.

Obviously, further considerations should be done by taking into account other factors, for example the number of lights (is it very large?) or a maximal allowed delay in newspapers delivering; anyway, no absolute superiority of the probabilistic approach may be drawn from a practical point of view (nor the converse!).

By a mathematical standpoint, however, we are faced with the problem of describing a function $t_n : V_n \to N$, or $t^*_n : S_n \to N$, in some concise manner.

A first way is to give its Range, specified by $\max_{V_n} t_n$, but further information is collected by assuming a distribution over V_n and deriving the average or the "tail" Prob $(t_n \geq k)$.

The complete knowledge of the random variable, i.e. $\mathrm{Prob}(t^*_n=k)$ for all k (which subsumes worst-case analysis via $\min\{k/\mathrm{Prob}(t^*_n=k)=0\}$) provides a good "snapshot" of the function, though not the function itself.

Section 3. A general Approach.

1. Probabilistic approaches to Combinatorial Optimization are quite recent; here, we show how they may be considered the "natural" extension of the detrministic one.

Referring to the naval battle as a simple example, now we consider a general combinatorial problem P defined as a set V_n of problem instances [2] and a <u>class</u> of algorithms A defined as a set S of algorithms.

To each couple $(s,v) \in S \times V$ we associate the value

$$T(s,v) = \text{time required by alg. } s \text{ on prob. inst. } v$$

Obviously, for a fixed s^o, $T(s^o,v)=t(v)$ [2] and for a fixed instance v^o, $T(s,v^o)=t^*(s)$.

Let Al and Pr be two players, the first one choosing in set S (of Algorithms) and the second one in set V (of Problem instances): if Al chooses s and Pr chooses v, then Al pays $T(s,v)$ to Pr

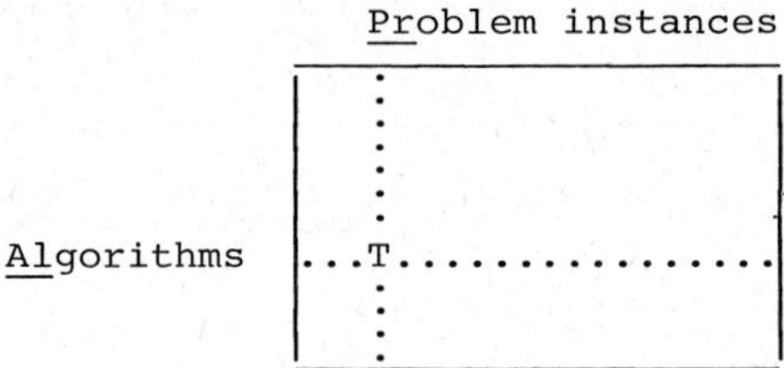

In classical game theory, we compute the two values

$$w(Al) = \min_{S} \max_{V} T(s,v)$$

$$w(Pr) = \max_{V} \min_{S} T(s,v)$$

For example, with $V=\{v/\Sigma v_j=m)\}$ and S the set of permutations in the naval battle, we get $w(Al)=n-m+1$ and $w(Pr)=1$ (and a large duality gap).

2. This minmax game has an interesting interpretation. When a Combinatorist performs worst-case analysis of an algorithm s^o for problem P, he finds (or simply bounds)

$$B(s^o)=\max_{V} T(s^o,v)$$

Further, for the same problem P, many algorithms are proposed in the aim of reducing the worst-case behaviour; $w(Al)$ is then the worst-case behaviour of the best algorithm in the class of studied algorithms A.

(Remark that, usually, class A is only implicitly considered in designing algorithms by analyzing only "intuitively good" procedures).

---

[2] Though we are really dealing with sequences of sets, functions and s.o., we omit the index n in the sequel, for simplicity.

In a similar way, we may interpret the role of Pr, but, due to a classical distinction between algorithm and problem, for a fixed problem instance v^o, the value

$$b(v^o) = \min_S T(s,v^o)$$

which measures the best performance of algorithms on v^o, is not a usual object in combinatorial analysis.

Nevertheless, to find $\max_V \min_S T(s,v) = w(Pr)$ is equivalent to show that an entire class of algorithms has a worst-case behaviour not better than w(Pr); indeed, by min max theory

$$w(Pr) = \max \min T \leq \min \max T = v(Al)$$

The kind of analysis which consists in looking for a problem instance v* maximizing b(v), (that is a v* for which all algorithms in S have a "bad" performance) is less common than the dual one consisting in the minimization of B(s) (that is the evaluation of worst-case for all algorithms in S).

However, an interesting example is given by Chvatal for the knapsack problem: he establishes a set of problem instances, which require exponential time to be solved by any of the algorithms belonging to a large well defined class.

3. Now, as usual in game theory, we introduce probabilities and make a probabilistic analysis of the game; this complicate the simple minimax criterion for selection of strategies, because many reasonable criterions exist for the two players.

As first, suppose that Al makes a probabilistic assumption about Pr's choice in V, then for any algorithm s, $T(s,\cdot)$ is a random variable; so Al must choose among random variables.

A possible criterion for Al is to fix k, then compute

$$\min_S \text{Prob}(T \geq k) \quad (3)$$

that is to find the algorithm with the best probabilistic behaviour, according to the Hungarian approach (random problem).

Suppose now, that Pr makes a probabilistic assumption about Al's choice in S, then for any problem instance v, $T(\cdot,v)$ is a random va-

---

(3) Similarly, we may take $\min_S$ Expect [T] and s.o.

riable, and a possible criterion for Pr is

$$\max_{V} \text{Prob}(T \geq k)$$

This value or an upper bound to the probability that the random algorithm takes more than k steps in solving problem *P* is obtained by a Cricket approach to *P*.

Remark, that the maximization is usually made <u>implicitly</u>.

4. Finally, our scheme of game suggests a straightforward generalization.

We make a probabilistic assumption on S and V altogether and study the random variable $T: S \times V \to N$; that is, we analyze a random algorithm behaviour on a random problem (Fully Randomized approach).

Though successfully used by some authors [Itai and Rodeh (1978) - Angluin and Valiant (1979)], this approach, has not been explicitly recognized in the literature. For example, Angluin and Valiant in the final part of their paper (1979) show how their random algorithm for the random Hamiltonian cycle problem may be "derandomized". Their technique is quite general and, in principle, very easy: the random algorithm S is replaced by a deterministic algorithm the steps of which depend on the input v, so, for random input, the random algorithm is simulated.

As an example, let us consider the naval battle game and the following algorithm for solving problem instance v:

step 1: From the last r^2 binary digits of v obtain r integers $z_1, \ldots, z_r$ in the range $\{1, \ldots, n-r\}$ ($r \approx \lg n$).

step 2: <u>For</u> $\ell = 1, \ldots, r$ <u>do</u> $s(\ell) \leftarrow z_\ell$

When this algorithm is used to solve a random problem, it simulates a Cricket type algorithm, indeed the z_ℓ's are then random, according to the distribution of v.

<u>PART 2</u>

<u>Section 1</u>. Hungarian Approach.

1. As an example of what we call Hungarian approach, we analyze a problem not defined on graphs, to show the generality of the underlying idea.

We consider a special case of the knapsack problem, for which the re sults contained in [d'Atri (1979)] are in their simpler form, so avoiding technicalties.

We have n <u>objects</u> with associated <u>weights</u> v_j and a knapsack of given <u>capacity</u> u: the problem is to choose a subset of objects in order to maximize total weight without exceeding the given capacity.

It may be formulated as the integer linear program

$$\text{(PP)} \qquad \begin{aligned} &\text{Max } \Sigma v_j x_j \\ &\Sigma v_j x_j \leq u \\ &x_j = 0 \text{ or } 1 \end{aligned}$$

where the coefficients are supposed positive integers.

(PP) is related to the solution of the following diophantine equation

$$\text{(E)} \qquad \Sigma v_j x_j = u \quad , \quad x_j = 0 \text{ or } 1$$

indeed, a solution of (E) is an optimal solution of (PP) (but not the contrary!).

(PP) and (E) are NP-complete [Lawler], but, if the coefficients are bounded by a function c(n), then a dynamical programming algorithm [Lawler] solves it in $O(n^2 \cdot c(n))$ steps.

Further, in linear time we can solve the continuous relaxation -i.e. $0 \leq x_j \leq 1$ instead of $x_j = 0$ or 1- and find a "good" approximate solution, as it is showed by the following trivial algorithm

GOLOSO (italian for greedy)

step 1: <u>If</u> $\Sigma v_j < u$ <u>then</u> $\overline{x} \leftarrow (1,1,\ldots,1)$ <u>and</u> STOP	Trivial case
step 2: $S \leftarrow 0$; $i \leftarrow 1$; $\overline{x} \leftarrow (0,0,\ldots,0)$	Initialize
step 3: <u>If</u> $S+v_i \leq u$ <u>then</u> $S \leftarrow S+v_i$ $\overline{x}_i \leftarrow 1$ $i \leftarrow i+1$ <u>and</u> GO TO 3	Sequential Filling
step 4: $\alpha \leftarrow u-S$	Residual Capacity

Algorithm GOLOSO computes a feasible solution $\overline{x}$ and a <u>residual</u> α which is an upper bound to the difference between the value (total weight) of $\overline{x}$ and that of an optimal solution, whose value is obviously less than u.

If the residual is zero then GOLOSO solves equation (E) and finds an optimal solution to (PP); but, in the worst-case it provides a solution differing by the optimal one by as much as $\max_j v_j - 1$. Obviously, such pessimistic situation only occurs for a special selections of coefficients (e.g., $u = \sum_1^r v_j+1$ and $v_j=K$ for $j>r$, where $K=\max_j v_j$).

2. Let $v_1,\ldots,v_n$ and u be random integer variables: all quantities defined through algorithm GOLOSO, as α or $\overline{x}$, are random objects whose distribution is related to that of weights and capacity.

Now we can correctly speak of average value of the approximation, which is the mean of the random variable α, or compute the probability of events such as "GOLOSO solves (PP)", which is the probability of the set of problem instances solved by GOLOSO.

We use the following probabilistic model

(a): $v_1,\ldots,v_n$ are uniformly distributed over $\{1,\ldots,c(n)\}$

(b): u is uniformly distributed over $\{1,\ldots,n\cdot c(n)\}$

(c): $v_1,\ldots,v_n$ and u are mutually independent random variables.

We derive the

LEMMA 1: Putting $\alpha = +\infty$ when the trivial case occurs,

$$\text{Prob}(\alpha=h) = \frac{c-h}{c^2} \quad \text{for } h=0,1,\ldots,c-1$$

$$\text{Prob}(\alpha=+\infty) = \frac{1}{2}\,\frac{c-1}{c}$$

Proof: First, we define the "pivot" $\gamma = \min\{i/ \sum_1^i v_j \geq u\}$ ($\gamma=n+1$ for $\alpha=+\infty$), and we remark that when $\alpha=0$ then the pivot object is taken ($\overline{x}_\gamma=1$) otherwise it is the first excluded one ($\overline{x}_\gamma=0$), i.e. $\gamma=i$ when the algorithm terminates, excepted case $\alpha=0$.

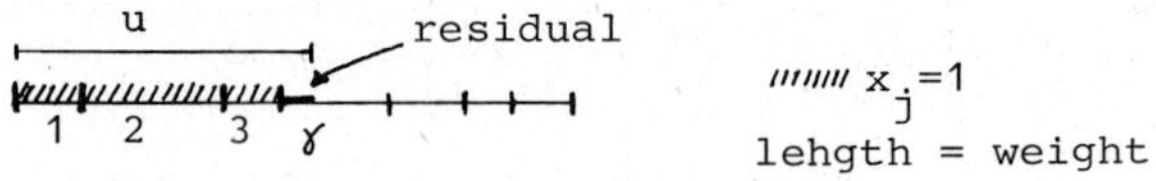

u

residual

1 2 3 γ

$x_j=1$

lehgth = weight

Set $S_k = \sum_1^{k-1} v_j$, $k=2,\ldots,n$ and $S_1=0$, then for $h=0,1,\ldots,c-1$ and $j=1,\ldots,n$:

$$\begin{aligned}
\text{Prob}(\alpha=h \text{ and } \gamma=j) &= \text{Prob}(S_j+h = u \text{ and } v_j > h)\\
&= \sum_{r=1}^{nc} \text{Prob}(u=r \text{ and } S_j+h=r \text{ and } v_j > h)\\
&= \sum_{r=1}^{nc} \text{Prob}(u=r)\cdot\text{Prob}(S_j+h=r \text{ and } v_j > h) && \text{- by hyp. (c)}\\
&= \frac{1}{nc}\sum_{r=0}^{nc-h} \text{Prob}(S_j=r-h \text{ and } v_j > h) && \text{- by hyp. (b)}
\end{aligned}$$

$$= \frac{1}{nc} \sum_{r=0}^{nc-h} \text{Prob}(S_j = r \text{ and } v_j > h)$$

$$= \frac{1}{nc} \text{Prob}(v_j > h) \quad \text{- when } v_j > h \text{, the range of } S_j \text{ is } 0,\dots,nc-h$$

$$= \frac{1}{nc}\,\frac{c-h}{c} \quad \text{- by hyp. (a)}$$

Finally, $\text{Prob}(\alpha=h) = \sum_{j=1}^{n} \text{prob}(\alpha=h \text{ and } \gamma=j) = n\cdot\frac{1}{nc}\cdot\frac{c-h}{c}$ and

$\text{Prob}(\alpha=+\infty) = 1-\sum_{h}^{c-1}\text{Prob}(\alpha=h)$.

Remark : $\text{Prob(GOLOSO solves (PP))} \geq \text{Prob}(\alpha=0) = \frac{1}{c}$

$\text{Prob(Trivial case occurs)} = \text{Prob}(\alpha=+\infty) \approx \frac{1}{2}$.

3. Now, we improve our algorithm by adding four further steps in which an object k such that $v_k = \max_{j>\gamma} \{v_j / v_j \leq \alpha\}$ is found, if it exists.

Let us denote by GOLOSO*, algorithm GOLOSO plus the following steps

step 5: $\bar{V} \leftarrow 0$

step 6: For j=i+1,...,n Do

If $\bar{v} \leq v_j \leq \alpha$ then $k \leftarrow j$ and $\bar{v} \leftarrow v_j$

step 7: If $\bar{v} \neq 0$ then $\bar{x}_k \leftarrow 1$

step 8: $\beta \leftarrow \alpha-\bar{v}$

The worst case analysis of GOLOSO* is similar to the pervious one (β as a measure of the approximation), but, under the probabilistic assumptions, we obtain the

THEOREM 1: For h=1,...,c-1

$$\text{Prob}(\beta \geq h) = \frac{1}{2}\,\frac{c}{n}\,\frac{1-(1-1/c)^n}{h}\left(1+\frac{1}{c}\left(\frac{c-2ch+h^2-h}{c}\right)\right).$$

From which, setting h=1 and computing $\text{Prob}(\beta=0)$

COROLLARY 1: Set $\rho=c(n)/n$ and suppose $c(n)=o(n)$, then

$$\text{Prob(GOLOSO* solves (PP))} \approx 1-\frac{1}{2}\rho.$$

Proof: For simplicity's sake we only prove the corollary.

Set $Q_{h,i} = \{\alpha=h \text{ and } \gamma=i\}$, $K_{h,i} = \{\nexists j, j\ i \text{ s.t. } v_j=h\}$, then

$$\text{Prob}(\beta\geq 1) = \sum_{h=1}^{c}\sum_{i=1}^{n}\text{Prob}(\beta\geq 1 \text{ and } Q_{h,i}) \quad \text{by decomposition.}$$

Now, for $\alpha=h$ and $\bar{\gamma}=i$, we have the following chain of equivalences

$$\beta \geq 1 \leftrightarrow \beta > 0 \leftrightarrow h > 0 \text{ and } \bar{v} \neq h \leftrightarrow \max_{j>i}\{v_j/v_j \leq \alpha\} \neq h \leftrightarrow h \notin \{v_j/j>i\},$$

thus

$$\begin{aligned} \mathrm{Prob}(\beta\geq 1) &= \sum_{h=1}^{c}\sum_{i=1}^{n} \mathrm{Prob}(K_{h,i} \text{ and } Q_{h,i}) \\ &= \sum_{h=1}^{c}\sum_{i=1}^{n} \mathrm{Prob}(Q_{h,i})\cdot\mathrm{Prob}(K_{h,i}) \quad \text{by hyp. (a)} \\ &= \mathrm{Prob}(\gamma=i)\cdot \sum_{i=1}^{n}(1-1/c)^{n-i} = \frac{c-1}{nc}\cdot\frac{1-(1-1/c)^{n}}{1/c} \end{aligned}$$

the result follows from the lemma and summing up; indeed, $\mathrm{Prob}(\gamma=1)+\ldots+\mathrm{Prob}(\gamma=n) = 1-\mathrm{Prob}(\gamma=n+1)$ and $\mathrm{Prob}(\gamma=i)$ is constant for $i=1,\ldots,n$.

4. We should design an improved version of our algorithm, say GOLOSO**, by inserting an initial step:

step 0: <u>Sort</u> $v_1,\ldots,v_n$ by nonincreasing value.

The aim of step 0 should be to increase the chance that in the second phase (steps 5 through 8) the algorithm finds an object of weight equal to α.

Indeed, while in the natural order objects with $v_j \leq \alpha$ may be placed before the pivot, i.e. $j < i$, after reordering, all objects with $v_j \leq \alpha$ are necessarly placed in position $j > i$, because $\alpha < v_i$ (see steps 3 and 4).

We have, now, $K_{h,i} = \{\nexists j, j>i / v_j=h\} = \{\nexists j, j\neq i / v_j=h\}$ and $\mathrm{Prob}\ (K_{h,i}) = (1-1/c)^{n-1}$.

Using this new value in the proof of corollary 1, we should drastically reduce $\mathrm{Prob}(\beta>0)$ to $\approx 1/2\ e^{-1/\rho}$.

But this simple substitution isn't possible, because the independence of $Q_{h,i}$ and $K_{h,i}$ is no more trivial; in fact, with respect to natural ordering, $Q_{h,i}$ and $K_{h,i}$ are defined by no common variable (u, $v_1,\ldots,v_i$ and $v_{i+1},\ldots,v_n$, respectively), but, due to sorting, all random variables play a role in the construction of the two events as defined by GOLOSO**.

To avoid technicalties, we refer to [d'Atri (1979)] for the proof of independence; nevertheless, we remark here the kind of difficulties which arise in probabilistic analysis of combinatorial algorithms, where each step affects the behaviour of the others.

Lastly, we note that corollary 1 may be converted in a non algori-

thmic number theory result as
"Prob (equation (E) has a 0/1 solution) $\rightarrow$ 1 as n $\rightarrow +\infty$".

Section 2: Cricket approach.

1. As an example of the Cricket approach, we analyze a problem strictly related to Rabin's one [Rabin (1976)], but stated here in the language of graph theory in order to display the usefulness of the technique for combinatorial optimization.
Before giving the general formulation let us describe its practical counterpart, a problem arising in the study of learning processes (obviously oversimplified).

We consider n individuals learning a set of k specific actions (i.e. replies to stimuli).
After the population has learned the k actions, it is treated with a biochemical substance inhibiting part of the actions: we don't know which ones.
So, we have to discover inhibited actions by experiments on the individuals; the task isn't trivial (test one animal per action) because only a fraction of them just remembers each action at the moment of the experiment: a lack of reply to one stimulus by one individual cannot be interpreted as an effect of the treatement.

An appropriate measure for experimental cost is the number of tests "individual-action" to be performed; assuming that a lower bound p to the fraction of individuals remembering each action is known (e.g. by previous studies) or postulated, we need $\approx k\cdot(1-p)\cdot n$ tests in the worst-case (we may be sure that an activity A has been inhibited only when at least n-pn+1 individuals aren't able to reply to the stimulus).

2. The problem is easily formulated in terms of a zero-degree subset problem (ZD) in a bipartite graph (V_1,V_2,E), $E \subset V_1 X V_2$, specified by its incidence kXn matrix $A=(a_{ij})$, where $a_{ij}=1$ iff vertex $i \in V_1$ is linked to vertex $j \in V_2$.
(ZD) consists in finding $S^{o} = \{i \in V_1 / deg(i) = 0\}$, where $deg(i)= \sum_{j \in V_2} a_{ij}$.

The following straightforward algorithm has complexity O(n) for constant k and $\min_{V_1} \{deg(i) \neq 0\} \geq pn$

step 1: $i \leftarrow 1$; $S \leftarrow \emptyset$

step 2: <u>If</u> ($a_{ij}=0$ for $j=1,\dots,n-pn+1$) <u>then</u> $S \leftarrow S \cup \{i\}$
<u>otherwise</u> $i \leftarrow i+1$

step 3: <u>If</u> $i \leq k$ GO TO 2

3. Now, we replace step 2 by a random step, obtaining algorithm GRILLO (italian for Cricket).

step 1: $i \leftarrow 1$; $S \leftarrow \emptyset$

step 2a: let $r_1,\dots,r_k$ be randomly chosen in V_2

2b: <u>If</u> ($a_{ir_j}=0$ for $j=1,\dots,\ell$) <u>then</u> $S \leftarrow S \cup \{i\}$
<u>otherwise</u> $i \leftarrow i+1$

step 3: If $i \leq k$ GO TO 2

GRILLO requires $O(\ell)$ steps for constant k, but, due to step 2a, it may fail to derive S^o, unless $\ell \geq n-pn+1$, by putting in S some vertex of degree $\neq 0$.

Nevertheless, a simple probabilistic analysis shows that this fact occurs "rarely".

First let's consider the meaning of the random selection step 2a: we consider a family of independent random variables R_{ij}, $i=1,\dots,k$, $j=1,\dots,\ell$ uniformly distributed on V_2, and then we consider the <u>random algorithm</u> GRILLO

step 2: <u>If</u> ($a_{ir_{ij}}=0$ for $j=1,\dots,\ell$) <u>then</u> $S \leftarrow S \cup \{i\}$
<u>otherwise</u> $i \leftarrow i+1$

Obviously, for any input graph, S is a random set.

We may establish the following chain of facts:

<u>First fact</u>: $S \supset S^o$ (S^o is not random),

indeed, if i is 0-degree then for no j $a_{ij}=1$ and i is added to S in step 2b.

<u>Second fact</u>: For a vertex i^o of non zero-degree $\mathrm{Prob}(i^o \in S) \leq (1-p)^{\ell}$,

indeed, $\text{Prob}(i^o \varepsilon S) = \text{Prob}(a_{iR_{ij}} \neq 1 \text{ for } j=1,\ldots,\ell) =$

$= \prod_1^{\ell} \text{Prob}(a_{iR_{ij}} \neq 1) =$

$= (1-\deg(i^o)/n)^{\ell} \leq (1-pn/n)^{\ell}$

Third fact: $\text{Prob}(S \neq S^o) \leq (1-p)^{\ell} \cdot (k-|S^o|)$

indeed, $\text{Prob}(S=S^o) = \text{Prob}$ (some non-zero degree i^o belongs to S) $\leq (k-|S^o|)\cdot\text{Prob}(i^o \varepsilon S)$

and

THEOREM 2: Algorithm GRILLO finds the set of 0-degree vertices in V_1 with probability $\geq 1-k(1-p)^{\ell}$, where $p=1/n\cdot\min\{\deg(i)\neq 0\}$.

Corollary: For constant p,k

$\text{Prob}(\text{GRILLO solves (ZD) in } 0(\lg n) \text{ steps}) \to 1$

Proof: $\text{Prob}(S=S^o) = 1-\text{Prob}(S\neq S^o) \geq 1-k(1-p)^{\ell}$.

Setting $\ell = \lg n$ the result follows.

Remark that the theorem has been obtained for a specific problem, i.e. for a particular graph G (non-random); from it we may derive a "probabilistic worst-case" behaviour on all graphs with $p \geq \bar{p}$, taking

$$\max_G \text{Prob(GRILLO doesn't solve (ZD) in G)} \leq k(1-\bar{p})^{\ell}.$$

4. Let us, now, show how Rabin's test for primality fits our scheme, by converting it in a graph problem.

The method is based on a test, TEST (n,b) on two integers n and b<n: if TEST = yes then n is composite (irrespective of the b-value). Further, for any composite number n there exist at least $3/4\cdot(n-1)$ numbers $b < n$ such that TEST(n,b) = yes: Rabin calls "witnesses" such numbers. Obviously, a prime number has no witness.

To find all prime numbers between n^o+1 and n^o+k (very large numbers) we should use the test $\approx \frac{k.n^o}{4}$ times at least; (for $k \ll n^o$).

Now, define the following (ZD) problem:

$V_1 = \{n^o+1,\ldots,n^o+k\}$, $V_2 = \{1,\ldots,n^o\}$ and $a_{ij}=1$ iff TEST(i,j) = yes.

Applying theorem 2 and noting that parameter $p \approx 3/4$, we obtain: "GRILLO finds the set of prime numbers in $\{n^o+1,\ldots,n^o+k\}$ with probability $< k4^{-\ell}$".

Remark: If k=1 and the number of steps is large enough, this reduces to verify, with a risk error α, whether n^o+1 is prime or not.

Section 3: Fully Randomized Approach.

1. In this section we show an application in which the Fully Randomized approach naturally arises as a development of probabilistic analysis.

But first, let us introduce the notion of random graph, as introduced by Erdös and Renyi.

We consider the set $G_{n,m}$ of all labeled graphs on n vertices with m edges, and suppose that the $\binom{\binom{n}{2}}{m}$ distinct graphs are equally likely; this implies that each edge has the same probability $m/\binom{n}{2}$ of belonging to the random graph (1).

The simplified version of the main result of our cencern is

Theorem 3 (Erdös-Renyi): *For $m = \frac{1}{2}(1+\varepsilon)n\cdot\lg n$, $\varepsilon>0$, random graph $G \in G_{n,m}$ satisfies Prob(G is connected) $\to 1$, as $n \to \infty$.*

2. Let us now study the minimum cost spanning tree problem (MCST) in a valuated graph, i.e. a graph whose edges are assigned a cost: it consists in finding a spanning tree of minimal cost.

This problem is solved by the so-called "greedy" algorithm which sequentially builds the tree, after reordering the edges by nondecreasing cost, and by adding a new edge only if no cycle is formed. The algorithm may be improved to yield an $O(m \lg n)$ time bound for "sparse" graphs or $O(n^2)$ for the complete one.

We call MINTREE the $O(m \lg n)$ implementation, without entering the details, of no interest here.

Set G(k) the partial graph constituted by those edges of G with cost $\leq k$ and denote by m(k) the number of such edges; we analyze the following algorithm by "standard" Hungarian approach.

QUICKTREE

step 1: Set $k = \lceil(1+\delta)\cdot c(n)\cdot\lg n/n\rceil$, $\delta > 0$

step.2: Apply MINTREE to graph G_k

(1) A similar model considers for each edge probability $m/\binom{n}{2}$ of belonging to G and, further, mutual independence.

3. Let G be complete and the costs be random independent variables, uniformly distributed over $\{1,\dots,c(n)\}$, then we have

<u>Theorem 4</u>: <u>Prob(QUICKTREE solves (MCST)) $\to 1$, as $n \to \infty$.</u>

The theorem derives from two lemmas.

<u>Lemma 1</u>: A minimum cost spanning tree in G_k is a m.c.s.t. in G.

<u>Proof</u>: A spanning tree of G_k is a spanning tree of G. In solving (MCST) in G, the greedy algorithms first examines edges of G_k and, then, those of $G-G_k$; if a m.c.s.t. is constructed using n-1 edges of G_k, no more edge is added and it is an m.c.s.t. of G.

<u>Lemma 2</u>: Let $k = \lceil (1+\delta)\cdot c(n)\cdot \lg n/n \rceil$, $\delta > 0$, then for some ε

$$\text{Prob}(m(k) \geq \frac{1}{2}(1+\varepsilon) n \lg n \to 1, \text{ as } n \to \infty.$$

<u>Proof</u>: k is such that, for each edge e, $\text{Prob}(e \in G_k) =$
$= \text{Prob}(\text{cost of } e \leq k) = k/c(n) \geq (1+\delta)\lg n/n$.
So, m(k) is a binomial random variable of mean $((1+\delta)\lg n/n)\cdot\binom{n}{2} = \frac{(1+\delta)}{2}(n-1)\lg n$; the result follows, taking $\varepsilon < \delta$.

4. The time required by QUICKTREE is $O(m(k)\lg n)$, which gives $O(n(\lg n)^2)$ with probability approaching 1, if $m(k)=O(n \lg n)$ with prob $\to 1$.

This happens for $k \to \infty$, that is $c(n)\cdot \lg n/n \to \infty$, but not for c(n) constant, which yields k=1 for large n; in such case $m(k)=O(n^2)$ with probability $\to 1$, and the algorithm performs poorly ($O(n^2 \lg n)$).

To deal with c(n) constant, we use the following <u>random</u> algorithm, maintaining the probabilistic hypothesis about the input.

RANTREE

step 1: Randomly choose $\frac{1+\varepsilon}{2}\, n\cdot\lg n$ edges out of G_1

step 2: Apply MINTREE to graph G_1' so obtained.

The analysis is quite simple:

<u>First fact</u>: $\text{Prob}(m(1) \geq \frac{1+\varepsilon}{2}\, n\cdot\lg n) \to 1$, for c(n) constant.

<u>Second fact</u>: G_1' is a random graph with $\frac{1+\varepsilon}{2}\, n\cdot\lg n$ edges at least [(3)].

<u>Third fact</u>: <u>Any</u> spanning tree in G_1 is optimal.

---

(3) If G_1 has $\frac{1+\varepsilon}{2}\, n \lg n$ edges at least.

Theorem 5: a) Prob(RANTREE solves (MCST)) $\to 1$, as $n \to \infty$

b) RANTREE requires $O(n(\lg n)^2)$ steps.

Proof: (with probability $\to 1$) G_1 has at least $\frac{1+\varepsilon}{2}$ $n \cdot \lg n$ edges, so G_1' is a random graph to which theorem 3 applies, it is connected and MINTREE finds a tree of cost n-1.

Section 4: Practice

1. As we noticed in the introduction, probabilistic approaches to Combinatorial Optimization constitute the formalization of certain techniques, widely used in the design and analysis of algorithms, as statistical analysis of experimental behaviour and tie-break instructions.

But, this aspect has only theoretical relevance in the comprehension of Combinatorial problems; while such methods will have, in next future, a great impact in the design and validation of algorithms. Indeed, almost all researches in the field provide a final result of the following kind:

"Under appropriate probabilistic assumptions (on input or algorithm) algorithm A (fast, simple and without special features) solves problem P with high probability".

This means that, if we are interested in solving P, we don't need to spend much time in realizing very complicated computer codes. However this cannot be interpreted as a general-purpose statement, because the particular result applies whenever the probabilistic assumption is justified: we have simply to use the right result for the right problem.

2. One may think that, from this point of view, the Cricket approach is more useful because we are able to simulate a random algorithm while we don't control input data ; but,this isn't true.

Indeed, in the validation of algorithms by experimental testing, we simulate data, too.

However, for the moment, there are only a few analysis of random algorithms [(4)] : a very interesting result obtained by Rabin in number theory states that $2^{100} - 593$ is prime with a probability of

[(4)] mainly, under the Fully Randomized Approach.

error less than 10^{-18}.
This number has been discovered in a few minutes run of a program searching for prime numbers along the lines of section 2; while non-random algorithms require $0(n^{1/4})$ steps at least, this one is $0((\lg n)^2)$.
It scarcely needs to be remarked that automatic computers have a fault probability greater than 10^{-18}: so that in some sense the above result can be considered deterministic or at least no less sure than one given by a deterministic algorithm.

3. The main shortcoming in practical applications of results on random problems is their asymptotical nature; this lack of precision may be eliminated by better evaluations of bounds on probabilities or by experimental simulation for parameter estimation.

The powerfulness of the probabilistic approach is well shown by the results we have obtained about the knapsack problem [d'Atri (1979)] in its general formulation:

(KP) $\max \sum_1^n c_j x_j, \quad \sum_1^n v_j x_j \leq u, \quad x_j = 0 \text{ or } 1.$

For (KP) many algorithms have been proposed in the literature, all of which claim a good performance; however, the most astonishing behaviour is reported by Laurière (1978):

n. of vars.	100	500	1.000	5.000	10.000	20.000	30.000	60.000
mean time in s.	.19	.92	1.05	3.1	6.00	11.12	16.78	30.06

(an almost linear time!).
The experiments have been performed on data set, randomly generated according to the uniform distribution on $\{1,\ldots,100\}$ for coefficients and on $\{1,\ldots,\sum_1^n v_j\}$ for the capacity u.
A probabilistic analysis similar to that of section 1, shows that a (simple) linear-time algorithm solves (KP) with a probability of failure $\approx e^{-n/c^2} \approx e^{-n/10.000}$ (non-asymptotical).
This algorithm does not include all special features of Laurière's algorithm.

Another important measure of the algorithmic performance is the number of variables left after the application of a reduction test (exploiting linear relations among coefficients): for example, if a 1.000 variables problem is reduced to a 100 vars. one, we consider 100 as a more appropriate measure of the size.

Example (Korsh-Ingargiola)

n. vars.	50	100	500	1.000	5.000
mean n. vars. left	7.7	10.1	25.2	38.6	66.3

The table shows the relation $m \cong \sqrt{n}$ (m=vars. left), and the used probabilistic model is similar to the previous one .

In [d'Atri (1979)] we give a good estimate for the distribution of m, which provides $m \leq n^{1/2+\varepsilon}$ with high probability (for little ε, $n^{1/2+\varepsilon} \approx \sqrt{n}$), under the assuption of uniform distribution of the coefficients on a given integer interval.

REFERENCES

D. Angluin, L.G. Valiant. "Fast Probabilistic Algorithms for Hamiltonian Circuits and Matchings". J. of Comp. System Sc. 18 (1979) April pp. 155-193.

V. Chvàtal. "Hard Knapsack Problems". Operation Res. (to app.).

G. d'Atri. "Analyse Probabiliste du Problème du Sac-à-dos". Thèse de l'Universtité Paris VI, Informatique (1979 Jun.).

P. Erdös, A. Renyi. "On Random Graphs". Publ. Math. Debrecen 6 (1959) pp. 290-297.

M.R. Garey, D.S. Johnson. "Computers and Intractability". Freeman and C., San Fzancisco (1979).

R.M. Karp. "The Probabilistic Analysis of some Combinatorial Search Algorithms". In Algorithms and Complexity. Tramb J.F. ed., Academic Press N.Y. (1976) pp. 1-16.

D.E. Knuth. "The Art of Computer Programming, vol. 3: Sorting and Searching". Addison-Wesley, Reading Mass. (1973).

A. Itai, M. Rodeh. "Finding a Minimum Circuit in a Graph". SIAM J. of Comp.; 7 (1978) December pp. 413-423.

M. Lauriere. "An Algorithm for the 0/1 Kanpsack Problem". Math. Prog. 14 (1978) vol. 1, pp. 1-10.

E.L. Lawler. "Combinatorial Optimization: Networks and Matroids". Holt, Rinehart and Wihston (1976).

E.L. Lawler. "Fast Approximation Algorithms for Knapsack Problems". Maths of O.R. 4, (1979) vol. 4, pp. 339-356.

M.O. Rabin. "Probabilistic Algorithms". In Algorithms and Complexity Tramb J.F. ed., Ac. Press N.Y. (1976) pp. 21-40.

Numerical Techniques for Stochastic Systems
F. Archetti and M. Cugiani (eds.)
© North-Holland Publishing Company, 1980

THE SIMPLE BAYESIAN ALGORITHM FOR THE MULTIDIMENSIONAL GLOBAL OPTIMIZATION

J. Mockus

Institute of Mathematics and Cybernetics
Academy of Sciences of the Lithuanian SSR
Vilnius

A stochastic model is proposed which allows to simplify calculations in Bayesian methods for global optimization problems. The concept of most "influential" obsservation is introduced and an actual algorithm relying on it is given. The performance of such algorithm is finally shown in comparison with other global optimiza tion algorithms.

Introduction

There exists some split between theory and practice in the field of global optimization. Mathematical methods usually are of the "worst case"type while the more efficient engineering methods [Mockus (1967)] often have no mathematical justification.
In order to make this split as small as possible the so called Bayesian approach was developed [Mockus (1972) and Mockus (1975)].
By the term "Bayesian methods of optimization"we mean the search pro cedures which minimize not the maximal (as in the "worst case" analysis) but the average deviation from the global minimum.
The deviation from the global minimum is defined as the difference:

$$\delta(f) = f(x_N) - f^*$$

where $f^* = \min_{x \in A} f(x)$, A -the search domain- is a compact subset of R^m, $f(.)$ is a continuous function, x_N is the final decision made by the optimization algorithm and N is the number of observations.
By observation we mean the computation of the objective function at some point in A.
The average deviation is defined mathematically as the integral:

$$(2) \qquad \int_C \delta(f)\; P\;(df)$$

where C is the set of all continuos functions.

P is some measure which represents the a priori information about the likelihood and importance of different subsets of C.

The Bayesian methods should minimize the average deviation.

The proper choice of an priori measure P is a difficult problem of the Bayesian approach.

It has been shown, under very general conditions, that there exists a unique stochastic function which represents any a priori information about the likelihood that a function f belongs to some subsets of C.

This means that a function $f \in C$ can be represented as a sample path of some stochastic function.

The conditions have been defined in Mockus (1978) under which the Bayesian methods, which are optimal in that they minimize the average deviation, can be also shown to converge to the global minimum of any continous function.

It has been also proved [Mockus (1977)] that the very reasonable assumptions of homogeneity, independence of the m-th differences and continuity of the sample paths are satisfied when the a priori measure P corresponds to a gaussian stochastic function which can be regarded as some generalization of the Wiener field.

The parameters of this stochastic function (expected value and variance) are usually unknown and should be evaluated using the available observations.

This means that the a priori measure P is updated using the available observations.

This can be regarded as some generalization of the classical Bayesian techniques where the a priori distribution is usually maintaned unchanged.

It follows that in the case of the one-step Bayesian approach [Mockus (1972)] the best point of the next observation $x_n + 1$ is:

$$(3) \quad x_{n+1} \in \arg\min_{x \in A} \frac{1}{\sigma_x^n} \int_{-\infty}^{+\infty} \min\,(y, C_n)\; e^{-\frac{1}{2}\left(\frac{y - \mu_x^n}{\sigma_x^n}\right)^2} dy.$$

Here μ_x^n and σ_x^n are the espected value and standard deviation conditioned by the observations already performed $f(x_1)$, $f(x_2)$, ... $f(x_n)$, $n = 0, 1, \ldots, N$,

$$C_n = \min_{x \varepsilon A} \mu_x^n - \varepsilon_n \quad , \qquad \varepsilon_n > 0$$

$$\text{usually} \quad \min_{x \varepsilon A} \mu_x^n = \min_{1 \leq i \leq n} \{f(x_i)\}$$

Unfortunately the computation of the integral in (3) and consequently of x_{n+1} happens to be rather complicated, especially in the multidimensional case.

No more than 100-200 observations of $f(x)$ can be actually handled when the expressions for μ_x^n and σ_x^n correspond to the usual multidimensional gaussian distribution. The stochastic model given by the Wiener field is a special case of this.

1. The simple Bayesian model.

It was clearly necessary to develope some alternative stochastic models which could be more convenient for computational purposes.

In order to develope those simpler stochastic models we assume that the conditional expectation μ_x^n of $f(x)$ is equal to the observed value y_i of the most influential observation x_i.

The conditional standard deviation is assumed to be an increasing function of the distance from this observation. Then for this simple model the expectation and the standard deviation, conditioned by the observations already performed can be expressed as

$$\mu_x^n = y_i \tag{5}$$

$$\sigma_x^n = \Delta_i \tag{6}$$

where $y_i = f(x_i)$ and $\Delta_i = \Delta(d_i)$ is a strictly increasing nonnegative function of $d_i = \| x-x_i \|$

$$i = \arg \min_{1 \leq j \leq n} \gamma_j \tag{7}$$

and $\gamma_j = \gamma_j(d_j)$ is a strictly increasing nonnegative function of d_j.

Theorem 1.

If μ_x^n and σ_x^n are defined by (5), (6) and (7) than the condition (3) can be reduced to the following condition:

$$(8) \qquad x_{n+1} \varepsilon \text{ arg max } \frac{\Delta_i}{\delta_i}$$

where n=0,1,...,N-1 and $\delta_i = y_i - C_n$, with the index i derived from condition (7).

Proof: The expression in (3) can be written as

$$x_{n+1} \varepsilon \underset{x\varepsilon A}{\text{arg max}} \frac{1}{\sigma_x^n} \int_{-\infty}^{C_n} (C_n - y) e^{-\frac{1}{2}\left(\frac{y-\mu_x^n}{\sigma_x^n}\right)^2} dy.$$

Some simple calculations yield:

$$x_{n+1} \varepsilon \underset{x\varepsilon A}{\text{arg min}} \phi(a_i) \text{ where:}$$

$$\phi(a_i) = \Delta_i \int_{-\infty}^{a_i} (u-a_i) e^{-\frac{1}{2}u^2} du$$

$a_i = -\frac{\delta_i}{\Delta_i}$ and $\delta_i > 0$ by the definition of y_i and C_n.

It will be shown that $\phi(a_i)$ is a decreasing function of a_i.
To do this the derivative of ϕ is calculated:

$$\frac{d\phi}{da_i} = \frac{\partial\phi}{\partial a_i} + \frac{\partial\phi}{\partial\Delta_i}\frac{\partial\Delta_i}{\partial a_i},$$

from the expression for $\phi(a_i)$ we can write:

$$\frac{\partial\phi}{\partial a_i} = -\Delta_i \int_{-\infty}^{a_i} e^{-\frac{1}{2}u^2} du .$$

This is nonpositive because $\Delta_i \geq 0$ and $a_i \leq 0$.

$$\frac{\partial\phi}{\partial\Delta_i} = \int_{-\infty}^{a_i} (u-a_i) e^{-\frac{1}{2}u^2} du .$$

This is nonpositive too because $a_i \leq 0$.

$$\frac{\partial\Delta_i}{\partial a_i} = \frac{\delta i}{a_i^2}.$$

This is nonnegative because $\delta_i \geq 0$.

It follows that $\frac{d\phi}{da_i} \leq 0$.

This means that the minimum of $\phi(a_i)$ and hence the minimum in (3), is reached when a_i is maximal. Hence

$$x_{n+1} \varepsilon \underset{x\varepsilon A}{\text{arg max}}\, a_i = \underset{x\varepsilon A}{\text{arg max}}\left(-\frac{\delta_i}{\Delta_i}\right) = \underset{x\varepsilon A}{\text{arg max}} \frac{\Delta_i}{\delta_i}$$

which completes the proof.

The simplest expression of the algorithm (8) is:

$$(9) \qquad \Delta_i = d_i^2$$

and

$$(10) \qquad \gamma_j = \frac{d_j^2}{\delta_j}$$

It follows from (7),(8),(9), (10) that

$$(11) \qquad x_{n+1} \in \arg\max_{x \in A} \min_{1 \leq j \leq n} \frac{d_j^2}{\delta_j}, \quad n=0,1,\ldots,N.$$

Since the exact maximization in (11) is obviously not necessary it can be performed using simple methods of the type of uniform search. No improvement was observed in the simulation of the two-dimensional case when the number of uniformly distributed evaluations of the expression

$$\min_{1 \leq j \leq n} \frac{d_j^2}{\delta_j}$$

was increased to more than 50.

The algorithm (11) can be generalized to the case of "noisy" observations using some smoothing technique when the mean value of the M most "influential" observations is substituted into the expression (11)

$$(12) \qquad x_{n+1} \in \arg\max_{x \in A} \frac{\bar{d}_i^2}{\bar{\delta}_i^2}$$

where

$$\bar{d}_i^2 = \frac{1}{M} \sum_{k=1}^{M} d_{i_k}^2, \qquad 1<M<N$$

$$\bar{\delta}_i = \bar{y}_i - C_n, \qquad n=M,\ldots,N$$

$$\bar{y}_i = \frac{1}{M} \sum_{k=1}^{M} y_{i_k}$$

$$i_k = \arg\min_{\substack{1 \leq j \leq n \\ j \neq (1,\ldots,k-1)}} \frac{d_j^2}{\delta_j}.$$

The algorithm (12) can be derived in a similar way to the algorithm (11) from the stochastic model in which

(13) $$\mu_x^n = \frac{1}{M} \sum_{k=1}^{M} y_{i_k}$$

(14) $$\sigma_x^n = \frac{1}{M} \sum_{k=1}^{M} d_{i_k}^2$$

It can be easily checked that the convergence conditions stated in Mockus (1978) hold for the stochastic models defined by (5),(6) and (13),(14).

Hence the algorithms (8),(11),(12) should converge to the global minimum of any continous function. In the case of noisy observations the convergence is with probability 1 in the probabilistic space of "noise".

2. The combination with local methods.

The stochastic models (5),(6) and (13), (14) seem fairly suitable when the results of observations y_i can be considered as the results of some local minimizations.

Since the local minimization is rather expensive it is assumed that the local optimization will be performed after N observations starting from the most "promising" points L which can be defined by the expression

(15) $$L = \{x_i : f(x_i) \le f(x_j),\ x_j \in R_i\}$$

where R_i is the set of R observations nearest to the point x_i.

The expression (15) is similar to the usual definition of local minimum and can be regarded as its discrete analogue.

If the integer R = 1 then the local minimization will be performed from any point x_i, i=1,...,N. If R=N than usually only one local optimization will be made from the best observed point.

The setting R=N can be recommended if it can be expected that the region of the global minimum is larger than or at least as large as that of any local minimum. Otherwise it is advisible to set R < N. For example R = m+1 or R = 2m.

The value of y_i can also be taken into account, excluding for example such points x_i where y_i is more than the average observed value.

In the case of noisy observations usually $L = \emptyset$ which means that no local optimization is performed.

3. The results of simulation.

The numerical performance of the algorithm (11) was investigated using the family of 2-dimensional functions

$$(16) \qquad f(x^{(1)},x^{(2)}) =$$

$$= \{[\sum_{i,j} (a_{ij}\sin(\pi i x^{(1)})\sin(\pi j x^{(2)})+b_{ij}\cos(\pi j x^{(1)})\cos(\pi j x^{(2)})]^2+$$

$$+ [\sum_{i,j} (c_{ij}\sin(\pi i x^{(1)})\sin(\pi j x^{(2)})-d_{ij}\cos(\pi j x^{(1)})\cos(\pi j x^{(2)})]^2\}^{1/2}$$

$i,j = 1,\dots,7$

$a_{ij},\ b_{ij},\ c_{ij},\ d_{ij}\ \varepsilon\ (0,1).$

Accordingly to Grishagin (1978) this is some approximation of the stress function in a thin elastic square plate under a cross sectional load.

Fifty test functions of the class (16) were considered, corresponding to a random uniform distribution of the parameters a_{ij}, b_{ij}, c_{ij} and $d_{ij}\ \varepsilon\ (0,1)$.

The value of R was set equal to N so that only one local optimization was made using about 40 evaluations of f by the simplex method of Nelder and Mead.

In the table 1 and figure 1 the relation is given between the percentage of cases in which the global minimum was found and the total number of observations

$$N_t = N + N_l$$

where N_l is the number of observations for the local search.

The index 1 corresponds to the simple Bayesian algorithm (11) when $\varepsilon_n = 10^{-6}$.

The index 2 denotes the usual Bayesian algorithm [Mockus (1975)].

The index 3 is related to Strongin's one-dimensional algorithm together with Peano type transformation [Strongin (1978)].

The index 4 corresponds to the uniform random search.

The index 5 shows the results of the uniform deterministic search.

The index 6 is present only in figure 1 and corresponds to the interactive search performed by an expert in the field of optimization V. Saltenis.

The simulation results using the family of functions (16) shows that the simple Bayesian algorithm 1 is better than the usual Bayesian algorithm 2.

However the preliminary results of simulation using the set of functions considered in [Mockus, Tiesis & Zilinskas (1978)] are rather opposite.

The programming and computation were carried out by J. Valevicene using the package of FORTRAN programs for Bayesian optimization [Tiesis (1979)] .

References

1) J.B. Mockus. Multiextremal problems in design. Moscow, 1967 (Russian).

2) J.B. Mockus. On the Bayesian methods for seeking the extremum. Automation and computers, N. 3, 1972, 53-62 (Russian).

3) J.B. Mockus. On bayesian methods of optimization. Towards Global Optimization. North-Holland, Amsterdam, 1975, 166-181.

4) J.B. Mockus. The sufficient conditions for the convergency of the Bayesian methods to the global minimum of any continuos function. The Optimal Decision Theory. Vol. 4, Vilnius, 1978, 67-80 (Russian).

5) J.B. Mockus. On Bayesian methods for seeking the extremum and their applications. Information Processing 77, North-Holland, 1977, 195-200.

6) V.A. Grishagin. The operative characteristics of some algorithms of global search. Problems of Random Search. Vol. 7, Riga, 1978, 198-208 (russian).

7) R.G. Strongin. Numerical methods in multiextremal problems. Moscow, 1978 (Russian).

8) J. Mockus, V. Tiesis, A. Zilinskas. The application of Bayesian method for seeking the extremum. Towards Global Optimization 2. North-Holland, Amsterdam, 1978, 117-129.

9) V.A. Tiesis. The package for nonlinear programming. Proceedings of the Conference on Computers. Kaunas, 1979 (Russian).

TABLE I

Nr of alg. \ Nt	80	90	100	105	125	140	200	240	340	370	440
I	46	60			80	88		96			
2	46		56	62	72	86					
3				56		68	82			94	100
4	30		38			44	52				78
5	26			44		68		84	92		94

FIGURE I

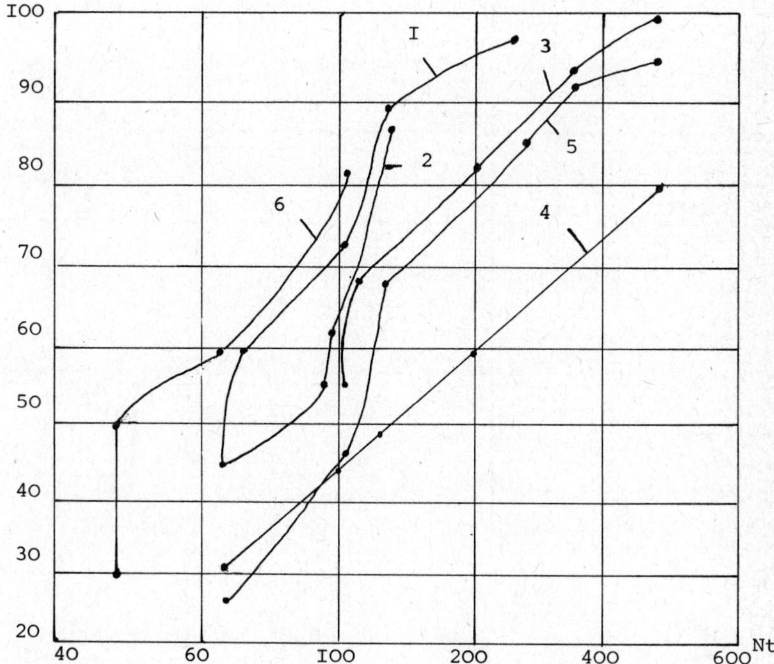

Numerical Techniques for Stochastic Systems
F. Archetti and M. Cugiani (eds.)
© North-Holland Publishing Company, 1980

GLOBAL STOCHASTIC APPROXIMATION: A REVIEW OF RESULTS AND SOME OPEN PROBLEMS

Ryszard Zielinski

INTRODUCTION

Problem: given a set χ and a real-valued function F on it find a point in χ at which the function F achieves its maximum.

We shall consider the problem under some circumstances motivated by practical applications, viz.we shall assume that the function F is not known except that its value at every point $x \in \chi$ can be observed, usually with a random error of the observation. The result of the observation of F at any given point $x \in \chi$ is then formally represented by a random variable Y_x and we shall assume that the expected value of Y_x is equal to F(x), i.e. the mean value of the error is equal to zero. A more precise statement of our original problem is now as follows: given a set χ and a family of random variables $\{Y_x : x \in \chi\}$ find a point in χ at which the regression function $F(x) = EY_x$ achieves its maximum.

Well-known processes of looking for (local) maxima of regression functions are those of Kiefer-Wolfowitz type [e.g. Kiefer and Wolfowitz 1952, Fabian 1971, Schmetterer 1978].

Well-known processes of looking for global maxima of (deterministic) functions are those of random seeking type [e.g. Brooks 1958, Kharlamov 1965, Rastrigin 1974]. A combination of this two ideas gives us a method of global stochastic approximation [Zielinski 1977], shortly g.s.a. A similar approach for a Hilbert space χ and twice differentiable functions F has been developed in [Vaisbord and Yudin, 1968]. In what follows we shall give a short review of results of [Zielinski 1977] and some extensions towards non-homogeneous procedures, e.g. for adaptive ones. We shall also formulate some open problems of a great practical interest.

My opinion is that there is no use in presenting all technical details in a conference lecture so I shall omit almost all of them. During our lecture we can treat χ as a domain in a finite-dimensional space of real numbers and consequently to consider densities of all probability measures on χ as those with respect to the usual Lebesgue measure. For example, Q is the uniform distribution on χ if χ is bounded and Q(A), $A \subset \chi$, is the volume of A divided by the volume of χ . All results are also valid in a more abstract setup presented in the paper [Zielinski 1977].

A review of results. If $F(x)$ can be computed without any (random) error the general structure of well-known random seeking processes is as follows. Choose at random a point $X_1 \in \chi$ and calculate $F(X_1)$. Sample a point $\xi_1 \in \chi$ according to a probability distribution over χ, which may depend on X_1, say Q_{X_1}. If $F(\xi_1) > F(X_1)$ put $X_2 = \xi_1$; put $X_2 = X_1$ otherwise. In general

$$(1) \quad X_{n+1} = \begin{cases} \xi_n, & \text{if } F(\xi_n) > F(X_n) \\ X_n, & \text{otherwise} \end{cases}$$

where ξ_n is a random point in χ distributed according to a probability measure Q_{X_n}.

A natural modification of the above procedure for the case in which $F(x)$ is estimated by a random variable Y_x is as follows

$$(2) \quad X_{n+1} = \begin{cases} \xi_n, & \text{if } Y_{\xi_n} > Y_{X_n} \\ X_n, & \text{otherwise} \end{cases}$$

It is easy to see that due to the random error of observation it could happen that $Y_{\xi_n} > Y_{X_n}$ although $F(\xi_n) < F(X_n)$ so that the consecutive approximation X_{n+1} would be worse than the previous X_n. Being aware of existence of the random error of observation we should hesitate in replacing X_n by ξ_n simply because the inequality $Y_{\xi_n} > Y_{X_n}$ holds, especially when Y_{X_n} has a high value. This can be formalized by introducing an increasing function $w : R^1 \to (0,1)$ and modifying the procedure (2) in the following way

$$X_{n+1} = \begin{cases} \xi_n, & \text{if } U_n > w(Y_{X_n}) \text{ and } Y_{\xi_n} > Y_{X_n} \\ X_n, & \text{otherwise} \end{cases} \quad (3)$$

where U_n, $n = 1,2\ldots$, are independent random variables distributed uniformly in the interval $(0,1)$. The procedure (2) is a special case of (3) for $w \equiv 0$. Some results concerning procedures like (2) and (3) are as follows.

1. Under appropriate distributions Q_x, for every (measurable) subset $S \subset \chi$

$$P\ \{X_n \in S \mid X_1 = x\} \to \bar{P}(S) \quad \text{as} \quad n \to \infty$$

and the limit distribution $\bar{P}$ does not depend on the initial point X_1.

2. If $Q_x = Q$ are uniform distributions then the limit distribution $\bar{P}$ has a density $\bar{p}(x)$ which is monotonic in $F(x)$. This means that in the limit the points with higher values of F are more probable than those with smaller values of the regression function :if $F(x_1) > F(x_2)$, $x_1, x_2 \in \chi$, then $\bar{p}(x_1) \geqslant \bar{p}(x_2)$ and the picture is like that in Fig.1.

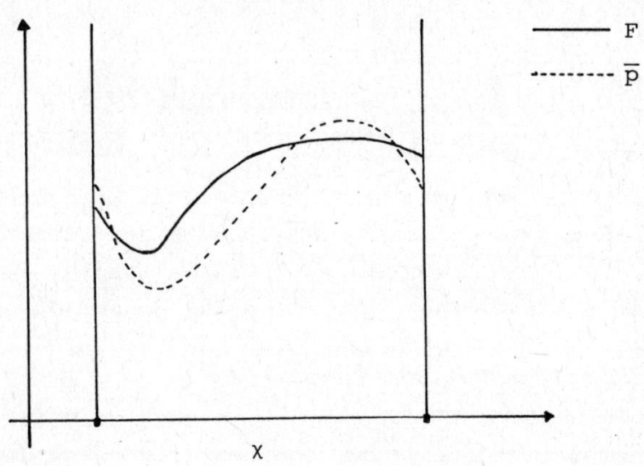

Fig. 1

3. The limit distribution of (3) with a strictly increasing function w is more concentrated at the points x with high values of F than that of (2). More precisely : let $\bar{p}$ and $\bar{p}_w$ be densities of the limit distributions in (2) and (3), respectively. There exists a number f such that $F(x) > f$ implies $\bar{p}_w(x) > \bar{p}(x)$. The picture is like that in Fig. 2.

4. If the error is bounded then X_n given by (3) [or (2)] converges with probability one to the point of global maximum of the function F. More precisely. Let F_0 = ess sup F and let ${}_\varepsilon\chi_0 = \{x: F(x) > F_0 - \varepsilon\}$, $\varepsilon > 0$. With probability one there exists an integer N such that $X_n \in {}_\varepsilon\chi_0$ for all $n \geqslant N$.

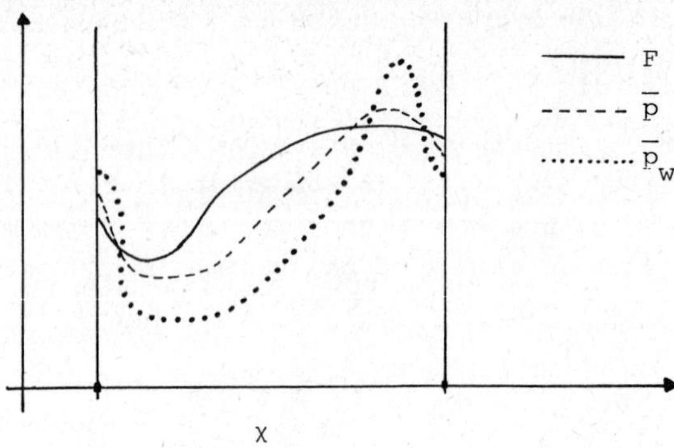

Fig. 2

All the above results deal with the case of homogeneous (in time) procedures, viz. the probability measures Q_x according to which the consecutive points are chosen as well as the function w in the randomized procedure (3) do not depend on n and do not change in the course of computation. The procedures Q_x and w changing with n will be considered later on.

A remarkable feature of the results is that they are quite different in two situations, namely when the error is unbounded we can only prove the convergence in distribution limit density being monotonic in F(x), and when the error is bounded we are able to prove the convergence with probability one. My opinion is that it is impossible to get a stronger result in the former case and that the above difference in results is due to the very nature of the two situations. To see why it is so consider a sequence $(X_n, n = 1,2,\dots)$ generated by (3). We obviously have $Y_{X_{n+1}} > Y_{X_n}$, $n = 1,2\dots$, and $Y_{X_n} \to \infty$ as $n \to \infty$. In the former case with positive probability we can get $F(X_{n+1}) < F(X_n)$ or even $F(X_{n+1}) < \min\{F(X_1),\dots,F(X_n)\}$ whereas in the latter case we have $1 = P\{Y_{X_{n+1}} > Y_{X_n}\} = P\{F(X_{n+1}) + \text{error} \geqslant Y_{X_{n+1}} > Y_{X_n}\} \leqslant P\{F(X_{n+1}) + e > Y_{X_n}\}$ so that with probability one $F(X_{n+1}) > Y_{X_n} - e$, where $e > 0$ is the maximal value of the error of observation.

<u>Non-homogeneous procedures</u>. We shall confine ourselves to the case of unbounded error and consider procedures in which both Q_x and w may change with n . The distribution Q_x at the n-th step will be de-

noted by $Q_x^{(n)}$ and the function w at that step by w_n.

Theorem 1. Assume that (i) P_1 and $Q_x = Q$ are uniform distributions over χ; (ii) $\inf_{x \in \chi} F(x) > -\infty$; (iii) $(w_n, n = 1,2,\ldots)$ is a sequence of non-decreasing functions from R^1 into $[0,1-\alpha]$, $\alpha \in (0,1)$ may depend on w_n.
Let $P_m\{X_n \in S \mid X_1 = x\}$ be the probability of the event $\{X_n \in S\}$ in the procedure (3) with w_m instead of w, when $X_1 = x$.
Then the limit distribution $\bar{P}_m(S) = \lim_{n\to\infty} P_m\{X_n \in S|X_1 = x\}$ exists and does not depend on x. If the limit $\bar{P}(S) = \lim_{m\to\infty} \bar{P}_m(S)$ exists then for the non-homogeneous procedure (3) with the function w_n at the n-th step we have.

$$\lim_{n\to\infty} P_n\{X_n \in S|X_1 = x\} = \bar{P}(S) \tag{4}$$

Proof. The first assertion is exactly the same as the first assertion of Theorem 3 in [Zielinski 1977].
To prove the second assertion consider the proof of Theorem 1 in that paper. For each procedure (3) with a fixed w_m we have

$$|P_m\{X_n \in S|X_1 = x\} - \bar{P}_m(S)| \leqslant (1-\varepsilon_m)^{n-1}$$

where due to the assumption (i), ε_m does not depend on m .
It follows that $P_m\{X_n \in S|X_1 = x\}$ tends to $\bar{P}_m(S)$ as $n \to \infty$, uniformly in m, and the Theorem follows.

Theorem 2. Suppose $w_r \equiv w$ is a given function. Let $q_x^{(m)}(.)$ be a density of the probability measure $Q_x^{(m)}$ and assume that $q^{(m)} = \inf_{x\in\chi} \inf_{y\in\chi} q_x^{(m)}(y)$ is positive.
Let $P_m\{X_n \in S|X_1 = x\}$ be the probability of the event $\{X_n \in S\}$ in the procedure (3) with $Q_x^{(n)}$ instead of Q_x.
Then the limit $\bar{P}_m(S) = \lim_{n\to\infty} P_m\{X_n \in S|X_1 = x\}$ exists and does not depend on x. If $q^{(m)} \geqslant c_m$ where (c_m) is a sequence of positive numbers such that $c_m = O(m^{-\gamma})$ with a positive $\gamma < 1$ and the limit $\bar{P}(S) = \lim_{m\to\infty} \bar{P}_m(S)$ exists, then (4) holds.

Proof. The first assertion is similar to that of Theorem 1. To prove the second one we have

$$|P_n\{X_n \in S|X_1 = x\} - \bar{P}(S)| \leqslant$$

$$\leq |P_n\{X_n \in S | X_1 = x\} - \overline{P}_n(S)| + |\overline{P}_n(S) - \overline{P}(S)| \leq$$

$$\leq (1-\varepsilon_n)^{n-1} + |\overline{P}_n(S) - \overline{P}(S)|$$

By the assumption concerning $q^{(m)}$ we have $(1-\varepsilon_n)^{n-1} \to 0$ as $n \to \infty$ and the theorem follows.

Remark. Both proofs presented here may be difficult to follow without knowing details of the proof of Theorem 1 in [Zielinski 1977] .

Theorems 1 and 2 enable us to change the function w and/or the probability measures Q_x in the course of computations. One of the methods of changing w is suggested in the original paper [Zielinski 1977, sec. 4.5] : it has been showed there that if $Q_x=Q$ is the uniform distribution and w is a non-decreasing function from R^1 into $[0, 1-\alpha]$, $\alpha \in (0, 1)$, then the procedure with $v=(b/a)w$, $b \in (a, 1)$, $a=\sup_{x \in R^1} w(x)$, is better than that with w in the sense that

$$\int F(x)\overline{P}_v(dx) > \int F(x)\overline{P}_w(x) \tag{5}$$

where $\overline{P}_v$ and $\overline{P}_w$ are the limit distributions in (3) with v and w, respectively.

Another method which leads to a non-homogeneous procedure consists in changing of probability measures Q_x. For example let χ be a bounded domain in R^k, $k \geq 1$, and let $Q_x^{(n)}$ be the normal distribution $N(x, \sigma_n^2 \Sigma)$, Σ being a fixed matrix. It is easy to see that the hypothesis of Theorem 2 is satisfied if σ_n^2 tends to zero not too rapidly.

Some open problems. It is obvious that looking for global maximum we should maintain the process (3) in such a way that, roughly speaking, each point $x \in \chi$ could be sampled with positive probability. If the error of observation is bounded this gives us the convergence to global maximum with probability one.

The first open problem is

1. How to choose Q_x and/or w to make the convergence as speedy as possible?

   and

2. How to modify $Q_x^{(n)}$ and/or w_n in non-homogeneous procedure to get a high speed of convergence? Is the non-homogeneous procedure really better than homogeneous one?

Consider the case where the error is unbounded. It has been proved that under some mild conditions the limit distribution exists and has a positive density. If $Q_x = Q$ is the uniform distribution then the limit density $\overline{p}(x)$ is monotonically increasing in $F(x)$. If $\overline{P}_1$ and $\overline{P}_2$ are two limit distributions we shall say that $\overline{P}_1$ is better than $\overline{P}_2$ if $\int F(x)\overline{P}_1(dx) > \int F(x)\overline{P}_2(dx)$ [see (5)]. Let $\mathcal{P}$ be the set of all (limit) distributions having positive density. It is easy to see that the best distribution in $\mathcal{P}$ does not exist. To prove this take a distribution $P \in \mathcal{P}$ with the density p and a number f such that $\pi = P\{x: F(x) > f\} > 0$. Let λ be a number from the interval $(1, \pi^{-1})$. If P' is the probability distribution with the density

$$p'(x) = \begin{cases} \lambda p(x) & , \text{ if } F(x) > f \\ \dfrac{1-\lambda\pi}{1-\lambda\pi}\, p(x) & , \text{ if } F(x) \leq f \end{cases}$$

then $\int F dP' > \int F dP$. It follows that for every distribution $P \in \mathcal{P}$ we can find a better distribution P' so that the best distribution does not exist. On the other hand we have

$$\sup_{P \in \mathcal{P}} \int F dP = F_0 \tag{6}$$

where F_0 is the global maximum of the function F. It means that for every $\varepsilon > 0$ there exists an ε-optimal distribution $\overline{P}_\varepsilon$ in the sense that

$$\int F\, d\overline{P}_\varepsilon > F_0 - \varepsilon \tag{7}$$

Now let $\tilde{\mathcal{P}}$ be a class of all limit distributions in the procedure (3) when $Q_x = Q$ is the uniform distribution and w is a non-decreasing function from R^1 into $[0, 1-\alpha]$ with a positive α (which may depend on w). An open problem is

3. Given $\varepsilon > 0$, does there exist an ε-optimal distribution in $\tilde{\mathcal{P}}$? or more generally

4. Given $\varepsilon > 0$, does there exist a sequence (w_n) such that non-homogeneous procedure (3) with $Q_x = Q$ (uniform distribution) gives us in the limit an ε-optimal distribution in the sense of (7)?

The above problems are in fact those of optimal construction of (randomized) processes of global stochastic approximation.

REFERENCES

1. Brooks, S.H. (1958): A discussion of random methods for seeking maxima. Opns. 6, 244-251.
2. Fabian, V. (1971): Stochastic approximation. In: Optimizing methods in statistics, Ed. J.S. Rustagi, Academic Press.
3. Kharlamov, B.P. (1965): On an algorithm of stochastic search of maximum in a deterministic field. Works of the Math. Institute V.A. Steklov LXXIX, 71-75 (In Russian).
4. Kiefer, J. and Wolfowitz, J. (1952): Stochastic estimation of the maximum of a regression function. Ann. Math. Statist. 23, 462-466.
5. Rastrigin, L.A. (1974): Systems of extremal control. "Nauka", Moscow (In Russian).
6. Schmetterer , L. (1978): From stochastic approximation to the stochastic theory of optimization. Forum Lecture at the EMS in Oslo.
7. Vaisbord, E.M. and Yudin, D.B. (1968): Stochastic approximation for extremal problems in Hilbert space. DAN, SSSR, Vol. 181, n.5 1034-1037 (In Russian).
8. Zielinski, R. (1977): Global stochastic approximation. Dissertationes MAthematicae CXLII.

Numerical Techniques for Stochastic Systems
F. Archetti and M. Cugiani (eds.)
© North-Holland Publishing Company, 1980

GLOBAL OPTIMIZATION: A STOCHASTIC APPROACH

C.G.E. Boender - A.H.G. Rinnooy Kan - L. Stougie - G.T. Timmer

Econometric Institute, Erasmus University Rotterdam

ABSTRACT

A stochastic method to solve the global optimization problem is briefly described. The results of some computational experiments are presented as well.

1. INTRODUCTION

In this contribution, we provide a brief description of a stochastic method for global optimization. Full details can be found in [1]. We also describe some of the computational results that were obtained with the latest version of the algorithm on a series of standard test problems.

The description of the algorithm is contained in Section 2. The computational experiments are the subject of Section 3 and some concluding remarks are contained in Section 4.

2. A DESCRIPTION OF THE ALGORITHM

Let f be a real valued function defined on $\mathbb{R}^n$. The global optimization problem is to find

$$y_* = \min_{x \in S} \{f(x)\} \tag{1}$$

where S is assumed chosen in such a way that the global minimum occurs in its interior.

As is the case with all stochastic methods to solve the global optimization problem (cf.[2]), our method is a mixture of global and local search. As part of the global search samples are drawn from a uniform distribution over S; during the local search, these samples are manipulated to yield ultimately a candidate solution y* to (1).

Under very mild conditions on f [2], the probability that the true global minimum will be found by such a method approaches 1 as the sample size increases. The goal of our algorithm is to find all the relevant local minima of f; y* is then defined to be the best of these minima. The minima themselves will be found with the aid of a *local search procedure*, based on a standard nonlinear programming routine.

For reasons of computational efficiency however, we would like to use this local search procedure no more than once for each local minimum. Put differently: if we define the *region of attraction* of a local minimum x* to be the set of all points in S starting from which the local search procedure will arrive in x*, we would like to initiate the local search procedure exactly once in each relevant region of attraction.

In order to find these regions of attraction, the algorithm invokes a *clustering procedure* (cf.[6]). Before every application of this procedure, the current sample is *transformed* by temporarily eliminating a fixed percentage of points with relatively high function values and by performing a single steepest descent step from the remaining ones. These steps cause the formation of groups of relatively close points, each of which corresponds to a promising local minimum. In a way yet to be described, the clustering procedure identifies these groups and the corresponding local minima by growing *clusters* around appropriately chosen *seed points*, until all points of the transformed current sample have been allocated to such a cluster. If in the course of doing so, one or more local minima are found that had not been discovered before, the sample is increased by drawing an additional fixed number of points. The new, extended sample is transformed and the clustering procedure is applied anew. Otherwise, the algorithm terminates by declaring the best local minimum y* that has been found, to be the candidate solution to (1). Finally, the user is presented with a range of *confidence intervals*, expressing probabilistic information on the likelihood of further improvement on this value. As demonstrated in [1] (cf.[5]), such a range of *level p asymptotic confidence intervals* is given, for every p larger than a certain threshold value p_o, by

$$[\underline{y}^{(1)} - \frac{\underline{y}^{(2)} - \underline{y}^{(1)}}{p^{-2/n} - 1}, \underline{y}^{*}] \qquad (2)$$

Here, $\underline{y}^{(1)}$ and $\underline{y}^{(2)}$ are the two smallest function values found in the entire sample.

It only remains to describe the clustering procedure in more detail. Prior to every application of this procedure to the transformed sample, we have available a set X^* of local minima found so far, and a set $X^{(1)}$ of sample points to which the local search procedure has been applied unsuccessfully, in the sense that it produced a local minimum that was known already. Initially both sets are empty. As far as the choice of seed points for the current clustering phase is concerned, we start by growing clusters from the local minima in X^*; if any points in the transformed sample remain unclustered, we start to use the points in $X^{(1)}$ as seed points; and if any points still remain unclustered, we choose the point $x^{(1)}$ with lowest function value among those, apply the local search procedure to $x^{(1)}$ to find a local minimum x^* and grow a cluster with either $x^{(1)}$ or x^* as a seed point, depending on whether or not x^* was already known to be a member of X^*. In the former case, $x^{(1)}$ is of course added to $X^{(1)}$.

The growing of a cluster around a given seed point itself is achieved by a modification of the well-known *single linkage clustering* technique [4]. Given a current cluster C, we calculate the distance d(x,C) of an unclustered point x to C as

$$d(x,C) = \min_{x' \in C}\{||x-x'||\} \qquad (3)$$

The unclustered point x for which d(x,C) is minimal is then added to C, unless its distance to C exceeds a certain *threshold value*. In the latter case the growth of the current cluster is terminated and a new one is started around another seed point. We should add that before a point is added to a cluster, we verify as an extra criterion of admittance if the gradient in that point roughly points in the direction of the local minimum corresponding to the cluster.

A proper threshold value is obtained by ignoring the effect of the steepest descent step and assuming that the points within each cluster still satisfy the original uniform distribution. If the current sample size (before transformation) is N, the probability that a point is at least at distance d from all the others can then be approximated for small d by

$$\left(1 - \frac{\pi^{n/2} d^n}{\Gamma(1+n/2)\, m(S)}\right)^{N-1} \tag{4}$$

where m(S) is the Lebesgue measure of S. We choose our threshold value to be equal to a distance d that is sufficiently large to make the probability (4) that the point is inaccurately not added to the cluster, equal to a prespecified constant α. The threshold value is therefore equal to

$$\left[\frac{\Gamma(1+n/2).m(S)}{\pi^{n/2}} . \left(1-\alpha^{1/(N-1)}\right)\right]^{1/n} \tag{5}$$

A related clustering technique, based on a generalization of *density clustering*, has been investigated as well, but was rejected as being marginally less accurate (see [1]).

3. COMPUTATIONAL RESULTS

The algorithm described in section 2 has been coded in FORTRAN and run on the DEC-20 computer of the Erasmus University Rotterdam. The test functions used are the standard ones available for global optimization,as presented in [3]; they are listed in Table 1.

The performance of the algorithm on each test function has been measured in two ways: in terms of the number of function evaluations required and in terms of the number of *standard time units* of computation, where this latter unit is defined [3] as the average time required for the evaluation of a certain test function. Both these measurements depend to a certain extent on the particularities of the sample at hand, and therefore the numbers given in Tables 2 and 3 below all represent the average outcome of four independent runs.

In the specific implementation of the algorithm tested, the parameter α corresponding to the termination criterion (5) of the single linkage clustering method was put equal to 0.002. Points were sampled in groups of 50, and the best ten percent was retained for clustering and local search. The optimum was found in all cases. An increase of the group size to 100 led to an even more reliable algorithm, in the sense that all local minima were found at the expense of roughly a fifty percent increase in both performance measurements.

In Tables 2 and 3 the results for our algorithm are compared to those of a few leading contenders, as reported in [3]. The outcome is quite satisfying. Yet it remains difficult to obtain a fair comparison between stochastic optimization methods, inasmuch as these methods all have the property that the user's confidence in their outcome can always be increased by enlarging the size of the sample. A fair comparison between the various methods would involve a comparison between the costs involved in achieving a certain level of confidence. Most algorithms, however, do not provide any such confidence information, and even our own confidence intervals hardly capture the impact of the extensive local search that is carried out to find the candidate solution. Moreover, the interpretation of the confidence interval is problematic until units of measurement for the objective function are specified. None the less, to give an idea of the size of these intervals, we present some typical examples in Table 4. The confidence intervals at least do not seem so large as to preclude their practical applicability.

Table 1 Test Functions (cf. [3])

GP	Goldstein & Price
BR	Branin (RCOS)
H3	Hartman 3
H6	Hartman 6
S5	Shekel 5
S7	Shekel 7
S10	Shekel 10

Table 2 Number of function evaluations

Method \ Function	GP	BR	H3	H6	S5	S7	S10
Törn	2499	1558	2584	3447	3679	3606	3874
De Biase/Frontini	378	597	732	806	620	788	1160
Price	2500	1800	2400	7600	3800	4900	4400
Branin	-*	-*	-*	-*	5500	5020	4860
New algorithm	362	242	346	555	554	661	768

*No results available

Table 3 Number of standard time units

Method \ Function	GP	BR	H3	H6	S5	S7	S10
Törn	4	4	8	16	10	13	15
De Biase/Frontini	15	14	16	21	23	20	30
Price	3	4	8	46	14	20	20
Branin	-*	-*	-*	-*	9	8.5	9.5
New algorithm	1.3	1	2.5	5	3	5	8

*No results available

Table 4 Confidence intervals

	Minimum value	Pr_1	L_1	Pr_2	L_2	Pr_3	L_3	Pr_4	L_4	Pr_5	L_5
GP	3.000	0.79	33	0.83	50	0.88	75	0.92	123	0.96	273
BR	0.3978	0.79	0.75	0.83	1	0.88	1.5	0.92	2.5	0.96	5.5
H3	-3.862	0.79	2	0.83	3	0.88	4	0.92	6.5	0.96	13
H6	-3.322	0.79	2	0.83	3	0.88	5	0.92	9	0.96	20
S5	-10.15	0.975	2	0.98	5	0.985	10	0.99	20	0.995	50
S7	-10.40	0.975	2	0.98	5	0.985	10	0.99	20	0.995	50
S10	-10.53	0.94	2	0.95	5	0.96	10	0.98	20	0.99	50

Pr_j: the probability corresponding to the j - th confidence interval (j=1,...,5)

L_j : the length of the j - th confidence interval (j=1,...,5)

4. CONCLUDING REMARKS

In the preceding sections we have provided a brief description of a stochastic method for global optimization and a summary of our computational results.

We suspect that the computational performane of the algorithm can be further improved by a more careful implementation of the local search.
We are currently using the VA10AD variable metric routine made available by the Numerical Optimization Centre in Hatfield, but we intend to undertake some further testing in this area.

A proper continuation of the method if, for instance, the final confidence interval is deemed unsatisfactory is very much a matter for future research.
The same is true for possible extensions to constrained global optimization and to combinatorial optimization.

ACKNOWLEDGEMENTS

We gratefully acknowledge the fruitful discussions with G. van der Hoek, L. de Biase and L. de Haan. This research was partially supported by NATO Special Research Grant 9.2.02 (SRG.7) and by the A.A. van Beek Fund in Rotterdam.

REFERENCES

[1] Boender, C.G.E., Rinnoy Kan, A.H.G., Stougie, L., Timmer, G.T., 'A stochastic method for global optimization', to appear.

[2] Dixon, L.C.W., Szegö, G.P. (1975), *Towards global optimization*, North-Holland, Amsterdam.

[3] Dixon, L.C.W., Szegö G.P. (1978), *Towards global optimization 2*, North-Holland, Amsterdam.

[4] Everitt, B. (1974), *Cluster analysis*, Heinemann, London.

[5] de Haan, L. (1979), 'Estimation of the minimum of a function using order statistics', Report 7902/S, Econometric Institute, Erasmus University, Rotterdam.

[6] Törn, A.A. (1978), 'A search-clustering approach to global optimization', in: Dixon, L.C.W., Szegö, G.P. (eds.), *Towards global optimization 2*, North-Holland, Amsterdam.

Numerical Techniques for Stochastic Systems
F. Archetti and M. Cugiani (eds.)
© North-Holland Publishing Company, 1980

SOME REMARKS ON STOCHASTIC APPROXIMATION METHODS

Jacek Koronacki

Institute of Mathematics, Polish Academy of Sciences
Warsaw, Poland

1. Introduction.

In these remarks we shall be concerned with the problem of finding a point Θ of minimum of a (regression) function f defined on the k-dimensional Euclidean space R^k when the only information available is that we can observe unbiased estimates of function values of f. In other words, for each $x \in R^k$, we can only observe noise-corrupted estimates, say V(x), of f(x), where $V(x)=f(x)+\varepsilon(x)$ and the expectation of the random variable $\varepsilon(x)$ equals zero.

Recursive procedures for solving the problem just described are well-known as those of stochastic approximation (SA). Extensive reviews on the subject are given in Fabian (1971), Schmetterer (1969), (1978); see also Dupač (1977), Fabian (1977), Khas'minskij (1977) (in what follows we shall often refer to Fabian (1971) without explicit mention). Let us note here that all convergence arguments given in these reviews (as well as our remarks below) rest heavily on the so-called martingale difference assumption and hence, at least as far as the convergence with probability one (w.p.1) is considered, on the martingale convergence theorem. This is not the case in Kushner and Clark (1978), where the martingale difference assumption is not required and conditions on the noise are considerably weakened (see e.g. examples in Sec.2.2 there; cf. also Ljung (1974) and (1978) where some similar results are obtained in a different way).

Let a sequence $(X_n)_1^\infty$ approximating Θ be given recursively,

$$X_{n+1} = X_n - a_n Y_n, \tag{1}$$

where a_n are positive numbers, X_n, Y_n are random k-dimensional vectors and, at each step, Y_n is an estimator of the gradient of f at X_n. In particular, one might take for Y_n the classical Kiefer-Wolfovitz (KW) estimator

$$Y_n = \frac{1}{2c_n} \begin{bmatrix} V(X_n+c_n e_1)-V(X_n-c_n e_1) \\ \vdots \\ V(X_n+c_n e_k)-V(X_n-c_n e_k) \end{bmatrix} = \frac{1}{2c_n} \begin{bmatrix} f(X_n+c_n e_1)-f(X_n-c_n e_1)+\varepsilon_n^{(1)} \\ \vdots \\ f(X_n+c_n e_k)-f(X_n-c_n e_k)+\varepsilon_n^{(k)} \end{bmatrix}, \quad (2)$$

where c_n are posivite numbers, e_i, $i=1,\ldots,k$, denotes the unit vector in the i-th coordinate direction and $\varepsilon_n^{(1)},\ldots,\varepsilon_n^{(k)}$ denote random variables due to noisy observations of the values of f at $X_n \pm c_n e_i$, $i=1,\ldots,k$, respectively (all random variables are assumed to be defined on a probability space $(\Omega,\mathcal{F},P)$ and relations between random variables are meant w.p. 1).

It is well-known that under the conditions similar to those given in Sec. 2 the sequence (X_n) defined by (1) and (2) converges to Θ w.p. 1 and, moreover, asymptotic distribution of $n^{1/3}(X_n-\Theta)$ is normal (procedures with the asymptotic normal distribution of the vectors $n^{\beta}(X_n-\Theta)$ will be referred to as the procedures with the speed of convergence of order $0(n^{-\beta})$).

In the sequel we shall comment on the asymptotic speed of randomized SA procedures, on steepest descent-like procedures and on stopping rules ıor recursive SA-type estimators. These comments will be apparently minor from the theoretical point of view but, perhaps, will prove to be of some value for the reader interested in practical applications of the SA methods.

2. The randomized SA procedures.

In complex problems of local minimization, notably when dealing with nonconvex regression functions of many variables, it is often recommended to apply a randomized procedure (in general, such procedures are known as random search procedures).

Let us begin with the following assumptions (with a few exceptions notation is borrowed from Fabian (1971)).

(A1) $f: R^k \to R$ attains its absolute minimum at a point $\Theta \in R^k$. The Hessian f_{xx} of f is continuous and bounded on R^k.
$\inf\{f(x)-f(\Theta); x \notin S(\varepsilon)\} > 0$, $\inf\{\|f_x(x)\| ; x \notin S(\varepsilon)\} > 0$,

for every $\varepsilon > 0$, where $S(\varepsilon)$ is the open ε-sphere around Θ, $\|.\|$ denotes the Euclidean norm and $f_x(x)$ denotes the gradient of f at x. ∎

(A2) $\{\xi_{p,n}, p = 1,\dots,m, m \geq 1, n = 1,2,\dots\}$ is a family of independent random vectors of dimension k; each vector is distributed uniformly over the surface of the unit sphere in R^k (around 0). ∎

(A3) $\mathcal{F}_n$ and $\mathcal{B}_n$ are the smallest σ-fields that measure, respectively, $(X_1,\dots,X_n)$ and $(X_1,\dots,X_n,\xi_{1,n},\dots,\xi_{m,n})$, σ^2 and c_n are positive numbers. With $E_{\mathcal{C}}$ denoting the conditional expectation given $\mathcal{C}$ and with

$$Z_n = Y_n - E_{\mathcal{F}_n} Y_n \tag{3}$$

we have

$$E_{\mathcal{F}_n} \|Z_n\|^2 < c_n^{-2} \sigma^2 \tag{4}$$

and

$$E_{\mathcal{B}_n} Y_n = c_n^{-1} \sum_{p=1}^{m} \xi_{p,n} [f(X_n + c_n \xi_{p,n}) - f(X_n)]. \; ∎ \tag{5}$$

(A4) f is thrice continuounsly differentiable in a neighbourhood of Θ. $f_{xx}(\Theta)$ is positive definite with the smallest eigenvalue λ.

There is a matrix Σ such that, for $W_n = c_n Z_n$,

$$\lim E_{\mathcal{F}_n} W_n W_n^T \to \Sigma \tag{6}$$

and, with χ the indicator function,

$$\lim n^{-1} \sum_{j=1}^{n} E(\chi\{\|W_j\|^2 > rj\}\|W_j\|^2) \to 0 \text{ for every } r>0. \blacksquare \quad (7)$$

Note that the procedure given by (1) and assumptions (A3) and (A4) is indeed a random version of the original KW procedure with one-sided difference estimator of the gradient (instead of the central difference estimator as in (2)). In order to see this it suffices to replace in (5) B_n by F_n, $\xi_{p,n}$ by e_p and to assume m=k and $E_{F_n}\varepsilon_n = 0$ where $\varepsilon_n = [\varepsilon_n^{(1)} \ldots \varepsilon_n^{(k)}]^T$.

Once the asymptotic properties of the original SA procedures are known the analogous results for their random versions are easy to obtain. The w.p. 1 convergence of the randomized KW procedure as well as of other procedures considered in the sequel was investigated in Koronacki (1975), under weaker assumptions on f and $(\xi_{n,p})$.

Now, conditions (4), (6) and (7) can be readily verified and using Fabian's Asymptotic Normality Theorem (see Fabian (1968) or Fabian (1971) - Th. 3.6) one can get almost immediately that, under (1) and (A1) to (A4) and with

$$a_n = an^{-1}, \quad c_n = cn^{-1/6}, \quad a\lambda\frac{m}{k} > 1/3,$$

the asymptotic distribution of $n^{1/3}(X_n - \Theta)$ is normal with mean

$$- ac^2 \left(\frac{m}{k} a\, f_{xx}(\Theta) - \frac{1}{3} I\right)^{-1} N_o$$

and covariance matrix $P\,M\,P^T$, where I denotes the identity matrix,

$$N_o^{(v)} = \frac{m}{2\,k\,(k+2)} \sum_{i=1}^{k} f_{x^{(i)}x^{(i)}x^{(v)}}(\Theta), \quad v=1,\ldots,k,$$

$f_{x^{(i)}x^{(i)}x^{(v)}}(\Theta)$ denotes the respective third order partial derivative of f at Θ, P is an orthogonal matrix such that $\Lambda = P^T f_{xx}(\Theta)\, P$ is diagonal and

$$M^{(ij)} = a^2 c^{-2} \frac{(P^T \Sigma P)^{(ij)}}{a\frac{m}{k}\Delta^{(ii)} + a\frac{m}{k}\Delta^{(jj)} - 2/3}, \quad i,j=1,\ldots,k.$$

Moreover, if

$$E_{B_n} \tilde{\mathcal{E}}_n \tilde{\mathcal{E}}_n^T \to \tilde{\Sigma},$$

where $\tilde{\Sigma}$ is a matrix and $\tilde{\mathcal{E}}_n = c_n (Y_n - E_{B_n} Y_n)$, then $\Sigma = \frac{m}{k}(\mathrm{tr}\,\tilde{\Sigma})\,I$ ($\tilde{\mathcal{E}}_n$ here plays the some role as $\mathcal{E}_n$ in (2)). Exact formulae for the moments of asymptotic distribution have been given above for illustrative purposes. In what follows we shall focus our attention on the order of the speed only.

Let us note that if f is quadratic, $c_n \equiv c$ and the remaining assumptions still hold, then the asymptotic distribution of $n^{1/2}\cdot(X_n - \Theta)$ is normal (in this case $\Sigma = \frac{m}{k}\cdot(\mathrm{tr}\,\tilde{\Sigma})\cdot I$ only if one-sided differences in (5) are replaced by the central ones, requiring thus 2m observations of function values per step instead of m+1; otherwise Σ is "increased" by $E\sum_{p=1}^{m}(\xi_{p,1}^T f_{xx}(\Theta)\xi_{p,1})^2 \xi_{p,1}\xi_{p,1}^T$).

As it is already seen, suitable randomization of directions along which the function values are observed should in general preserve the order of the speed and, at the same time, should lead to a reduction in number of observations required at each step (let us recall that in (5) m is arbitrary, $m \geqslant 1$, and only m+1 observations are required per step). In fact this is exactly the case with the randomized versions of the estimators of Fabian (1971) and Zielinski (1979). Furthermore, the same holds true for another Fabian's estimator, the random version of which is the following (see Koronacki (1975))*):

$$Y_n = c_n^{-1} \sum_{p=1}^{m} \xi_{p,n} [\mathrm{sgn}\,(V(X_n + c_n \xi_{p,n}) - V(X_n))]. \qquad (8)$$

---

*) Convergence w.p. 1 of the randomized KW procedure and of the procedure with Y_n given by (8) as well as the asymptotic speed of the former have been studied under different assumptions (and for m=1) in Kushner and Clark (1978).

In this last case (5) in (A3) must be replaced by Assumption 4 of Koronacki (1975). It should be noted however that the speed obtained for the sequence (X_n) defined by (1) and (8) is of order $0(n^{-1/4})$ only, unless a distribution function G in Koronacki (1975) is assumed to have bounded and continuous second order derivative (then the speed is already $0(n^{-1/3})$).

All in all the asymptotic properties of the randomized SA procedures seem appealing. One must not however forget that they hold asymptotically, when n tends to infinity. Only the limit value of the variance $E_{F_n} W_n W_n^T$ does affect the results considered. Beside that, it is only the bias of an estimator Y_n what counts in calculating the order of the speed. Due to their simplicity and nice asymptotic properties the randomized procedures should indeed be recommended when standard SA methods are hardly applicable, but by no means they can be claimed to be superior to the latter ones. Let us conclude this remark with the following curious example.

Suppose that f is strictly convex and the point of minimum θ is known to lie in a cube,

$|\theta^{(i)}| < A$, $i = 1, \ldots, k$, for certain $A>0$. Let

$$E_{B_n} Y_n = c_n^{-1} \xi_{1,n} f(X_n + c_n \xi_{1,n}) \cdot k \tag{9}$$

and

$$X_{n+1} = [X_n - a_n Y_n]_A , \tag{10}$$

where $X \equiv [\tilde{X}]_A$ is the projection of $\tilde{X}$ on the cube $[-A,A] \times \ldots \times [-A,A]$: $X^{(i)} = A$, if $\tilde{X}^{(i)} > A$, $X^{(i)} = -A$, if $\tilde{X}^{(i)} < -A$ and $X^{(i)} = \tilde{X}^{(i)}$ otherwise.

Under the usual assumptions (X_n) converges then to θ w.p. 1 and, in particular, if f is quadratic and $c_n \equiv c$ we get by the Asymptotic Normality Theorem that the speed of the procedure is $0(n^{-1/2})$! Note that for f quadratic the "strange" estimator (9), requiring only 1 observation per step, is unbiased! This example serves as a good illustration for the inherent limitations of possible inference from the asymptotic results (cf. also paragraph (1.12) in Fabian (1971)).

3. A steepest descent-like SA procedure.

Let in this Section sequence (X_n) be defined as follows.

$$X_{n+1} = X_n - \alpha_n Y_n, \tag{11}$$

where α_n are random variables and Y_n is any of the estimators (randomized or not) mentioned in Sec. 2.

Conditions guaranteeing the w.p. 1 convergence of SA procedures of the form (1) are well - known and have already been quoted, at least in part, in Sec. 2. But, in order to obtain the truly practical methods for stochastic optimization it is intuitively clear that, at each step and once X_n and Y_n are already given, a kind of one-dimensional minimization starting from X_n in direction $-Y_n$ should be performed (thus allowing the α_n to be random). This intuition has been confirmed by several experimenters (c.f. e.g. Janač (1971). Fabian (1960) proposed to determine α_n in the following way: given X_n and Y_n take observations V_i of $f(X_n - i a_n Y_n)$, where a_n is a positive real, until $V_1 > V_2 > \ldots > V_{i-1}$ and put $\alpha_n = i \cdot a_n$ if $V_1 > V_2 > \ldots > V_i \leq V_{i+1}$. In this procedure each one-dimensional search is therefore assumed to proceed with a step constant length $a_n \|Y_n\|$. Fabian correctly pointed out that "this assumption is not essential" and added that "removing it leads to complications of proofs or to results insufficiently general".

In this Section we shall consider a natural and straightforward generalization of the Fabian's procedure for the one-dimensional minimization, the convergence proof of which is only slightly more complicated and, at the same time, which seems to be quite reasonable.

Let, for each n, the one-dimensional minimization procedure consists of two stages. At the first stage it proceeds with a step of increasing length while at the second the length of steps is constant. For the sake of definitness we shall assume below that, at the first stage, each step is thrice as large as the preceding one.

Now, to be more precise, define the following sequence of random variables (with $\mu_n = [\log_3 \gamma n^{\mu}]$, where $[\nu]$ denotes the integer part of ν, γ and μ are positive numbers and μ is such that $\sum_n a_n^2 \cdot c_n^{-2} n^{2\mu} < \infty$):

$$V_{1,1}^n = V(X_n - a_n Y_n)$$

$$V_{2,1}^n = V(X_n - (a_n + 3a_n) Y_n)$$

$$V_{3,1}^n = V(X_n - (a_n + 3a_n + 9a_n) Y_n)$$

$$\vdots$$

$$V_{\mu_n+1,1}^n = V\left(X_n - \frac{a_n}{2}(3^{\mu_n+1} - 1) Y_n\right)$$

and let α_1^n be a random variable such that (with $\omega \in \Omega$)

$$\alpha_1^n(\omega) = (a_n/2)(3^{i_1^n} - 1)$$

for ω in the set (on the r.h.s. we omit superscript n in i_1^n)

$$A_{i_1^n,1} = \begin{cases} \{V_{1,1}^n > V_{2,1}^n > \ldots > V_{i_1,1}^n \leqslant V_{i_1+1,1}^n\}, & 1 \leqslant i_1 \leqslant \mu_n \\ \{V_{1,1}^n > V_{2,1}^n > \ldots > V_{i_1,1}^n\}, & i_1 = \mu_n + 1. \end{cases}$$

Further, define the sequence

$$V_{0,2}^n = V(X_n - \alpha_1^n Y_n)$$

$$V_{1,2}^n = V(X_n - (\alpha_1^n + a_n) Y_n)$$

$$V_{2,2}^n = V(X_n - (\alpha_1^n + 2a_n) Y_n) \quad \text{etc.}$$

and, finally, put in (11)

$$\alpha_n(\omega) = \alpha_1^n(\omega) + i_2^n a_n$$

for ω in the set

$$A_{i_2^n,2} = \{\omega \in A_{i_1^n,1} : V_{0,2}^n > V_{1,2}^n > \ldots > V_{i_2^n,2}^n \leqslant V_{i_2^n+1,2}^n\}.$$

Observe that at the first stage no more than μ_n+1 steps can be performed. Suitably modyfying the proof of Fabian one can show that convergence properties of procedure (11) with α_n defined above are the same as those of the original Fabian's method (with the only exception that condition $\Sigma a_n c_n^{-2} < \infty$ should be replaced by $\Sigma a_n^2 c_n^{-2} n^{2\mu} < \infty$).

Let us recall that unimodality of f is not needed for the existence of a convergent (w.p. 1) subsequence of (X_n); cf Fabian (1960) - Sec's 3-5 or Koronacki (1975)*).

Remark 1. It is recommended, especially when the values of i_1^n are large, that $\alpha_1^n(\omega)$ be defined as follows

$$\alpha_1^n(\omega) = (a_n/2)\,(3^{i_1^n-1} - 1), \quad \omega \in A_{i_1^n,1}.$$

The reason is that, even if there is no random noise, the procedure can "overjump" the point of minimum (along $-Y_n$) at its last successful step. Clearly, such a change of α_n does not affect the convergence property.

Remark 2. At the second stage of a one-dimensional minimization procedure one could e.g. apply successive halving of the first unsuccesful step of the first stage.

Remark 3. One should not expect that using of one-dimensional minimization procedures can improve the order of the speed; cf. Révész (1974). It should however improve "efficiency" of the SA methods for n fixed, when X_n is "far" from a point of minimum of f.

4. Stopping rules for SA-type estimators.

In this Section we shall confine ourselves to the KW procedure (1), (2) with f defined on the real line (in what follows we assume tacitly that a suitable modification of (A3) holds). We shall discuss briefly the problem of finding such a stopping rule (s.r.) that a fixed - length confidence interval for Θ be attained. From the "asymptotic" point of view the s.r. in question could be constructed using the Asymptotic Normality Theorem (cf. e.g. Sielken (1973)), but we shall not follow this approach.

The problem of stopping SA-type estimators is striking for its relative absence in the literature. As far as we are aware, in a

*) Notation on p. 520 of Koronacki (1975) is misleading. Everywhere on this page a should be replaced with a_n and, in Assumption 2 and Th. 3, a_n should be replaced with $\tilde{a}_n$.

nonasymptotic set-up there have been proposed only two solutions to this problem. The first is due to Dvoretzky (1956) (see Sec.8there) who considered the (so-called Robbins-Monro) situation of finding a root of a regression function. Dvoretzky's solution can be adapted to the case of looking for a minimum of a regression function f, if f is continuons and "nearly" quadratic; i.e. if there exist constants C_o, K_1, K_2 such that for every $0<c\leqslant C_o$

$$2\,c\,K_1(x-\theta)^2\leqslant[f(x+c)-f(x-c)]\cdot(x-\theta)\leqslant 2\,c\,K_2\cdot(x-\theta)^2.$$

Dvoretzky's s.r. is obtained by using the Chebyshev's inequality and is deterministic (i.e. the s.r. obtained is not a random variable).

The second s.r. is due to Farrell (1962) and can be described in the following way. Let $\underline{\theta}\leqslant\theta\leqslant\overline{\theta}$, for some known $\underline{\theta}$ and $\overline{\theta}$. Let $(X_n^i)_{n=1}^{\infty}$, i=1,..., 2l, be 2l independent sequences of random variables obtained by the simultaneous use of 2l KW procedures, with $X_1^i = \underline{\theta}$ for i=1,..., l and $X_1^i = \overline{\theta}$ for i=l+1,...,2l. Finally, let $X_n' = \min\{X_n^1, \ldots, X_n^l\}$, $X_n'' = \max\{X_n^{l+1}, \ldots, X_n^{2l}\}$ and $1-\alpha$ be a confidence level required. Now, if (A1) is satisfied and

$$4\sigma^2\partial^{-2}\sum_{i=1}^{\infty}(a_i/c_i)^2 \leqslant(\alpha/2)^{1/l},\quad a_1K_o<1,\quad \sum_{i=1}^{\infty} a_ic_i \leqslant\partial/K_o, \tag{12}$$

where ∂ is a positive number, σ^2 and K_o are such that $E_{F_n}\,\varepsilon_n^2\leqslant\sigma^2$ and $|f_{xx}(x)|\leqslant K_o$, respectively, then, for any ε positive and any integers M,N such that $|X_N''-X_M'|\leqslant\varepsilon$, the interval $(-2\partial+X_M', 2\partial+X_N'')$ is the confidence interval for θ, of length $\leqslant\varepsilon+4\partial$ and with confidence level $1-\alpha$. In particular, one can define M,N to be such integers n_1, n_2 that $|X_{n_1}''-X_{n_2}'|\leqslant\varepsilon$ and the sum n_1+n_2 assumes the least value. (Farrell considered in fact the Robbins-Monro situation and assumed l=1. An extension of the Farrell's result to the minimization problem (and l>1) is a trivial excercise; c.f. Koronacki (1979).)

Let us conclude this Section with the remark that we have aimed here at calling the reader's attention to the stopping problem of SA estimators as to this very problem which so far remains nearly unsolved, at least from the practical point of view (note that e.g.

conditions (12) are very, hopefully unnecessarily, stringent).

REFERENCES

[1] DUPAČ, V. (1977), Math. Operationsforsch Stat., ser.Statistics, 8, 107 - 117.

[2] DVORETZKY, A. (1956), Proc. Third Berkeley Symp. Math. Statist. Prob., vol. 1, 39-55.

[3] FABIAN, V. (1960), Czech. Math. J, 10, 123 - 159.

[4] FABIAN, V. (1968), Ann. Math. Statist. 39,1327 - 1332.

[5] FABIAN, V. (1971), Optimizing methods in statistics (J.S.Rustagi, ed.), Ac. Press, 439 - 470.

[6] FABIAN, V. (1977), Proc. Int. Statist. Inst. Symp., Calcutta.

[7] FARRELL, R.H. (1962) Amm. Math. Statist, 33, 237- 247.

[8] JANÁČ, K. (1971), Simulation, 16, 51 - 58.

[9] KHAS'MINSKIJ, R.Z. (1977), Math. Operationsforsch Stat., ser. Statistics, 8.

[10] KORONACKI, J. (1975), Int. J. Control, 21, 517 - 527.

[11] KORONACKI, J. (1979), Banach Center Publications, vol. 6, Polish Scientific Publishers.

[12] KUSHNER, H.J. and Cark, D.S. (1978), Stochastic approximation methods for constrained and unconstrained systems, Springer - Verlag.

[13] LJUNG, L. (1974), Prepr. Stoch. Control Symp., Budapest

[14] LJUNG, L. (1978), Ann. Statist. 6.

[15] RÉVÉSZ, P. (1974), Studia Sci. Math. Hung., 9, 453 - 460.

[16] SCHMETTERER, L. (1969), Multivariate Analysis II, Proc. 2 nd Int. Symp., Dayton. Ohio, Ac. Press, 443 - 460.

[17] SCHMETTERER, L. (1978), Proc. European Meeting of Statisticians, Oslo.

[18] SIELKEN, R.L., Jr. (1973), Z. Wahrsch. verv. Geb., 26, 67 - 75.

[19] ZIELINSKI, R. (1979), Proc. 7th Prague Conf. Inf. Th., Dec. Functions, Random Processes, Academia.

10